Measurement, Instrumentation and Experiment Design in Physics and Engineering

Measurement, Instrumentation and Experiment Design in Physics and Engineering

MICHAEL SAYER, P. Eng.
Professor of Physics
Queen's University at Kingston
Ontario, Canada

ABHAI MANSINGH
Professor of Physics, and
Director, South Campus
University of Delhi

Prentice-Hall of India Private Limited
NEW DELHI - 110 001
2000

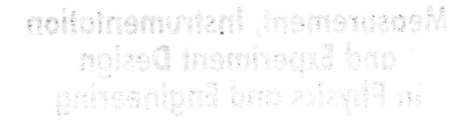

Rs. 195.00

MEASUREMENT, INSTRUMENTATION AND EXPERIMENT DESIGN
IN PHYSICS AND ENGINEERING
by Michael Sayer and Abhai Mansingh

ISBN-81-203-1269-4

The export rights of this book are vested solely with the publisher.

Published by Asoke K. Ghosh, Prentice-Hall of India Private Limited, M-97, Connaught Circus, New Delhi-110001 and Printed by Syndicate Binders, B-167, Okhla Industrial Area, Phase I, New Delhi-110020.

To

Anne Sayer and Sydney Southcot

Contents

6. VACUUM TECHNIQUES 169–195

7. OPTICAL INSTRUMENTS 196–226

8. X-RAY MEASUREMENTS 227–252

9. RADIOACTIVITY AND MATTER 253–274

10. RADIATION DETECTION AND MEASUREMENT 275–304

11. ANALYTICAL INSTRUMENTATION 305–321

12. OCCUPATIONAL HEALTH AND SAFETY 322–334

Preface

The art of the physicist combines a broad theoretical understanding of phenomena with an ability to make careful measurements. Physicists and Engineering Physicists are adept as problem solvers, bringing novel theoretical and experimental insights to problems encountered in other areas of science and technology.

The objectives of this book are to demonstrate the principles of experimental practice in physics and physics-related engineering, to show how measurement, experiment design, signal processing and modern instrumentation can be used most effectively, and to encourage the creative use of experimental and theoretical physics in areas which may be unfamiliar. A knowledge of basic electricity and magnetism and circuit theory is assumed along with some introduction to aspects of semi-conductor and quantum physics. Extensive opportunities are provided to use spreadsheet methods in experimental design and evaluation.

The important topics dealt in the text are experiment design, signal to noise enhancement and the use of electronics, operational and phase sensitive amplifiers for the acquisition and processing of data. This includes computer-based instrumentation system with a particular emphasis on standard interfaces such as the IEEE488.

Primarily addressed to students at the advanced undergraduate and introductory graduate level in physics, applied and engineering physics, and engineering, the book will serve as background material for a wide range of experimental physics and engineering. It is not a compendium of all types of transducers and experimental measurements. The emphasis has been to review the background physics and experimental techniques in important areas of application so that a reader develops his or her own insight and knowledge to work with any instrument and its manual. Questions are provided throughout to assist in this end. Since most of the laboratory practices involve some aspects of temperature measurement, optical techniques, vacuum practice, electrical measurements and nuclear instrumentation, these areas are covered in detail.

In a world increasingly conscious of the potential effects of science and technology on the environment, laboratory safety and the safety of the public at large are of importance. International and national aspects of these matters are reviewed in Chapter 12.

Measurement and instrumentation is not a passive and isolated subject that can be defined in terms of specific experiments. Real experiments involve complex systems, considered approximations, and compromises between competing requirements such as cost, accuracy, speed of measurement and ease of operation. In each chapter, design examples and problems are presented—many of them have arisen from real life situations.

The authors wish to thank all students at Queen's University who have taken this course and who have contributed to the development of the material. The important contributions of colleagues who have taught within the course and/or whose knowledge and insights have provided material and questions are recognised and acknowledged. This is particularly the case for the sections on nuclear techniques. Contributors include D.A. Hutchins, G.F. Lynch, Hay-Boon Mak, J.P. Harrison, H.C. Evans, J.D. MacArthur and G.T. Ewan. The efforts of G.T. Ewan, S.A. Langstaff, A.M. Sayer, J.L.Whitton and A. Kapoor in reviewing the manuscript and in making suggestions for improvements and additions are appreciated. Secretarial assistance in early drafts by D. Robertson and M. Reid was of great value.

Michael Sayer
Abhai Mansingh

1
Physical Measurement

Experimental science can be represented by a cycle of activities.

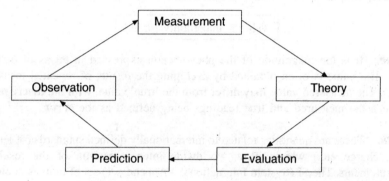

Fig. 1.1 The cycle of experimental science.

The scientific process normally starts with an **observation** which is quantified formally or informally by some type of **measurement**. A curious onlooker will then seek to relate these facts to other background knowledge to find a reason or **theory** for the observation. The theory is tested by evaluating whether it fits the recognized facts, and it is **confirmed** if it predicts new observations.

An important part of this cycle is how observations are quantified, **measured** and assessed. The achievements of Newton and Faraday in quantifying the concepts of mechanics or electric fields are remarkable especially at the time when accurate clocks, electrometers, voltmeters or ammeters did not exist. In modern science and engineering, the basic measurement methods are combined with techniques to enhance the accuracy of measurement for creating sensitive **instrumentation**. **Design** is the creative act which makes best use of these instruments, avoids errors in measurement or interpretation, and applies physical insight to achieve defined objectives.

1.1 MEASUREMENT

Measurement is the process by which the natural world is quantified. Good measurements are required to describe a phenomenon, to compare with theory, and to make an engineering design.

1

In *physics* a good experiment is considered to be one in which all pertinent variables of a phenomenon except one are held constant and the phenomenon is measured as the single variable is changed. Any observed variations are then a function only of the parameter under test. The quality of the information is indicated by the quoted **uncertainty** in the data [Ref. 1, 2].

In *engineering or life sciences,* experimental systems are often too complex to allow this procedure, and measurements have to be made under conditions where one or more variables interact, or uncontrollable changes in the system occur. The behaviour is then inferred from the analysis·of a sequence of measurements using statistical methods [Ref. 3, 4].

1.2   THE RESULT OF A MEASUREMENT

Without exception, the result of a measurement has three parts:

> Value ± uncertainty [units]

Value.   It is the magnitude of the phenomenon expressed in terms of defined units. The value is often obtained by averaging the results of repeated measurements. The measured value may differ from the 'true' value, with the **discrepancy** between the measured and true readings being defined as the **error.**

Units.   These are quantities related to internationally defined standards (of length, time, charge etc.) which allow worldwide intercomparison of the result of measurements. The SI (System International—Appendix 1) set of units is in almost universal use for science and engineering.

Uncertainty.   It can be defined as an estimate of the degree to which the value of a measurement can be repeated, or is expected to conform to the 'true' value.

> There is no magic formula for the uncertainty. It is a **professional** judgment made by an experimenter. The number of significant figures quoted in a result should be consistent with its uncertainty.

Significant Figures

When a result is quoted as $M \pm m$ (units), this implies that the value of M is estimated to within 1 part in M/m. The uncertainty in M or m is indicated by the number of non-zero digits. The number of significant figures in both the value and the uncertainty should be consistent with common sense.

1.893

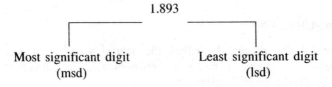

Most significant digit   Least significant digit
(msd)   (lsd)

The following points are worth considering in this regard:

1. If there is a decimal point, the rightmost digit is the least significant digit even if it is a zero.

2. When it is necessary to roundoff significant figures, the insignificant figures should be treated as a fraction.

   - If the fraction is greater than 1/2, the least significant digit is incremented.

   - If the fraction is less than 1/2, the least significant figure is not incremented.

   - If the fraction equals 1/2, the least significant digit is incremented *only if it is odd.*

3. When quoting an experimental result in intermediate calculations, the number of significant figures should be approximately 1 more than that dictated by the experimental uncertainty. For computer analysis of data, it is generally advisable to retain all available digits in intermediate calculations and round only the final or quoted intermediate results.

Question (i)   (a)   How many significant figures are in these numbers?

$$11.64 \quad 1020 \quad 1.020 \times 10^3 \quad 4000 \quad 0.004 \quad 400.$$

              (b)   How should the final result be quoted?

                 (i) $L = 1.979 \pm 0.012$m (1 part in 164)

                (ii) $L = 1.979 \pm 0.082$m (1 part in 24)

1.3 SOURCES OF UNCERTAINTY AND EXPERIMENTAL ERROR

Uncertainty and experimental error arise from two sources:

Systematic error. This is the degree to which a measured value differs from the 'true' value because of errors inherent in the measurement. This may be due to an incorrect scale, a wrong calibration or an erroneous assumption. In instrumentation, changes in the original calibration of an instrument over time, or if the instrument is used under abnormal conditions are major sources of systematic error. Systematic errors are generally not statistical in nature and may be corrected by measurement of a **standard**.

Random errors. These are uncontrolled variations in the measurement which define the precision of the result. They may arise from difficulties in making a measurement, or from inherent fluctuations in the processes concerned—e.g. arising in the counting of charges collected by an electrode or in the probability of radioactive decay by a nucleus. The results of repeated measurements are **distributed** about a mean value. In instrumentation these random fluctuations are termed **noise**.

The **accuracy** of a measurement is the degree to which the quoted value reflects the **true** value. The **precision** is the degree to which a repeated measurement gives the same result. Figure 1.2(a) shows a measurement which is inaccurate but highly precise. Figure 1.2(b) shows one which is accurate but imprecise [Ref. 1].

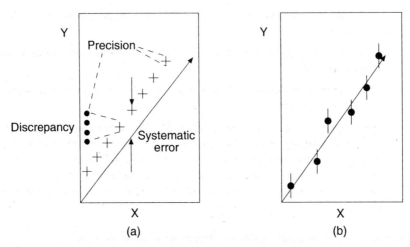

Fig. 1.2   (a) An experiment which has high precision but low accuracy, (b) an experiment which has low precision but high accuracy. The line is the 'true' value. The error bars represent the range over which readings occur; adapted from [Ref. 1].

1.4 SYSTEMATIC ERROR

Calibration and Standards

In much instrumentation, the existence of systematic error is most effectively checked by the measurement of **standards** or of samples having known properties. This process is known as **calibration**. It is important to use standards with values across the entire range of parameters of the proposed measurement.

> No matter how simple or complex a measuring tool may be, **always** check it against something you know.

If necessary, the reading of an instrument may be corrected by a **calibration factor.** The correction is often a simple multiplication or addition. The uncertainty in the measured value of a test sample is then determined by combining the uncertainty to which the standard is known, the uncertainty with which the standard is measured by the instrument, and the uncertainty with which the sample under test is measured [see Design Problem 1.2]. In practice, a standard is usually used to adjust the output of an instrument to read to a stated accuracy. Periodic calibrations are then made against the standard.

Systematic Errors Arising from Experimental Design

Systematic errors can arise from incorrect assumptions made during a measurement. These can be reduced by recognition of the problem and correct experimental design. For example, consider the measurement of a resistance by a current/voltage method using the circuit shown in Fig. 1.3(a). Representative data points are shown in Fig. 1.3(b).

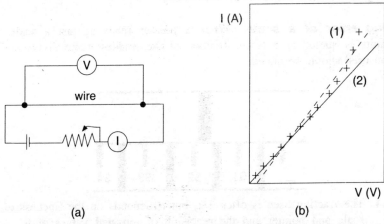

(a)   (b)

Fig. 1.3   (a) An experimental measurement of resistance R, (b) line 1 has a constant slope but does not pass through zero, line 2 suggests that the experimental slope increases at higher voltages.

If it is assumed that Ohm's Law $R = \Delta V / \Delta I$ applies and that R is constant during the measurement, line 1 may be taken to be an appropriate fit. The discrepancy at the origin could be an error in the zero setting of the voltmeter. However, a more careful examination of the data (line 2) suggests that the slope of the line increases as the voltage increases. Since the resistance of a metal usually depends on temperature as $[R(T) = R(0)[1 + \alpha (T - T_0)]$, where α is a constant], if the wire is heated by the passage of current, the temperature rise $(T - T_0)$ is likely related to the dissipation of electrical power $\approx V^2/R$ or I^2R. Thus $R(V) \approx \Delta V / \Delta I$. $[1 + k'V^2]$ where k' is a constant, and line 2 extrapolated to low applied voltage is likely to give a better value of the room temperature resistance.

In this case, good design suggests that an experimenter do one or more of: (i) limit measurements to a voltage range where heating is negligible, (ii) fit the function $R(V)$ to the data, and (iii) use a flowing coolant about the wire to maintain it at a constant temperature.

> It is important to examine all data sets critically by graphical plots and to assess the physics of a measurement. A blind computer fit of a linear function to data can lead to substantial errors of interpretation.

1.5 RANDOM ERROR

Inherent variations in a measured parameter are directly assessed by making **repeated** measurements to examine the distribution of values obtained. Certain practices are commonly used to estimate the uncertainty expected in a **single** measurement.

Reading Error and Calibration Accuracy

Reading error of a scale. When a pointer reads against a scale, the uncertainty is quoted as ± (some fraction of the smallest scale division). The criterion used should be stated

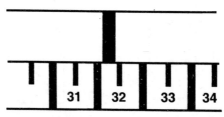

Fig. 1.4 The fraction used is often 1/2, but it depends on the fineness of the scale and pointer and the precision of repeated values of the same object.

Mechanical linkages. Gears and mechanical connections between a knob and the system it moves inevitably involve some looseness or 'backlash'. This means that the system may not be in exactly the same position when a reading is approached from one direction (high) as in the other (low). The uncertainty due to backlash should always be checked by experiment. It is a good practice to always approach a reading from the same direction as long as all calibrations are carried out using the same procedure.

Thermal or mechanical drift. Slow thermal or mechanical changes may create unidirectional shifts in a measured value. Careful experimental design is required to minimize such drifts. Where appropriate, the quoted uncertainty must include the possibility of the drift.

Instrumentation. The accuracy of a measurement is stated in the specification for the instrument either as ± (defined value) or ± (% full scale). This is often scale dependent. In general, there is a correlation between the guaranteed accuracy of an instrument and its cost.

Reading Error versus Statistical Analysis

The uncertainty estimated in a single reading by the above methods should be consistent with the distribution of values obtained by repeated measurements. If there is a substantial difference, the origin of the difference should be examined.

DC VOLTS		ACCURACY** ± (% value + counts)	
RANGE	RESOLUTION	24 hr* 22–24°C	1 yr 18–28°C
200 mV	1 μV	0.007 + 2	0.016 + 3
2 V	10 μV	0.005 + 2	0.011 + 2
20 V	100 μV	0.006 + 2	0.015 + 2
200 V	1 mV	0.006 + 2	0.015 + 2
1000 V	10 mV	0.007 + 2	0.015 + 2

* Relative to calibration accuracy. ** When properly zeroed.

Fig. 1.5 The specification reflects both the initial calibration and the stability of the calibration with time and operating conditions.

For example, the reading error of a scale reflects the best precision that can be achieved. If the range (highest to lowest) of repeated measurements is larger than ± (reading error) it is probable that backlash or some complexity in the measurement exists.

> Always repeat an initial measurement a number of times to assess the origin of uncertainty.

The **range of values** observed allows the precision of the measurement and the primary origin of error to be assessed. If the range is consistent with or greater than the reading error, the mean and an analysis of the range (see Sec. 1.7) defines the precision of the result. The reading error is then ignored.

The **calibration accuracy** of a meter and of an instrument should always be recorded so that the probability of systematic error is known. Whether the uncertainty arising from the calibration is included in the final quoted uncertainty depends on the experimental precision of the measurement. If the experimental precision is comparable with the calibration accuracy, both factors must be considered. If the range of observations far exceeds the calibration accuracy, the latter is ignored.

Question (ii)   Repeated voltage readings using the DC voltmeter specified in Fig. 1.5 on the 20 V scale 1 year after purchase were 1.2340, 1.2505, 1.2155, 1.2200 V. What is the value of the voltage?

1.6 DEFINITION OF THE UNCERTAINTY

The nature of the quoted uncertainty should always be defined. Does it refer to a single measurement, a mean value for repeated measurements on the same sample, or is it a characteristic of a set of samples? This progression is shown in Fig. 1.6.

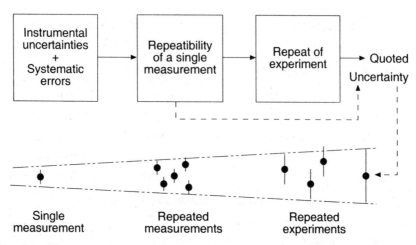

Fig. 1.6 The progression in the uncertainty from that of a single measurement to that for a system.

Suppose the resistivity ρ of a material is determined by measuring the resistance $R \pm \Delta R$ ohms of a sample of length $L \pm \Delta L$ m and area $A \pm \Delta A$ m² followed by calculation of $\rho \pm \Delta \rho$ from RA/L. The estimate of uncertainty $\pm \Delta \rho$ is that for a single **measurement.**

Using the same or a similar sample, one might repeat the measurement with different applied voltages (which may affect heating of the sample), or reduced length (which may change the effect of electrodes). Averaging the results now gives $\rho_{sample} \pm \Delta \rho_{sample}$ for this type of **sample.**

New samples from fresh batches of material may now be prepared and measured to check the reproducibility of the material. Averaging the results gives $\rho_{material} \pm \Delta \rho_{material}$. This is the true result for the **resistivity** of the **material.**

1.7 THE ANALYSIS OF REPEATED MEASUREMENTS

Repeated measurements of a quantity are invariably distributed about a mean or average value. This may be due to:

- experimental difficulties in repeating a measurement, e.g. roughness of the end of a bar, differences in contact resistance, or reading error.

- variations in the medium in which an experiment is performed, e.g. twinkling of starlight as it passes through the earth's upper atmosphere.

- inherent fluctuations in a quantity, e.g. collection of charge at an electrode, detection of radioactive decay products from a nucleus.

The results are evaluated by taking a set of N values of a quantity x, calculating the average value $\bar{x}$, and considering how the values are distributed about $\bar{x}$. Such a set of N values is

$$[x_1, x_2, x_3 \ldots x_i \ldots x_n]$$

Average value:
$$\bar{x} = \frac{1}{N} \Sigma (x_1 + x_2 + x_3 + \ldots + x_i + \ldots + x_n) = \frac{1}{N} \sum_{i=1}^{n} x_i$$

where Σ refers to a sum over N readings, and i goes from 1 to n.

The **distribution** of the sample of N readings about the mean value can be assessed in several ways.

Mean deviation:
$$d = \frac{1}{N} \Sigma |(x_i - \bar{x})|$$

where $|(x_i - \bar{x})|$ is the absolute value of the deviation of each reading from the average x.

Variance:
$$s^2 = \frac{1}{(N-1)} \Sigma (x_i - \bar{x})^2$$

This is the **sample variance** for a finite number of N readings. The denominator $(N - 1)$ is used to account for the fact that one piece of information is used to estimate the mean value $\bar{x}$. If $N \to \infty$, $(N - 1) \to N$ and the **standard deviation** $\sigma = \sqrt{s^2}$ is obtained. The variance and standard deviation reflect the *distribution* of values about the mean.

Variance of the Mean Value

Since the *value* of the mean $\bar{x}$ is calculated from the result of all N measurements, it is intuitive that this **mean** value is known to greater certainty than that of a **single** reading. The variance of the **sample mean** s_m^2 is defined as

$$s_m^2 = \frac{1}{N(N-1)} \Sigma (x_i - \bar{x})^2$$

This result is justified in Sec. 1.9.

Question (iii)   Ten repeated measurements of a light intensity are: 10.3, 12.6, 11.5, 14.3, 15.2, 13.6, 12.3, 14.5, 12.9, 10.8 W/m². Calculate the mean value, mean deviation, sample variance and variance of the sample mean. Estimate the standard deviation of the mean value σ_m from the square root of the variance of the sample mean and use it to state the average value for the light intensity.

Distribution Functions

The distribution of readings about $\bar{x}$ can be assessed by considering the number of readings which occur in a particular interval defined by x and $x + \Delta x$. The size of the 'bin' or interval Δx is arbitrary and is chosen for convenience. Such a plot is shown in Fig. 1.7.

Figure 1.7(a) is a plot of the number of readings (which fall in a particular interval x to $x + \Delta x$) versus the value of x. There are a total of N readings in the distribution. The ith bin is defined as x_i. For a finite number of readings this graph is a histogram of 'boxes' (shown dotted) called the **sample** distribution. Figures 1.7(b) and 1.7(c) show two types of anomalous distribution which may arise

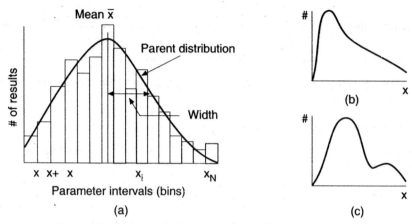

Fig. 1.7 (a) A frequency distribution for a set of data. The number of
measurements which fall between x and $x + \Delta x$ are plotted against x to
form a histogram of boxes. The average line through the boxes is that
which would be expected if N is large and Δx is small, (b) and (c)
represent non-symmetrical distributions due to experimental artifacts.

from experiment. For example, there may be a localised 'bump' on the end of a
bar.

The Parent Distribution

If the number of readings is very large ($N \rightarrow \infty$), the width of the boxes can be
made very small while still retaining a reasonable number of points in each box.
The histogram then approaches a smooth line known as the. **parent** distribution.
This line gives the probability $P(x, n)$ that the number of values in the range x_i
to $x_i + \Delta x$ is $n(x_i)$. The **mean** of the parent distribution is μ, while the variance is
σ^2. The parameter σ is known as the **standard deviation.** For a finite number of
measurements the values in the boxes at any x_i of the sample histogram are
scattered about the line defining the parent distribution.

A convenient technique for calculating the standard deviation is obtained by
expanding the expression for the variance with $N \rightarrow \infty$.

$$\sigma^2 = \frac{1}{N} \Sigma (x - \mu)^2 = \frac{1}{N} \Sigma (x^2 - 2\mu x + \mu^2) = \frac{1}{N} \Sigma x^2 - \mu^2$$

since $\Sigma x = \mu N$.

> The standard deviation σ is the square root of the average of the
> squares minus the square of the average.

The **distribution function** $P(x)$ is the probability of a value occurring
at a particular value between x and $x + \Delta x$. $NP(x)$ implies that n out of
N values occur at x. The overall distribution must contain all N values of x. Thus
$\Sigma n(x) = N$, and $\Sigma P(x) \Delta x = 1$.

> A measured distribution function has no defined *a priori* shape and merely represents the experimental data.

A frequency plot gives valuable insights into a set of measurements and into the design of an experiment. For example, if the distribution of values is symmetrical about the mean value (Fig. 1.7(a)), a process involving purely random variations is suggested. If the symmetry of the distribution is distorted about the maximum value (Figs. 1.7(b) and (c)) other processes may be involved. These include the use of a limited number of samples, multiple effects which perturb the measurement, or an anomaly in the system. An example of an anomaly is a localized bump on a length of wood.

1.8 THE MATHEMATICAL DESCRIPTION OF DATA DISTRIBUTION FUNCTIONS

Analytical expressions which predict the distribution of results from an experiment are valuable to interpret data and to guide experimental design. The parameters are primarily the number of repeated experiments N and the probability p for which a particular result is expected. Some simplifications arise in limiting cases corresponding to known experimental situations. This gives rise to different types of **statistical distribution functions.** The expressions are listed, then their mathematical form is justified.

Binomial Distribution

$$P_B(x, n, p) = \frac{n!}{x!(n - x)!} p^x (1 - p)^{(n-x)}$$

This is the most general expression for predicting the outcome x of n repeated measurements in which the probability of a given result in a single experiment is p. No restrictions are placed on n and p.

This distribution predicts the number of heads (x) or tails $(n - x)$ obtained in n tosses of a coin $(p = 1/2)$; the number of, say 6's achieved in n throws of a dice $(p = 1/6)$; or the digits (0 or 1) of a digital number such as 101110 if the digits are selected sequentially at random $(n = 6, p = 1/2)$. The average number of 'expected' results is np, e.g. 500 heads out of 1000 coin throws.

Poisson Distribution

$$P_P(x_i, n, p) = \frac{\mu^x}{x!} e^{-\mu}$$

This is a limiting case of the Binomial distribution when the number of experiments or samples, n is large but the probability p of a result is small. The predicted value $\mu = np$ is small.

Such conditions occur in experiments involving counting in nuclear or atomic physics or in electron flow in electronic devices. The detection of gamma ray emission from a sample of nuclear material results from the small probability of decay of nuclei (*p* small) within a large starting population (*n* large). In electronics, *n* electrons may arrive at a collector at randomly spaced infrequent intervals (*p* small). The collected current will fluctuate with time.

Gaussian Distribution

$$P_G(x, \mu, \sigma) = \frac{1}{\sigma\sqrt{2\pi}} \exp\left[-\frac{1}{2}\left(\frac{x-\mu}{\sigma}\right)^2\right]$$

This is a **description** of data from experiments in which the results are distributed symmetrically about a mean value μ with a width given by the standard deviation σ. μ and σ are chosen as constants of the function. This distribution function is a convenient way of describing commonly observed experimental results on large sample populations. Compared to the Binomial and Poisson distributions, the Gaussian distribution is more of a convenient mathematical description of an observed experimental behaviour than a conclusion resulting from physics.

1.9   DERIVATION AND PROPERTIES OF THE DISTRIBUTION FUNCTIONS

Binomial Distribution $P_B(x, n, p)$

In tossing a coin, the probability of a head or a tail is $p = 1/2$. The probability of 2 heads in tossing two coins is $1/2 \cdot 1/2 = 1/4$, and by extension the probability of *n* heads in *n* coins is $(1/2)^n = p^n$. If large numbers of coins are thrown, it is intuitive that on average, roughly half the coins will be heads, and half, tails. Thus the average value $\bar{x} = np$.

To predict the probability of a distribution in which *x* coins are heads ($p = 1/2$) and ($n - x$) are tails ($p = 1 - 1/2$) we calculate:

Probability of selecting	* Probability that	* Probability that
x similar coins from *n* ·	*x* are heads	($n - x$) are tails

The number of ways to select *x* indistinguishable coins from *n* is

$$\frac{n * (n-1) * (n-2) * \ldots (n-x+1)}{x * (x-1) * (x-2) \ldots 1} = \frac{n!}{x!(n-x)!}$$

hence

$$P_B(x, n, p) = \frac{n!}{x!(n-x)!} p^x (1-p)^{(n-x)}$$

Question (iv)   1000 airborne particles per second cross a perpendicular plane in a turbulent airstream. The stream impinges on one or other of two rectangular apertures one of which is three times the width of the other (a) How many particles enter the small aperture?, (b) If 350 particles are observed in the small aperture, justify whether the flow pattern has changed.

Poisson Distribution $P_p(x, \mu)$

If n is large: $(n \rightarrow \infty)$, $n \gg x$ and $\mu = np$ is small, parts of the binomial equation can be approximated

$$\frac{n!}{(n - x)!} = n(n - 1)(n - 2)(n - 3) \ldots (n - x + 1) \rightarrow n^x$$

If p is small: $(p \rightarrow 0)$ and $n \gg x$ $(1 - p)^{(n - x)} \rightarrow (1 - p)^n$

Writing $(1 - p)^n = \left(1 - \frac{1}{1/p}\right)^{(1/p)np} = \left(\frac{1}{e}\right)^{np} = e^{-\mu}$

using the definition of the exponential function

$$e = \mathop{\mathrm{Lt}}_{x \rightarrow \infty} \left(1 + \frac{1}{x}\right)^x \text{ and setting } x = 1/p$$

The Binomial distribution is then approximated as the Poisson distribution

$$\boxed{P_p(x, \mu) = \frac{\mu^x e^{-\mu}}{x!}}$$

The form of the Binomial and Poisson Distributions for different values of n, p and np are shown in Fig. 1.8. If np is small, the distribution is a not symmetrical about its maximum value and the maximum value does not occur at np. The reason for this is that values of $x < 0$ are not allowed. For both distributions, the functions are symmetrical about $\mu = np$ if $\mu = np > 20$.

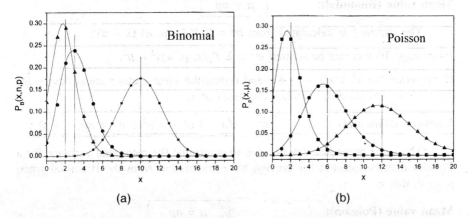

(a) (b)

Fig. 1.8 (a) Binomial distribution function for $n = 20$ and $p = 0.5$, 0.16 and 0.1.
(b) Poisson distribution function for $n = 200$ and $p = 0.01$, 0.03 and 0.06, and $\mu = np$.

If there are only 2 to 3 counts per hour for an experiment in nuclear physics, the mean value μ is also between 2 and 3 and the number of counts in successive hours will not be symmetric about μ. This may make the distribution hard to interpret. In the design of counting experiments, the counting time interval is often chosen so that the mean value or number of counts is > 10 and the form of the distribution approaches symmetry. Thus a better experimental design may be to count for a period of 24 hrs when the count rate is 48–72 counts per 24 hours. On the other hand, if it can be established in counting intervals of 1 hr that the data conforms to a Poisson Distribution, the analytical form can allow μ to be predicted with improved accuracy in the shorter time interval.

Mean and Variance for the Binomial and Poisson Distributions

The mean and variance of both the Binomial and the Poisson distributions can be calculated by a mathematical device. This is illustrated for the Binomial function. The mean μ is calculated by weighting a value of x by its probability to occur $P(x, n, p)$ and summing over all the values of x.

$$\mu = \Sigma \, x P_B(x, n, p) = \Sigma \, x \, \frac{n!}{x!(n-x)!} \, p^x \, (1-p)^{n-x}$$

Cancelling x and taking out a value of np gives

$$\mu = np \, \Sigma \, \frac{(n-1)!}{(x-1)!(n-x)!} \, p^{x-1} \, (1-p)^{n-x}$$

If it is noted that $(n-x) = (n-1) - (x-1)$, the second summation has the form of the binomial function for $(n-1)$ and $(x-1)$. Since all $(n-1)$ values lie within this distribution, the summation on the right hand side of this equation may also be equated to 1. Therefore, for a Binomial function

Mean value (Binomial): $\boxed{\mu = np}$

The variance is calculated from $\sigma^2 = \Sigma \, P_B(x, p, n) \, (x - \mu)^2$.

From page 10 this can be written $\sigma^2 = \Sigma \, P_B(x, p, n)x^2 - \mu^2$.

If we write that $x^2 = x(x-1) + x$ and utilize the same mathematical device for $x(x-1)$ as was used to calculate the mean of x,

Variance (Binomial): $\boxed{\sigma^2 = n(n-1)p^2 - n^2p^2 + np = np(1-p)}$

The same procedure can be used to determine the mean and variance for a Poisson distribution, but for brevity, by extrapolation from the Binomial when $p \to 0$, this is:

Mean value (Poisson): $\boxed{\mu = np}$

Variance (Poisson): $\boxed{\sigma^2 = p = \mu}$

This result is important in the theory of measurement and in signal processing. In a random process described by Poisson statistics with mean value μ, the width of the distribution estimated by the standard deviation is the square root of the mean ($\sigma = \sqrt{\mu}$). No other information is required to define the form of the distribution function. The effect of this relationship shows up strongly in the **relative error** when the data is plotted on a scale set by the mean value—Fig. 1.9.

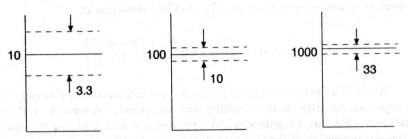

Fig. 1.9 Plots showing the distribution of values expected for means of 10, 100 and 1000 in a Poisson distribution for which $\sigma = \sqrt{\mu}$. Each graph is normalised at its mean value μ.

$$\text{Value} = \text{mean} \pm \text{relative error in mean} = \mu \pm \sigma/\mu$$

$$= \mu \pm \sqrt{\mu}/\mu = \mu \pm 1/\sqrt{\mu}$$

If $\mu = np$, the relative error is proportional to $1/\sqrt{n}$ and the relative certainty improves or the width of the distribution decreases with $1/\sqrt{n}$. Thus the accuracy improves with the square root of the number of 'counts' in a value. This may be accomplished by repeating the measurement of a quantity a large number (n) times and summing the results, or by taking measurements (counting) over a longer time interval.

> The certainty improves by 'root n'

This relationship has important consequences for the design of experiments. *However, experimental judgement must also be applied.* It is clearly of value to increase the number of measurements n, but the degree to which the result is improved is only a factor of $1/\sqrt{n}$. Since each experiment requires time and resources, a set of n experiments takes n times the time for one experiment. Equipment must be stable over this time and the inherent costs in labour or time must be worth the benefits.

> An experimental design based on repeated measurements is a compromise between the additional time or resources required and the improvement in uncertainty achieved.

Question (v)   In a nuclear counting experiment the count rate is about 5 counts/hr. How long do you have to count to establish the uncertainty in the count rate to better than 1%?

Gaussian Distribution $P_G(x, \mu, \sigma)$

Many experiments give rise to data which originate in large sample populations $n > 10$–20 and which lead to values of $x \gg 1$. Such data is distributed symmetrically about a mean value. The interest is in the width of the distribution, and in predictions which can be made once the distribution is known. The Gaussian distribution provides a symmetrical function mathematically described by two parameters representing a mean μ and a standard deviation σ.

$$P_G\,(x, \mu, \sigma) = \frac{1}{\sigma\sqrt{2\pi}}\,\exp\left[-\frac{1}{2}\left(\frac{x-\mu}{\sigma}\right)^2\right]$$

Values of μ and σ for a large population are estimated either directly from the experimental data or by assuming that the results conform to a Poisson distribution ($\sigma = \sqrt{\mu}$). **Conclusions** about the data are then drawn on the basis of the known properties of the Gaussian distribution function.

The importance of the Gaussian function is that its value (representing the probability of observing a particular reading), and its integral (representing the fraction of measurements which fall within a particular interval) are tabulated or can be computed as a function of $[(x-\mu)/\sigma]$. Important conclusions are listed in Fig. 1.10 for a Gaussian data set with mean μ and standard deviation σ.

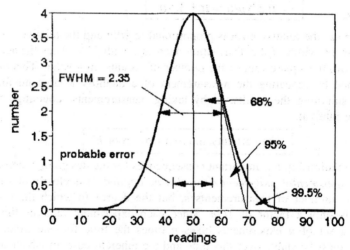

Fig. 1.10   Gaussian distribution of 100 readings with mean $\mu = 50$ and standard deviation $\sigma = 10$.

The percentage figures show the probability or **confidence limits** for a value to lie within σ, 2σ and 3σ of the mean value respectively. Thus 68% of the readings will lie within σ, 95% within 2σ, and 99.5% within 3σ. The **most probable value** of $\pm 0.675\sigma$ implies that 50% of the readings lie within this range and 50% outside. The distribution has a **Full Width at Half Maximum** (FWHM) of 2.35σ. The FWHM is extensively used in nuclear physics—see page 277.

Question (vi)   Commercial resistors of 2% accuracy have a range of values with a standard deviation of 2% nominal resistance. If a particular application requires a resistor value of $1000 \pm 40\ \Omega$, how many 2% resistors will be rejected out of a batch of 1000?

1.10 PROPAGATION OF ERROR

Most experiments involve the computation of a final result (u_i) from a combination of independent or unrelated measurements $(x_i, y_i, ...)$ such as

> volume = length × width × height
>
> count rate = sample rate − background rate
>
> activation energy for a semiconductor = $E = kT \ln (R_T/R_0)$

where k is Boltzmann's constant, T is absolute temperature, R_T and R_0 are the resistances of a sample at T and 0 K respectively.

The design of an experiment must take into account the relative error to be allowed in each of the parameters. The variance in the function $u = f(x, y, ...)$ is

$$\text{Variance } \sigma_i^2 = \lim_{n \to \infty} \Sigma \, (u_i - \bar{u})^2$$

On expanding $(u_i - \bar{u}) = f(x_i, y_i, ...)$ as a Taylor series and substituting,

$$\sigma^2 = \lim \Sigma \left[(x_i - \bar{x}) \frac{\partial u}{\partial x} + (y_i - \bar{y}) \frac{\partial u}{\partial y} + ... \right]^2$$

Multiplying out the term in brackets gives two types of terms

$$(x_i - \bar{x})^2 \left(\frac{\partial u}{\partial x} \right)^2 = \sigma^2 \left(\frac{\partial u}{\partial x} \right)^2$$

$$2(x_i - \bar{x})(y_i - \bar{y}) \left(\frac{\partial u}{\partial x} \frac{\partial u}{\partial y} \right)$$

The second term is called the **covariance.** If the measurements are independent and unrelated, variations in x and y may be randomly positive and negative and the terms in the covariance will sum to **zero.** The **error propagation formula** is then,

$$\sigma_u^2 = \sigma_x^2 \left(\frac{\partial u}{\partial x} \right)^2 + \sigma_y^2 \left(\frac{\partial u}{\partial y} \right)^2 + ...$$

Two examples of the use of this expression are given, while Table 1.1 lists its conclusions for specific functions.

Example 1. Variance of a sum or difference $u = \bar{x} \pm \bar{y}$

$$x = \bar{x} \pm \sigma_x \qquad y = \bar{y} \pm \sigma_y \qquad \frac{\partial u}{\partial x} = 1 \qquad \frac{\partial u}{\partial y} = \pm 1$$

$$\sigma_u^2 = \sigma_x^2 (1)^2 + \sigma_y^2 (\pm 1)^2 \qquad \sigma = \sqrt{\sigma_x^2 + \sigma_y^2}$$

Example 2. Variance of a mean $u = \bar{x} = \dfrac{1}{(N)} \Sigma \, (x_1 + x_2 + \dots + x_n)$

$$\frac{\partial u}{\partial x_1} = \frac{1}{N} \dots \text{etc}$$

$$\sigma_u^2 = \frac{1}{N^2} \, (\sigma_1^2 + \sigma_2^2 + \sigma_3^2 + \dots + \sigma_N^2)$$

$$\sigma_u^2 = \frac{1}{N^2} \, N\sigma^2 = \frac{\sigma^2}{N} \quad \text{if the standard deviation in each reading is the same.}$$

Table 1.1 The Standard Deviation σ_u for Important Functional Forms

u	σ_u or $\dfrac{\sigma_u}{u}$	Comment
$x \pm y$	$\sigma_u = \sqrt{\sigma_x^2 + \sigma_y^2}$	Standard deviations add in **quadrature**
$ax \pm by$	$\sigma_u = \sqrt{a^2 \sigma_x^2 + b^2 \sigma_y^2}$	
$\pm\, axy$	$\dfrac{\sigma_u}{u} = \sqrt{\dfrac{\sigma_x^2}{x^2} + \dfrac{\sigma_y^2}{y^2}}$	**Relative** standard deviations add in quadrature
$\pm\, a\dfrac{x}{y}$	$\dfrac{\sigma_u}{u} = \sqrt{\dfrac{\sigma_x^2}{x^2} + \dfrac{\sigma_y^2}{y^2}}$	**Relative** standard deviations add in quadrature
$x^{\pm m}$	$\sigma_u / u = \pm\, m\, \sigma_x$	
$e^{\pm ax}$	$\sigma_u / u = \pm\, a\, \sigma_x$	
$a^{\pm bx}$	$\sigma_u / u = \pm\, (b \ln a)\, \sigma_x$	
$\ln (\pm bx)$	$\dfrac{\sigma_u}{u} = \dfrac{b\, \sigma_x}{x}$	
$\bar{x} = \dfrac{1}{N} \Sigma x_i$	$\sigma(\bar{x}) = \dfrac{\sigma}{\sqrt{N}}$	If all standard deviations are the same at σ
$\bar{x} = \dfrac{\Sigma(x_i/\sigma_{i2})}{\Sigma\,(1/\sigma_i^2)}$	$\sigma(\bar{x}) = \dfrac{1}{\Sigma(1/\sigma_i^2)}$	If each reading x_i has standard deviation σ_i

The result that the standard deviation of a mean is reduced by the square root of the number of readings is the same as that reached by consideration of the Poisson distribution in Fig. 1.9. The result also justifies the experimental definition of the variance of the sample mean s_m^2 discussed on page 9.

Although not proven formally, Table 1.1 also shows the mean and standard deviation for data in which each point is weighted by an individual standard deviation. This enables averages to be calculated when the data points have different uncertainty.

Question (vii)   (a) An X-ray diffraction peak is known to occur at a diffraction angle of 31.6 ± 0.1°. When measured by a spectrometer the value of the diffraction angle recorded for the peak is 31.1 ± 0.2°. If the difference is added to all readings as a calibration factor, what is the value and uncertainty of this factor?

(b) What is the value of a resistance R Ω if the voltage drop measured across it is 1.53 ± 0.01 V when a current of 1.25 ± 0.02 A is passed?

1.11 ANALYSIS OF DATA

Automated data acquisition from instrumentation produces copious information which is generally analyzed using a spreadsheet such as Excel, QuattroPro or Lotus 123. The mathematical actions need to calculate means, and variances can be readily accomplished. More comprehensive statistical analyses to fit data to functional forms using regression or least squares fits are part of such a spreadsheet. The principle on which such fits are made is illustrated in Fig. 1.11.

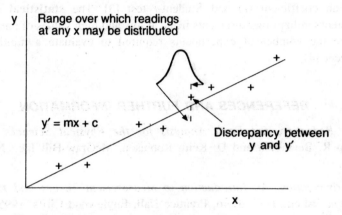

Fig. 1.11   The distribution of a set of data points about a proposed linear fit.

For each data point at x, repeated measurements will give values of y which are distributed about y with a standard deviation σ_y. It is desired to fit a set of data points (x, y) to a function $f(x, y)$. A linear function $y' = mx' + C$ may be chosen for simplicity of illustration. In making this fit, two factors are considered:

(a) the total discrepancy between the measured points and the fitted line should be a minimum.

(b) it is known that the individual data points have a statistical distribution about their 'average' values.

The degree to which any given data point y deviates from the proposed fit y' must therefore be weighted or compared with its statistical variation about y. At each x, this is accomplished by comparing the 'variance' of the discrepancy $(y - y')^2$ between the measured (y) and the fitted (y') value with the variance σ_y^2 describing the distribution of the experimental points about their mean value.

If σ_y is assumed to arise from a Poisson distribution, $\sigma_y^2 = y$. The 'best fit' for all points then occurs when the sum over all points:

$$\frac{\Sigma (y - y')^2}{\sigma_y^2} = \frac{\Sigma (y - y')^2}{y} \text{ is a minimum}$$

This procedure is the basis for a 'least squares' regression analysis to fit data to a function. Mathematical details of the fitting routine can be obtained from texts such as Bevington [1].

1.12 MULTI-PARAMETER EXPERIMENTS

Engineering and biological experiments often result in experiments which involve multiple or linked variables. A range of tests are then required to determine the significance of any given data set. This includes the analysis of variance (ANOVA), correlation coefficient (r) and Student-t test [3]. The **statistical design of experiments** utilizes measurements made at low and high values of parameters to minimize the number of experiments required to evaluate a multiparameter experiment [4].

REFERENCES AND FURTHER INFORMATION

1. *Data Reduction and Error Analysis for the Physical Sciences*, 2nd ed., Philip R. Bevington and D. Keith Robinson, McGraw-Hill Inc., New York, 1992.

2. *Experimentation: An Introduction to Measurement Theory and Experiment Design*, 3rd ed., D.C. Baird, Prentice Hall, Englewood Cliffs, 1995.

3. *Applied Statistics and Probability for Engineers*, D.C. Montgomery and G.C. Runger, John Wiley & Sons, Inc., New York, 1994.

4. *Statistics for Experimenters: An Introduction to Design, Data Analysis and Model Building*, G.E.P. Box, W.G. Hunter and J. Stuart Hunter, John Wiley & Sons, New York, 1978.

5. The evaluation of data using a spreadsheet such as QuattroPro or Excel is a major part of modern experimental science—particularly when data is acquired through a computer interface. Examine an available spreadsheet closely and assess the statistical functions available. In most cases, the spreadsheet assumes that the standard deviation on each data point is the same, although in some, allowance can be made for different standard deviations on any point, and for varying confidence levels. Often the most difficult task in using a spreadsheet is to decide on the number of significant figures to be used in the final quoted result. It is essential to make sure that the result makes **physical** sense.

ANSWERS TO QUESTIONS

(i) (a) 4, 3, 4, 1, 1, 3, (b) (i) 1.98 ± 0.01 m (ii) 1.98 ± 0.08 m

(ii) Since the calibration accuracy $\pm.00015$ V $+.0002$ V is small;
$V = 1.23 \pm 0.02$ V

(iii) 12.8 ± 0.5 W/m^2

(iv) Mean of 250/cm^3, probability of 350 unusual

(v) 2000 h

(vi) 50

(vii) (a) 0.5 ± 0.2, (b) 1.22 ± 0.02 Ω

DESIGN PROBLEMS

1.1 An experimenter uses an X-ray spectrometer which has a rotating drum scale and a pointer to measure angles. A view of this scale is shown.

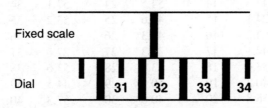

Fixed scale

Dial   31 32 33 34

(a) It is expected to measure an X-ray peak which occurs at an angle of around 31.6°. After viewing the scale and pointer, with what accuracy will you quote the result?

(b) A preliminary set of measurements are made in which little care is taken in making the measurements (e.g. not always approaching from the same direction, careless estimation of position etc). Five repeated measurements in degrees are as follows:

$$30.8, \ 31.3, \ 31.8, \ 31.7, \ 32.2$$

Is this acceptable?—if not, why not?

(c) The mechanical linkages in the spectrometer are adjusted to remove 'backlash', and care is taken to always approach the measurement in the same direction. Five repeated measurements in degrees are as follows:

$$31.2, \ 31.3, \ 31.7, \ 31.6, \ 31.7$$

Is this acceptable? If so what is the measured angle of the peak.

1.2 (a) Using the spectrometer in Problem 1 with a calibration peak which is known to occur at $31.6 \pm 0.1°$, a set of 10 repeated measurements give values in degrees:

$$30.8, \ 31.3, \ 31.2, \ 31.4, \ 31.3,$$
$$31.4, \ 30.9, \ 31.2, \ 31.1, \ 31.1$$

What is the systematic error in the scale setting of the spectrometer? The correction should be added to or subtracted from the actual scale reading.

(b) The spectrometer is used to determine the position of another peak which is measured at the following angles on repeated measurements in degrees

$$35.1, \ 35.3, \ 34.9, \ 35.5, \ 35.1,$$

$$35.6, \ 35.4, \ 35.5, \ 34.8, \ 35.1$$

What is the value of the corrected peak position?

1.3 The following set of 40 repeated data points is acquired during an experiment:

Point	Value	Point	Value	Point	Value	Point	Value
1	26.3	11	63.0	21	41.8	31	45.2
2	39.1	12	29.8	22	27.3	32	58.5
3	52.3	13	41.6	23	40.6	33	57.3
4	47.5	14	54.5	24	57.1	34	37.3
5	24.1	15	44.6	25	53.2	35	32.1
6	64.2	16	49.3	26	42.3	36	48.3
7	42.1	17	35.3	27	27.6	37	33.6
8	53.6	18	63.0	28	53.6	38	50.6
9	40.2	19	50.5	29	44.5	39	36.8
10	36.8	20	40.6	30	57.3	40	38.2

(a) Use a spreadsheet to calculate the experimental sample mean $\bar{x}$ and mean deviation $\langle x_i - \bar{x} \rangle$ for this distribution.

(b) Calculate the **sample** variance

$$s^2 = \frac{1}{N-1} \Sigma (x_i - \bar{x})^2$$

(c) Compare the value of s with the estimate of the standard deviation σ computed from

$$\sigma^2 = \frac{1}{N} \Sigma (x_i)^2 - \bar{x}^2$$

(d) Work out a frequency distribution for the data using bins of width 2 and plot the data on a bar graph.

1.4 Use the data given in Problem 1.3. If the results are binned with a bin spacing of 5 units from 20 to 70, compare the distribution of experimental points in the bin with a Gaussian distribution having parameters determined from the experimental data. Use a bar graph. Calculate the Gaussian at points at the value of the bin. Normalize the two fits by noting that the total number of points in both distributions must be the same (40). Multiply the Gaussian probability by a factor which results in the sum of the bin occupancy being close to 40.

1.5 The following observations (*V*) were made using a voltmeter:

15, 18.5, 14.5, 19.5, 18.4, 16.5, 17.3, 15.5, 13.3

For this sample set, calculate the (a) mean value, (b) standard deviation, (c) standard deviation of the mean value, (d) State the value of the quantity, (e) In a subsequent measurement, a value of 11.5 was observed. How likely is this reading to be correct, and not due to a mistake? Justify your answer quantitatively.

1.6 If you are in the apple business and sell bags of 20 apples for $2 with a guarantee that you will replace free of cost any bags that contain a bad apple, how much money will you make selling 2000 bags if your cost of a bag is $1.00. The probability of an apple being bad is 0.006. What is the average number of bad apples per bag?

1.7 A Neutrino Observatory expects to detect 2 solar neutrinos in a 24 hour period. What is the probability of detecting 8 neutrinos over this time period?

If you were concerned that such an observation was due to some spurious effect (e.g. ground based vibrations, or power fluctuations within the instrumentation, how might you go about assessing this probability? [The required answer is qualitative in nature].

1.8 The resistance $R = V/I$ of a wire of radius 0.5 ± 0.005 mm and length 2.00 ± 0.01 m is measured using the circuit shown in Fig. 1.3. The quoted accuracy for the digital meters used are ± 0.01% × scale reading. The measurements made during the experiment are as follows:

V (V)	0	0.8	2.1	2.5	3.2	3.4	4.1	4.8
I (mA)	0	4.8	9.8	12.0	14.7	18.2	19.9	25.0

Do you have to worry about the accuracy of a voltmeter or ammeter?

What is the value of the resistivity ρ (Ω-m) calculated from the expression $R = \rho L/A$?

1.9 The resistance of a semiconductor as a function of temperature varies as

$R = A \exp (E/kT)$

where E is an activation energy of 0.71 ± 0.05 eV,

k is Boltzmann's Constant (8.63×10^{-5} eV/K) and T is the absolute temperature in degrees K. If $A = 1.3 \pm 0.1$ Ω, what is the value of the resistance R when $T = 673 \pm 5$K? Neglect any error in the definition of k.

2

Instrumentation and System Design

2.1 EXPERIMENT DESIGN

Suppose we consider the factors which enter the measurement of a particular quantity, for example, to measure the spectral emission from a star, light intensity from an optical wave guide, or radio-tracer activity in a human patient.

The simplest experiment is:

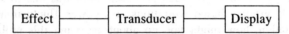

where the **display** or output normally is an electrical current or voltage generated by detection of the physical effect by a **transducer**.

Three steps are involved:

signal measurement
signal to noise enhancement
signal processing.

Signal Measurement

The transducer output—the **signal** (S), should be proportional to the quantity being measured, and of a magnitude which is large compared to that of spurious or random fluctuations which are introduced either in the signal itself, or through fluctuations associated with the transducer and the display device. These constitute a **noise** (N) in the system which may be represented as:

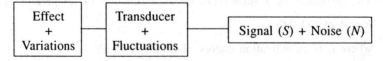

An example of noise inherent in a signal is the 'twinkling' of starlight as it passes through the earth's atmosphere, while resistance fluctuations generated in an insensitive, high resistance, or badly made photoresistor are a property of the transducer.

The **detectability** of a signal is defined by the **signal/noise ratio** (S/N). Calculation of this ratio is the first consideration for any measurement system. S and N may be defined in terms of power levels, voltages, currents, or any other parameters of a system.

Signal to Noise Enhancement

Simple amplification cannot improve the detectability of a signal since both signal and noise are amplified equally. Specialized techniques must be included to increase S with respect to N.

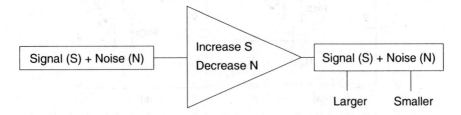

Fig. 2.1 The requirement for signal to noise enhancement.

This is accomplished by utilizing some characteristic which differs between the signal S and the noise N. This is often accomplished through differences in the frequency spectrum of the information in the two signals. This will be discussed in Chapter 4.

Signal Processing

Since physical relationships exist between measured quantities—e.g. power ~ (voltage)^2, it is often useful to convert the output from a voltage proportional to a signal to a voltage proportional to some **function** of the signal, e.g. the square, logarithm or inverse. This may be carried out by special amplifiers (hardware), or by computer based instrumentation (software). This can make complex graphs linear, or allow the direct calculation of experimental parameters during an experiment.

Electronic instrumentation is a significant part of signal processing. Processing can be analog (continuous in time) or digital (pulsed). An important aspect of instrumentation is the conversion of analog signals to digital data and vice versa.

2.2 TRANSDUCERS

A **transducer** is a passive circuit element (resistor, capacitor or inductor), or an active voltage or current source (transistor or diode) which, along with associated batteries, signal generators, load resistors or other circuit components, produces a voltage or a current of a magnitude related to the quantity being measured.

How the circuit is chosen can simplify or complicate the results, or produce an output in a form needed for calculation. This is particularly important if you wish to make a **direct readout** of the magnitude of the quantity being measured. For example, it is much more effective for a household thermometer to read temperature in °C on a digital display than to read the height of a glass thermometer column, or to measure a thermocouple voltage and then look up a calibration curve in tables. To be 'user-friendly' is the objective of much modern instrumentation. These considerations can be illustrated by two examples:

Example 1. Suppose the transducer is a resistor where $R(Q)$ varies as Q. Q may be light intensity for a photoresistor, or temperature for a heated metal. Possible circuits to measure $R(Q)$ are:

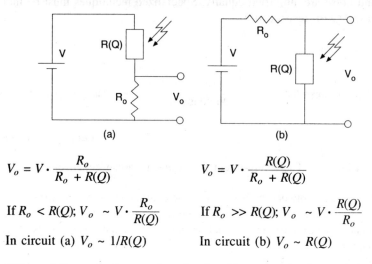

(a)   (b)

$$V_o = V \cdot \frac{R_o}{R_o + R(Q)} \qquad\qquad V_o = V \cdot \frac{R(Q)}{R_o + R(Q)}$$

If $R_o < R(Q)$; $V_o \sim V \cdot \dfrac{R_o}{R(Q)}$ If $R_o \gg R(Q)$; $V_o \sim V \cdot \dfrac{R(Q)}{R_o}$

In circuit (a) $V_o \sim 1/R(Q)$ In circuit (b) $V_o \sim R(Q)$

Either of these results may be of value. For example, for a photoresistor, the resistance $R(L)$ decreases when the light intensity (L) increases. In circuit (a), the output voltage (V_o) will increase when the light intensity (L) increases. For a metallic resistor, the resistance $R(T)$ increases when (T) rises. If circuit (b) is used, the output voltage also increases with (T).

Example 2. Suppose $R(Q)$ is a thermistor with a temperature dependent resistance of the form

$$R(T) = A \exp (E/kT)$$

where E is an activation energy in electron volts, k is Boltzmann's Constant $(1.38 \times 10^{-23}$ J/K), and T is temperature in K. A is a constant for the resistor. It is required to determine E from measurements of $R(T)$ versus T. Two ways of plotting the experimental data are shown below (Figs. 2.2(a) and (b)). These include (a) a simple plot of R vs T as a non-linear function and (b) a logarithmic plot which would allow analysis of the data as a linear graph:

$\ln (R(T)) = \ln A + E/kT$ as shown in Figs. 2.2(a) and (b) respectively.

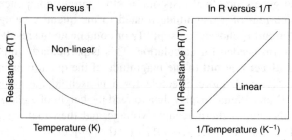

Fig. 2.2   (a) Non-linear graph, (b) Linear graph with a slope of (E/k).

Such graphs may be made either by data analysis **after** measurements of $R(T)$ and T or **during** data acquisition. Specialized amplifiers which can be used for this purpose are discussed in Sec. 5.2. They are connected as shown below:

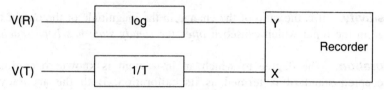

Fig. 2.3 Linearization of a complex relationship using specialized amplifiers.

Question (i) The resistance of a resistance thermometer changes from 100 Ω to 200 Ω when T changes from 0°C to 300°C. Calculate the corresponding voltage output from the test circuit.

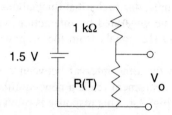

2.3 TRANSDUCER CHARACTERISTICS

The characteristics of a transducer are defined in terms of a number of parameters. Some or all of these quantities are defined in specification sheets.

Accuracy. It is the conformity of an indicated value to an accepted standard, or true value. It defines the limit the errors will not exceed when the instrument is used under stated operating conditions.

Resolution. The degree to which equal values of a quantity can be discriminated by an instrument is known as its resolution. For example, it corresponds to the last stable figure on a digital display.

Repeatability. The repeatability is the agreement among a number of consecutive measurements of the output for the same value of the input—under the same operating conditions and *when approached from the same direction.*

Reproducibility. It is known as the agreement among repeated measurements of the output for the same value of the input made over a period of time under the same operating conditions and *when approached from either direction.*

Hysteresis. The summation of all effects in which the output has different values when the same value of the input is applied first in an increasing, and then in a decreasing direction, is called hysteresis.

Linearity.   For successive equal increments of the input, the linearity is the deviation of the plotted transducer output from a straight line. This is often defined in terms of a percentage of the maximum or full scale output.

Sensitivity.   It is the ratio of the change in the magnitude of the output to the change in the input which caused it *after the steady state has been reached.*

Calibration.   The degree to which an instrument is known to conform to an accepted standard is termed as its calibration. Both the accuracy and reliability of an instrument depends on its construction and on how well it holds its calibration.

Noise.   Noise consists of signals generated within the transducer, independent of the input signal, which contribute to the output. Such signals may be intrinsic to the transducer (for example, due to thermal fluctuations of carrier concentrations in a semi-conductor), or be generated by interaction with the environment (for example, by pickup of 60 Hz signals from the ac power supply).

Response time.   It is the time interval between a change in the measured quantity and the time an instrument reads a new equilibrium value. The form of the response is shown in Fig. 2.4. This response is often defined in terms of three characteristic times:

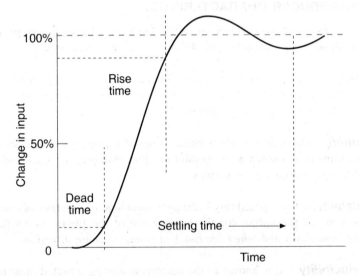

Fig. 2.4   The time response of an instrument after a change in the measured parameter.

Dead time.   A time during which a new signal or variation in a signal cannot be detected due to some physical characteristic of the system or the transducer, is known as dead time (τ_D).

Effects which lead to a dead time are the time taken for a variation in a process to reach the transducer (for example, for heat to penetrate an insulating blanket between a heater and a temperature sensor), or if the transducer is insensitive for a period after a measurement (for example, the dead time of a Geiger counter).

Rise time. The time taken by the instrument to respond to a step change in a measured quantity is known as the rise time. The change is often complex and τ_R is defined as the time taken to change from 10% to 90% of the final value.

Settling time. The time required for an instrument to attain a stable reading within a stated percentage of its equilibrium output is called settling time. This is τ_S, often taken to be the time to the first minimum.

2.4 SELECTION OF AN INSTRUMENTATION TRANSDUCER

The selection of an instrumentation transducer for a particular purpose or application is a complex process and is a matter of scientific and engineering judgement.

Transducer selection is based not only on a technical assessment of the measurement requirement and the system, but also on human factors such as the type of operating personnel and on the reliability and availability of suppliers of the equipment and spare parts. A checklist of the type used by a large industrial company in selecting transducers for its equipment for plant installation is shown in Table 2.1.

It is of interest to consider this list and to evaluate the relative importance of the items in several different applications, for example—a university laboratory, an industrial plant, or a telecommunications satellite. A poor choice of transducer and control circuitry can have disastrous consequences. In the Three Mile Island Nuclear Power Plant accident, operators made incorrect decisions based on faulty transducer information—partly due to poor layout of the instrumentation panel.

In the laboratory, equipment damage can occur if transducer failure leads to loss of power control. For example, if control circuitry is designed improperly, failure of a temperature sensor for a furnace can lead to full power being applied to the heating elements. Since this inevitably occurs in the middle of the night, the furnace and its contents overheat and are destroyed before the operator has opportunity to detect the transducer failure and correct the problem. The correct design should ensure that the furnace switches off when the transducer fails.

Table 2.1 Factors to be Considered in Selecting an Instrument Transducer

Input Variable	Transducer Input/Output	
Range (max. and min. values)	Accuracy	Hysteresis/backlash
Overload protection	Repeatability	Threshold/noise level
Frequency response	Linearity	Stability
– Transient response	Sensitivity	Zero drift
– Resonant frequency	Resolution	Loss of calibration
	Friction	

Measurement Reliability

Ease and speed of calibration and testing
Time available for calibration prior to and during use
Duration of operating mission
Stability against drift of zero point and proportionality constant
Vulnerability to sudden failure (mean time before failure)
Fail safety (will failure of the transducer mean system failure and will it invalidate
   data from other transducers)
Failure recognition (how will transducer failure be apparent)

Human and Environmental Factors

Operating environment (laboratory or industrial plant)
Operating personnel (Ph.D. or unskilled labourers)
Ease of access
Control panel layout
Access of operating personnel to calibration controls

Overall System

Output characteristics (voltage, current, other...)
Size and weight
Power requirements
Accessories needed
Mounting requirements
Environment of transducer location
Crosstalk between transducers
Effect of presence of transducer on measured quantity
Need for corrections dependent on other transducers

Purchase and Maintenance

Availability and delivery; is the item off-the-shelf?
Development necessary for operation
Availability of calibration and test data from manufacturer
Availability of spare parts and technical advice
Price
Previous experience with seller

2.5 THE TRANSDUCER AS AN ELECTRICAL ELEMENT

The purpose of a transducer is to carry out the conversion

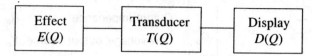

where $D(Q)$ is usually some form of electrical signal. Formally we can write:

$$D(Q) = T(Q) \cdot E(Q)$$

where $T(Q)$ is called the transfer function for the transducer. Q is some measured variable (frequency, temperature, light intensity, etc). The transducer is generally a resistance or an electrical source which can be regarded as either a voltage or a current generator. The first stage in designing any equipment involves the following three activities.

1. To **model** the transducer by a simple electrical circuit representing its circuit properties.

2. To **connect** the transducer to circuit elements representing attached components (cables, voltmeters, oscilloscopes etc.)

3. To **calculate** the output voltage using simple circuit theory.

Modelling a Transducer

A resistive transducer is a resistance with a value $R(Q)$ dependent on the quantity sensed. A voltage or current source can be regarded as a battery or source of emf $V(Q)$ with an appropriate **internal impedance** R_o.

In most cases the output of the transducer and the measurement circuit will have some dependence on frequency ω, where ω is the angular frequency $2\pi f$. If the measured quantity varies with a specific frequency, this frequency dependence is clear. However, even transducers designed to measure static phenomena—for example, the light level in a room, have to cope with switching the light on or off. This implies a frequency dependence.

A transducer may store **charge** and therefore act as a capacitor. It may have **internal** sources of fluctuating voltages independent of the signal and thereby act as a **noise** generator.

Under **small signal** or linear conditions, these components can be combined in some form of equivalent circuit. A major activity in instrumentation lies in devising suitable equivalent circuits for electrical, pneumatic, thermal and even vacuum systems.

In the electrical case, an equivalent circuit is defined for an element which generates a voltage $V(Q)$ as a result of some external stimulus (Q) as follows.

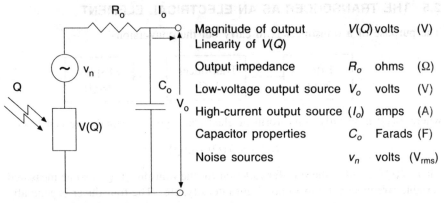

Magnitude of output	$V(Q)$ volts (V)
Linearity of $V(Q)$	
Output impedance	R_o ohms (Ω)
Low-voltage output source	V_o volts (V)
High-current output source	(I_o) amps (A)
Capacitor properties	C_o Farads (F)
Noise sources	v_n volts (V_{rms})

Fig. 2.5   The components of a transducer circuit model.

For a given transducer, not all these components will be present. However, **all** instrumentation design starts with a physical drawing and an equivalent circuit of the transducers.

2.6 MODELLING EXTERNAL CIRCUIT COMPONENTS

A typical experiment uses wires or coaxial cables to connect components, and connects such cables between plugs or BNC connectors. Transducers are driven at various frequencies by signal generators, and voltages are measured by amplifiers or digital voltmeters.

Instruments are expensive and complex, and it would appear difficult to describe them by a simple circuit element. However, a circuit only recognizes voltages and currents which appear at the terminals of components. This leads to immediate simplification.

Any instrument can be treated as a box which has **input** or **output** properties characteristic of the frequency range of interest. The internal complexities of the instrument can be lumped into a frequency dependent amplification factor $A(\omega)$ or voltage source $V(\omega)$. The instrument is reduced to the simplest circuits possible to describe the input and output properties respectively.

Setting up and using circuit models requires physical insight and common sense. The basic principles are simple:

Everything is an *R-C* circuit

Keep it simple for a given frequency range

Use elementary circuit calculations wherever possible. Use network theory only in the last resort

---

Circuit models for some important experimental elements will now be described.

Signal Generator

The output of most instruments can be regarded as a variable frequency voltage source. When current i is drawn from this source, the output voltage V_o at the terminals decreases. This can be modelled as a 'perfect' voltage source V having zero internal impedance connected in series with a resistance R_o equal to the **output impedance** of the generator.

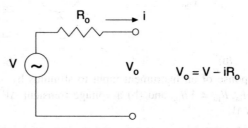

$$V_o = V - iR_o$$

Fig. 2.6 The equivalent circuit of a battery or other signal source.

The output resistance R_o for a signal generator is normally either 50 Ω or 600 Ω. Values of R_o lower than 50 Ω are possible, but require specialized amplifiers to achieve. Such amplifiers include the transistor emitter follower, or some types of operational amplifier (see Sec. 5.1).

Question (ii)   When a 500 Ω resistance is connected across the terminals of a signal generator, the output changes from 3 V to 2.75 V. What is the output impedance of the generator?

Amplifier or Voltmeter

A 'perfect' amplifier increases the magnitude of a signal, takes no current, and has the same amplification at all frequencies. The input of most instruments involves some form of amplifier.

Real amplifiers do take current, and have a frequency response partly defined by the input circuit, and partly determined by their internal electronic design. The amplifier is modelled as:

Input circuit	+	perfect amplifer with a frequency response set by its design

The equivalent circuit for the input can be devised by considering what happens if (i) a dc voltage source, and (ii) a rapid voltage transient is applied to the input terminals. The dc voltage probes the **low frequency** response; the transient examines the **high frequency** behaviour.

When combined, the input can be regarded as that of a 'perfect' amplifier which has an internal frequency response $A(f)$ set by the manufacturer's specifications. The input impedance is infinite so that no current flows into its terminals. An *R-C* network connected from the input terminal to ground then represents the input properties.

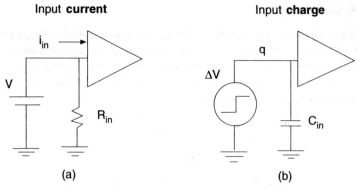

Fig. 2.7   The response of an instrument input to stimulus by: (a) a dc battery V, current i_{in}, $R_{in} = V/i_{in}$, and (b) a voltage transient ΔV, charge $q = \int i \, dt$, $C_{in} = q/\Delta V$.

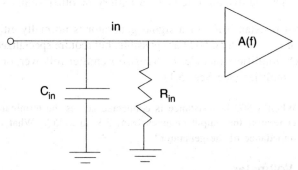

Fig. 2.8   The input circuit to a 'perfect' amplifier.

The physical origin of R_{in} is some combination of the actual input current to the first stage of the amplifier, and leakage current from the input terminal to ground. C_{in} arises mostly from the static capacitance associated with the input connector. For a well designed instrument, both of these quantities will vary little under different operating conditions. In most instruments, values for R_{in} and C_{in} are marked beside the input connector. For many oscilloscopes and general purpose instruments, a common value for R_{in} is 1 MΩ. Digital voltmeters have an input resistance of 10–100 MΩ. The input capacitance C_{in} is generally in the range of 30 pF.

Coaxial Cable

It is common practice to interconnect instruments by means of coaxial cable. This consists of a central copper wire, a dielectric sheath, and a braided metal external conductor which is normally connected to ground. The cable is protected by an outer plastic sheath, and connections to the conductors are made using carefully constructed connectors soldered and crimped to the inner and outer wires. Various systems of connectors for easy connection between instruments are in use. The BNC connector is very common and a typical assembly is illustrated below:

Fig. 2.9   Coaxial cable and BNC connectors.

The principle of the coaxial connector is that of the Faraday shield. The signal is propagated on a central copper wire which is enclosed by an outer grounded conductor. There is no electrical field due to outside sources within the grounded sheath, and interference or electrical 'pick-up' to the central wire is minimised. It is good practice to build all instrumentation systems within grounded metal boxes interconnected by well made coaxial cables.

Coaxial cables require care in use. Their effects should be carefully considered, otherwise they can be a source of electrical noise in their own right. A circuit model must take into account the following points:

• the resistance and inductance of the central wire.

• the capacitance of the dielectric sheath.

• the characteristic impedance Z_o.

The resistance and capacitance is **distributed** along the length of the cable, which therefore must be treated at high frequencies (GHz) as a **transmission line** of characteristic impedance Z_o. A circuit model of a typical cable is shown in Fig. 2.10, where C_c is 88 pF/m, L_c is 0.26 μH/m and R_c is 10 Ω/km. The

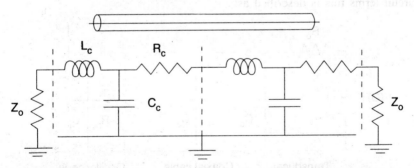

Fig. 2.10   Physical form and circuit model of a coaxial cable.

significance of the inductance depends on the frequency and on the impedance of the circuits to which the cable is connected. At 1 MHz, the inductive impedance $(Z = \omega L)$ and resistance of the central core per unit length is 1.6 ohm/m and 0.01 ohm/m respectively. This is negligible compared to the input and output impedances of most transducers and amplifiers at this frequency. However, the capacitance to ground of 88 pF/m is **always** present and can never be neglected.

Coaxial cables are a form of **transmission line**. On such a line, electro-magnetic waves propagate at a speed of

$$c_t = \frac{1}{\sqrt{L_c C_c}} \text{ m/s}$$

and the line has a characteristic impedance Z_o

$$Z_o = \sqrt{\frac{L_c}{C_c}} \ \Omega$$

When the line is terminated at each end by a resistance equal to the characteristic impedance, reflections of the signal from the ends of the line are minimized. Commonly available coaxial cable has a characteristic impedance of 54 ohms compatible with signal generators having a 50 Ω output impedance. The output voltage from the cable is often measured across a 50 Ω termination. Signals propagate within the cable at a speed of around 70% of the speed of light in air.

Problems with coaxial cables can be encountered in the measurement of very low signal voltages. Piezoelectric or triboelectric voltages generated by mechanical vibrations between the conductors and the dielectric can be a major source of unwanted noise. To reduce this problem special coaxial cables lubricated with carbon can be used, or alternatively pairs of wires are used as twisted pairs within a shielded box.

2.7   CIRCUIT CALCULATIONS

The voltage available for amplification by a measurement circuit results from the interconnection of a transducer, a coaxial cable, and a measuring system. In circuit terms this is described as:

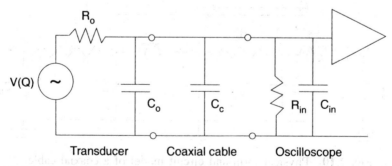

Fig. 2.11   A circuit diagram for a transducer-coaxial cable-instrument system.

The dc voltage available for amplification by the measurement circuit is the result of the potential divider set up by R_o and R_{in}. At high and intermediate frequencies, the frequency response of the circuit is also determined by the various capacitances associated with the transducers, connecting cables and the amplifier input. These must be incorporated into an equivalent circuit.

In order to evaluate the circuit response, it is of value to compute the output voltage using simple circuit theory. Network calculations are usually not very effective and should be avoided. A good practice is to draw a sequence of simple circuits which are resolved using simple circuit theorems. Such diagrams should be labelled with the **actual** value of the components used in the circuit.

Useful Circuit Theorems

Resistors and capacitors in series and parallel:

	Parallel	**Series**
Resistors	$1/R_t = 1/R_1 + 1/R_2$	$R_t = R_1 + R_2$
Capacitors	$C_t = C_1 + C_2$	$1/C_t = 1/C_1 + 1/C_2$

Potential divider

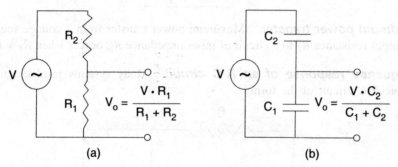

$$V_o = \frac{V \cdot R_1}{R_1 + R_2}$$

$$V_o = \frac{V \cdot C_2}{C_1 + C_2}$$

(a)   (b)

Fig. 2.12   (a) A resistive and (b) a capacitive potential divider.

The potential distribution along a series chain

$$\text{Voltage across element } x = \frac{\text{Impedance of element } x}{\text{Total impedance of chain}} \cdot \text{voltage across chain}$$

Thevenin equivalent

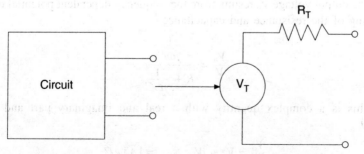

Fig. 2.13   Thevenin equivalent of a circuit treated as a 'black box'.

Any two terminal network can be replaced by a single source V_T and internal impedance R_T,

> V_T is the **open circuit** voltage measured looking back into the terminals
>
> R_T is the **impedance** measured between the terminals assuming that all voltage sources are set to zero.

Question (iii)   Determine the Thevenin equivalent circuit for the potential divider network shown. The output voltage is measured across the 1 kΩ resistor.

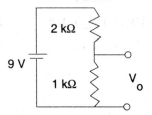

Maximum power transfer.   Maximum power transfer from a voltage source of output resistance R_T to a circuit of input impedance R_{in} occurs when $R_T = R_{in}$.

Frequency response of an R-C circuit.   Many systems resolve into a simple *R-C* circuit of the form:

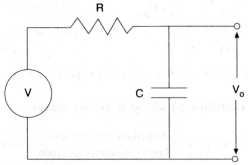

Fig. 2.14   An *R-C* potential divider.

The output voltage V_o results from the frequency dependent potential divider consisting of the resistance and capacitance.

$$\frac{V_o}{V} = \frac{\dfrac{1}{j\omega C}}{R + \dfrac{1}{j\omega C}}$$

This is a complex quantity with a real and imaginary part and phase angle θ

$$\frac{V_o}{V} = V_o + jV_{\text{imaginary}} = |A|\,e^{j\theta}$$

Using complex algebra to evaluate the magnitude $|A|$ and phase θ of V,

$$\text{Magnitude } A = \frac{V_o}{V} = \frac{1}{\sqrt{1 + \omega^2 C^2 R^2}}$$

$$\text{Phase } \theta = \tan^{-1}(\omega CR)$$

The frequency dependence of this circuit is very important in many areas of instrumentation. It can be seen most clearly if the graph is plotted on logarithmic axes ($\log_{10}(V_o/V)$ vs $\log_{10} f$). Such a graph is called a Bode plot.

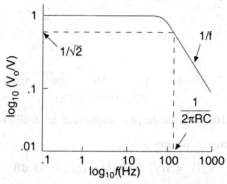

Fig. 2.15 Bode plot for an *R-C* circuit.

At **low frequency** the magnitude is independent of frequency.
At **high frequency** the magnitude varies or 'rolls off' as 1/frequency.
The **corner frequency** f_c, is defined by the frequency at which:

$$\omega CR = 1; \text{ i.e. } f = f_c = \frac{1}{2\pi RC}$$

$$\text{At } f_c \left| \begin{array}{l} \textbf{Magnitude } V_o/V = 1/\sqrt{2} \\[2mm] \textbf{Phase } \Theta = \tan^{-1} 1 = 45° \end{array} \right.$$

The frequency f_c is a convenient method of defining the frequency response of the circuit. The frequency interval from $0–f_c$ is termed the **frequency bandwidth**. Above f_c, $\omega RC \gg 1$, and the response varies as $V_o/V = 1/(\omega RC)$. This is called the '**roll-off**' of the circuit.

Decibel units. It is a common practice in instrumentation to use a logarithmic unit called the decibel (dB) to describe power (W), voltage (V), and current ratios. The unit is based on a relationship concerned with the measurement of power (Sec. 2.9)

$$\frac{\text{Power (1)}}{\text{Power (2)}} (dB) = 10 \log_{10} \frac{P_1}{P_2}$$

When this is applied to voltage or current ratios, since $P \propto V^2$, if the above frequency response curve is described in dB,

$$\frac{\text{Voltage (1)}}{\text{Voltage (2)}} \text{ (dB)} = 20 \log_{10} \frac{V_1}{V_2}$$

The frequency response of the Bode plot described in dB has characteristic features:

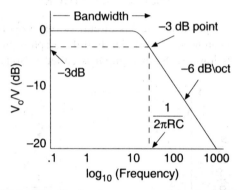

Fig. 2.16   The Bode plot expressed in decibels (dB).

The **cut-off frequency** f_c occurs at

$$V_o/V = 1/\sqrt{2} = 20 \log 1/\sqrt{2} = -3 \text{ dB}$$

At high frequencies, the 'roll off per octave', or the change in output if the frequency is doubled is 1/2

$$V_f/V_{2f} = 20 \log 1/2 = -6 \text{ dB for } f \rightarrow 2f$$

This is written as **–6 dB/octave**. For a decade change in frequency,

$$V_f/V_{10f} = 20 \log 1/10 = -20 \text{ dB/decade}$$

The –3 dB frequency and the 'roll-off' (normally –6 dB/octave) are commonly quoted in instrument specifications.

Question (iv)   Calculate the corner frequency f_c for an instrument having an input resistance of 1 MΩ and an input capacitance of 30 pF (1 pF = 1×10^{-12} F).

An Example of Circuit Calculations

Signal generator connected to an oscilloscope

System definition:

— Signal generator with 1 V output
Frequency range 0–1 MHz
Output impedance 600 Ω

— Oscilloscope with 0–100 MHz bandwidth
Input impedance 1 MΩ ‖ 30 pF

— 1 m of 54 Ω coaxial cable having C_c = 88 pF/m

Figure 2.17 (a)–(e) shows the steps in the calculation. It is highly recommended to draw the set of diagrams showing the progression of simplification. Each calculation then becomes simple and less subject to error.

(a) The physical diagram:

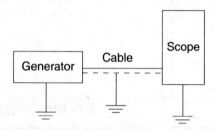

The terminals which are connected to ground are noted for **each** instrument.

(b) The equivalent circuit:

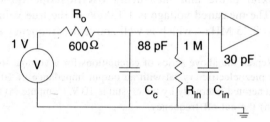

All components are drawn including the 1 m length of coaxial cable used to connect the two instruments.

(c) The circuit is rearranged to simplify the calculation:

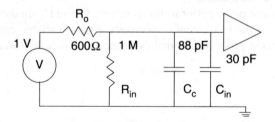

(d) The resistor network is simplified using Thevenin's theorem and the series/ parallel relationships:

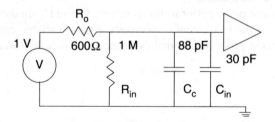

(e) The output voltage is computed:

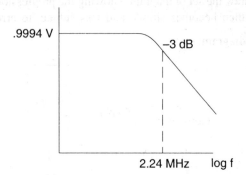

Fig. 2.17   (a)–(e) stages in the computation of the output voltage of a measurement circuit.

The coaxial cable and measuring oscilloscope do not perturb the measurement. The measured voltage is 1/1.0006 of the true value, while the cut-off frequency is 2.25 MHz—which is well above the frequency range of interest.

Question (v)   Repeat the above series of calculations for a voltage source consisting of a piezoelectric crystal with an output impedance of 10 MΩ. The voltage generated internally by the crystal is 10 V. Compute (a) the voltage output, (b) the cut-off frequency.

2.8 INSTRUMENT PROBES

It is often desirable to increase the input impedance and to reduce the input capacitance of an instrument.

An instrument probe performs this function. R_1 and C_1 are the input resistance and capacitance of the instrument respectively. R_2 and variable capacitance C_2 are built into a flexible probe which makes connection with the circuit to be tested.

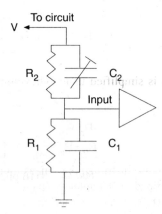

Fig. 2.18   Circuit arrangement for an *R-C* instrument probe for high frequency measurements.

The probe and input circuits form a resistive potential divider at low frequencies, and act as a capacitative potential divider at high frequencies. The circuit has an output which is independent of frequency if the low and high frequency responses are identical.

$$V_o = \frac{R_1}{R_1 + R_2} = \frac{C_2}{C_1 + C_2}$$

Rearranging, this condition implies that

$$R_1 C_1 = R_2 C_2$$

If $R_1 = 1$ MΩ and $C_1 = 30$ pF, $R_2 = 10$ MΩ, then $C_1 = 3$ pF. The resistance of the instrument seen by the circuit is $(R_1 + R_2) = 11$ MΩ, while the capacitance is the series combination of (C_1 and C_2) which is 2.72 pF. The voltage measured by the instrument is reduced by a factor of $R_1/(R_1 + R_2) = 1/11$.

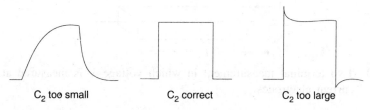

C_2 too small C_2 correct C_2 too large

Fig. 2.19   Stages in the adjustment of an instrument probe response to an input square wave.

It is not difficult to increase the amplification to make up the (× 1/11) attenuation of the probe. The large (and possibly variable) capacitative loading of a coaxial cable is reduced. High frequency capacitance measurements invariably use a probe. The value of C_2 is adjusted by examining a test square wave voltage. C_2 is adjusted until no distortion of the waveform is observed (Fig. 2.19).

2.9 POWER MEASUREMENTS

The electrical quantities normally measured in circuits are voltage V (V) and current I (A). However, power P (W) can be regarded as a more fundamental property

$$P = V I = I^2 R = V^2/R \text{ (W)}$$

Electrical power is dissipated as heat or is converted to another form of energy (mechanical motion, acoustic pressure, etc). Power measurements often have the advantage that they are independent of frequency (see page 71).

Average power can be measured by the temperature rise induced in a resistive or power dissipating element. Instruments to measure radiative and optical power by this means are described in Chapter 3 (page 71). Instantaneous electrical power is determined from simultaneous measurements of current and voltage by some form of watt meter. The measurement of ac power at frequency f in reactive circuits has to take into account phase shift between the current and voltage.

2.10 MEASUREMENT METHODS

In many cases a transducer can be treated as a simple resistance. Experimental methods and circuit techniques for resistance measurement are important in instrumentation and transducer design. They are often of interest in determining the electrical properties of materials.

Two Terminal Measurements

In Fig. 2.20, R (ohms) = V/I, $R = \rho\,L/A$ where ρ is the **resistivity** in (ohm-m). Errors may be introduced due to electrode resistances.

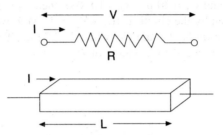

Fig. 2.20 Two terminal measurement in which voltage V is measured at the current electrodes.

Four Terminal Measurements

Contact and lead resistances can introduce significant errors. In a four-terminal measurement (Fig. 2.21), current is introduced through one set of leads, while the resultant voltage is measured across a second set using a *high impedance* voltmeter. Since negligible current flows through the resistive contact, electrode errors are minimized. The resistance measured is that between the voltage probes.

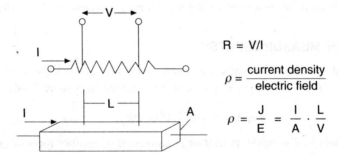

$$R = V/I$$

$$\rho = \frac{\text{current density}}{\text{electric field}}$$

$$\rho = \frac{J}{E} = \frac{I}{A} \cdot \frac{L}{V}$$

Fig. 2.21 Four terminal measurements using one set of current probes, and a second set of voltage probes. The voltage is measured by a high resistance voltmeter so that minimum current is drawn.

Sheet Resistivity Measurements using 4-Probe van der Pauw Methods

The sheet resistivity ρ_s Ω/square of thin sheets or coatings of low resistance material on insulating substrates may be measured by a four-probe measurement.

If the coating thickness is small (t) such that the electric field across the thickness is uniform, the sheet resistivity is related to the bulk resistivity by $\rho_b = \rho_s \cdot t$ Ω-m. The calculations are simplest if the sheet is infinite in extent, but a correction factor can be applied for smaller samples. Two types of geometry are shown in Figs. 2.22(a) and (b):

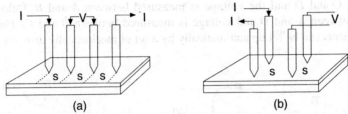

(a)   (b)

Fig. 2.22   Four terminal probes in (a) a linear and (b) a square configuration of spacing s. The sample size is a square of side d.

Current is passed through the outer probes or one set of side probes. Voltage is measured using the second set. The sheet resistivity is calculated by the following equations. For the square probe, the conductivity is corrected to take account of the size of the sample.

Linear probe (Ref. 2.5).   Sample dimensions >> probe spacing

$$\text{Sheet resistivity } \rho_s = \frac{V \pi}{I \ln 2} \text{ Ω/square}$$

Square probe (Ref. 2.6).   Probe spacing s, sample dimensions d, four probes are equally spaced on a square sample

$$\text{Sheet resistivity } \rho_s = \frac{V}{I} \cdot G(d/s) \text{ Ω/square}$$

where $G(d/s)$ is a correction factor calculated from the dimensions of the sample d in units of probe spacing s.

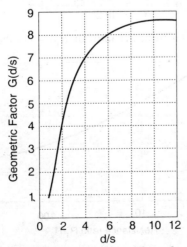

Fig. 2.23   Correction factor for sample size d for a four point probe of spacing s.

van der Pauw Measurements on Thick Samples of Arbitrary Shape

A very useful technique (L.J. van der Pauw, Philips Technical Review, 20, 220 (1958/59)) applies to a slab of material of arbitrary shape of thickness t. Four terminals (A, B, C, D) are prepared at arbitrary positions around the perimeter. Two successive measurements are made in which current I is passed through terminals C and D and the voltage is measured between A and B, followed by current between A and D and voltage is measured between B and C. These two measurements can be set up automatically by a set of mechanically driven switches.

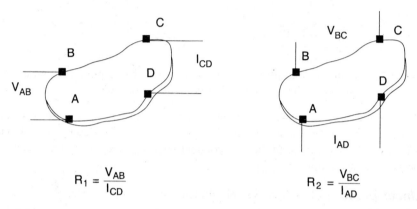

$$R_1 = \frac{V_{AB}}{I_{CD}} \qquad\qquad R_2 = \frac{V_{BC}}{I_{AD}}$$

Fig. 2.24 Successive electrode configurations for van der Pauw geometry measurements.

The resistivity $\rho = 1/\sigma$ is calculated from:

$$\exp(-\sigma\pi t R_1) + \exp(-\sigma\pi t R_2) = 1$$

This equation can be solved numerically, or an expression can be used:

$$\rho = \frac{1}{\sigma} = \frac{\pi t}{\ln 2} \frac{R_1 + R_2}{2} f$$

where f is a factor deduced from the ratio of R_1/R_2 via the graph:

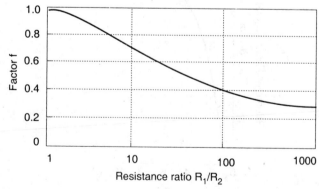

Fig. 2.25 Correction factor for van der Pauw geometry [Smits, 1958].

Question (vi) A square probe of spacing 2.5 mm is used to measure the sheet resistivity of a thin nickel film prepared on the surface of square glass plates 1cm on each side. The voltage measured for an input current of 10 mA is 15 mV. The film thickness is 700 nm. Calculate the bulk resistivity of the deposited nickel.

2.11 DC AND AC BRIDGE MEASUREMENTS

In physical measurements, particularly in electrical control circuits, it is of importance to measure a **change** in the resistance of a component about a preset value, or to compare one voltage level with another. A bridge measurement is a differential measurement which compares the output of two potential dividers.

$$V_1 = V \cdot \frac{R_1}{R_1 + R_2}$$

$$V_2 = V \cdot \frac{R_3}{R_3 + R_4}$$

I_g is the current through the detector.

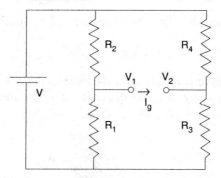

Fig. 2.26 A bridge circuit drawn as two resistive potential dividers.

As shown in Fig. 2.26, if $V_1 = V_2$ so that $I_g = 0$, the condition for a balanced Wheatstone Network is obtained.

$$\frac{R_1}{R_2} = \frac{R_3}{R_4}$$

In order to calculate the output voltage from a bridge, if the current drawn is small, it is convenient to derive the Thevenin equivalents for the two potential

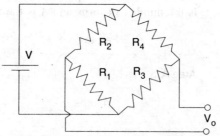

Fig. 2.27 A Wheatstone Bridge.

dividers. Knowing the equivalent voltage sources and internal impedances, the current which flows through the central arm can be readily calculated.

Question (vii)   If the battery voltage is V, what voltage (i.e. $V_1 - V_2$) will be measured (a) when all the resistors are equal at 100 Ω, (b) when $R_3 = 200$ Ω and the rest are 100 Ω. Compare the result to that derived in Question 2(i).

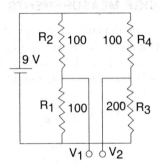

Impedance Measurements and the AC Bridge

At angular frequency $\omega = 2\pi f$, the impedance of a capacitor is $Z_C = 1/\omega C$.

$$C\,(F) = \frac{\varepsilon_r \varepsilon_o A}{d}$$

where ε_r is the relative permittivity of the material, $\varepsilon_o = 8.85 \times 10^{-12}$ F/m is the permittivity of free space, d is the thickness, A is the area.

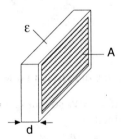

Fig. 2.28   Capacitor structure.

At angular frequency $\omega = 2\pi f$, the impedance of an inductance is $Z_L = \omega L$. For a coil wound on a core of relative permeability μ, $L = \mu \mu_o n A$, where $\mu_o = 4\pi \times 10^{-7}$ H/m, n is the number of turns and A is the area of cross-section.

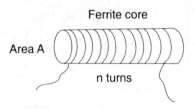

Fig. 2.29   Inductor structure.

Because energy is stored in the electric and magnetic fields associated with a capacitor and inductor respectively, no power is theoretically dissipated by these components. In practice, dielectric loss in capacitor materials and winding resistance in coils (page 115) does lead to power dissipation and a resistive component of the total impedance. The equivalent circuit includes a resistor in either series or parallel.

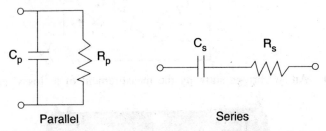

Parallel   Series

Fig. 2.30   The parallel and series circuit models for a dielectric.

The quality factor Q or dielectric loss $D = \tan \delta$ describes power loss in the component.

$$Q = \frac{1}{\tan \delta} = \frac{\omega L_s}{R_s}$$

$$Q = \frac{1}{\tan \delta} = \omega C_p R_p = \frac{1}{\omega C_s R_s}$$

The parallel and the series parameters are both equally valid descriptions of the properties of capacitors. If one parameter is measured, the other can be calculated. The relationships between circuit elements are as follows:

$$D = \frac{1}{Q} = \omega R_s C_s = \frac{1}{\omega R_p C_p}$$

$$R_s = \frac{D^2}{(1 + D^2)} R_p$$

$$C_s = (1 + D^2) C_P$$

Capacitors and inductors are generally measured using AC bridges (Fig. 2.31) in which one arm of a resistive Wheatstone Network is replaced by the appropriate RC network. The bridge is operated under AC excitation (normally 1000 Hz).

The 'balance' occurs when no signal is detected across the centre of the network. This requires adjustment of controls for both capacitance, and resistance. The bridge is often calibrated in terms of capacitance C and dissipation $D = \tan \delta = 1/\omega CR$. C is measured directly, R is computed from D and C measured at a known frequency.

The bridge shown in Figure 2.32 is manually balanced. More modern equipment uses the same principle but an automatic balance is achieved using electronic circuitry.

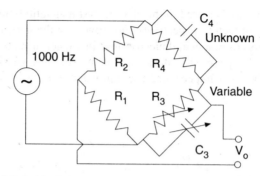

Fig. 2.31 An AC bridge showing the measurement of a 'lossy' capacitor.

Fig. 2.32 A manually balanced General Radio 1650 A impedance bridge operating at 1000 Hz. The right hand control adjusts the capacitance or inductance (C, L) the left hand control adjusts the resistive component (D). (General Radio Inc. Model 1650 A Instrument Manual)

Question (viii)   A capacitor is measured using a bridge operating at 1000 Hz to have a parallel capacitance C_p = 9.5 nF and dissipation D = tan δ = 0.03. Calculate the component values for the parallel and series equivalent circuits.

REFERENCES AND FURTHER INFORMATION

1. *Keithley Low Level Measurements Handbook*, 4th ed., Keithley Inc., 28775 Aurora Rd., Cleveland 44139, U.S.A..

2. *Introduction to Transducers for Instrumentation*, Honeywell Inc., Test Instruments Div., P.O. Box 5227, 4800 E. Dry Creek Road, Denver, Colo.

3. *Burr Brown Integrated Circuits Data Book*, Burr Brown Corporation, Tucson, Arizona.

4. *Amplifier Applications Guide* (1992), Analog Devices Inc., 1 Technology Way, P.O. Box 9106, Norwood, Massachusetts 02062-9106 (USA).

5. F.M. Smits, "Measurement of Sheet Resistivities with the Four-Point Probe", *Bell System Technical Journal,* **20,** 711 (1958).

6. F. Keywell and G. Dorosheskli, "Measurement of the Sheet Resistivity of a Square Wafer with a Square Four-Point Probe", *Review of Scientific Instruments,* **31,** 833 (1960).

ANSWERS TO QUESTIONS

(i) 0.14 V to 0.25 V (non-linear)

(ii) 45 Ω

(iii) $V_T = 3$ V, $R_T = 0.67$ kΩ

(iv) 5.3 kHz

(v) 0.91 V, 1.48 kHz

(vi) 7.35×10^{-8} Ω-m

(vii) 0, -1.5 V

(viii) $C_p = 9.5$ nF, $R_p = 560$ kΩ, $C_s = 9.5$ nF, $R_s = 500$ Ω

DESIGN PROBLEMS

2.1 Design a circuit to control the temperature of a water bath at a setpoint temperature of 75°C. Choose a resistive sensor and draw up a bridge circuit which will provide control around the setpoint. Suggest values for all components to be used. Heat can be supplied to a heater in the bath using a power amplifier which requires 0 to +0.5 V for zero to maximum power output.

2.2 Design a bridge or other circuit which will provide an output signal to control the intensity of light in a room at a pre-set (but adjustable) level. Assume that the power controller itself is available and that it produces power over a range from zero to maximum for an input voltage from 0 to V.

Describe the function and physical basis of any transducer you employ and suggest values for all circuit components.

2.3 A silicon bar is used as a magneto-resistor, i.e. its zero field resistance of 10 kΩ increases with an applied magnetic field B Tesla such that

$$R_B = 10(1+10 \text{ B}) \text{ k}\Omega$$

(a) Design a circuit which will give an output voltage proportional to B for small values of B ($B < 0.1T$). Choose appropriate batteries and resistors. What will be the voltage sensitivity of this circuit in V/T?

(b) It is proposed to connect this circuit to an oscilloscope having an input impedance of 1 MΩ in parallel with 55 pF. The cable length is 2 m and the cable capacitance is 85 pF/m. What is the highest frequency sinusoidal signal you might expect to measure without significant attenuation?

2.4 Design a frequency independent, high voltage probe to allow an oscilloscope with an input impedance of 1 MΩ, input capacitance of 30 pF, and full scale sensitivity of 100 V to read 1000 V full scale. Give component values, power and voltage ratings for all components used in the design.

2.5 In order to avoid reflections from the end of the line, coaxial cables should always be terminated at both ends by a resistance equal to the characteristic impedance Z_o. A circuit which ensures this for three 53 Ω cables serving different stations is shown. What is the value of R?

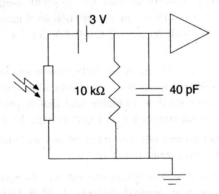

2.6 A photoconductive material is used in the circuit shown. The output voltage developed across a 10 kΩ resistance is measured by an oscilloscope with an input resistance of 10 MΩ. The capacitance associated with the connection cables and amplifier input is 40 pF.

The detector has a dark resistance of 100 kΩ and an illuminated resistance of 50 kΩ.

(a) What is the output voltage at low frequencies in the dark and light?

(b) What is the cut-off frequency of the circuit when the photoconductor is illuminated?

2.7 When light of a particular intensity illuminates a photoresistor, the illuminated resistance $R_L = 1500$ Ω. The sensitivity $(\Delta R/\Delta I)$ of this illuminated resistance R_L to variations in the light intensity I (mW/cm^2) is -200 Ω $-$ cm^2/mW, i.e. the resistance falls with increasing light intensity.

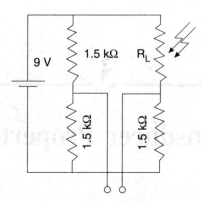

The photoresistor is included in a bridge circuit powered by a 9 V battery to make a sensitive measurement of ΔR through the voltage difference $(V_2 - V_1)$, and hence to detect small variations in the light intensity ΔI.

(a) What will be the output voltage for a change in light intensity of 0.1 mW/cm^2?

(b) If this voltage is measured by connection to an oscilloscope with an input impedance of 1 MΩ in paralled with 30 pF using a 0.5 m length of coaxial cable having a cable capacitance of 85 pF/m, determine whether the circuit can respond to light variations of frequency > 500 kHz without attenuation.

3

Transducer Properties

Many types of transducers exist. In this chapter representative types will be examined to demonstrate how the transducer parameters noted in Chapter 2 arise in practice. The areas which will be covered are: temperature, light and linear motion.

3.1 TEMPERATURE MEASUREMENTS

The range of temperature encountered in a laboratory or in industrial applications extends from millikelvin to over 3000 K.

This range of temperature covers physical phenomena in the realm of quantum physics—liquid helium and superconductivity, to classical radiation phenomena at the highest temperature at which materials exist (about 3300 K). Above that we enter the realm of plasma physics and optical spectroscopy. The production and measurement of temperature is summarized in Table 3.1.

Table 3.1 Production and Measurement of Temperature

0	4.2	77	194	273	600	1100	200	3000
Magnetic effects	Liquid He	Liquid nitrogen	Dry ice (CO_2)	Ice	Flames resistive	heating	Vacuum furnace	
^{3}He–^{4}He	Liquid gas	cooling			Nichrome	SiC	Tungsten	

Magnetic
  Resistance thermometer
    Au(Ge)_____platinum _____
      Thermocouples
      copper-constantan_____
      chromel-alumel_____
        platinum-13% rhodium-platinum_____
            rhenium-tungsten_____
        Semiconductor sensors
        Thermistors _____
         Optical sensing
          infrared emission_____
           optical pyrometers_____
           spectroscopy_____

Experimental methods to produce this temperature range include resistive heating and the use of liquid gases. Magnetic effects are required in the millikelvin range.

High temperature	**Low temperature**

Attainment: resistive heating

Liquid gases — liquid $N_2 \rightarrow$ 77K
— liquid $He_2 \rightarrow$ 4.2K
Below 4K — pumped He ($\rightarrow$ 1K)
^{3}He-^{4}He refrigerator
magnetic effects

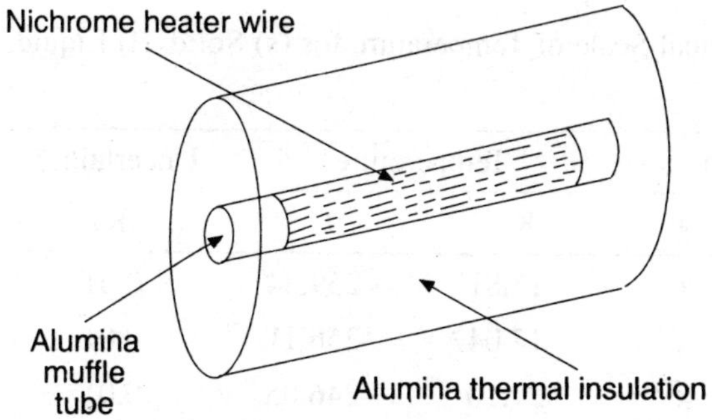

Nichrome heater wire

Alumina muffle tube

Alumina thermal insulation

Fig. 3.1 A tube furnace for measurements to 1100°C.

Accessible areas of science are those which can be attained by simple and economic means. Such means are often determined by technical factors which are unrelated to the experiment being carried out.

For example, for high temperatures, measurements are relatively easy up to 1100°C because resistive heating wires such as nichrome or chromel are stable in air to this temperature. Silicon carbide heating elements (more expensive to buy and operate) can be used to about 2000°C, but complex electric arcs and specialized techniques are required above this temperature. Relatively little science is done in this temperature range.

Below room temperature, liquid nitrogen with a boiling point of 77K and liquid helium at 4.2K provide easy access to these temperature ranges. Pumping on He can reduce the temperature to 1K. The millikelvin range requires specialized techniques such as adiabatic demagnetization and the use of thermodynamic phase transitions between the isotopes of helium ^{3}He and ^{4}He. The invention of a new technique such as an efficient liquid helium ^{3}He-^{4}He refrigerator makes new scientific fields widely accessible. The vertical lines in Table 3.1 indicate temperature regions where the techniques are similar.

3.2 DEFINITION OF TEMPERATURE

The thermodynamic parameter of temperature $\{T\}$ is defined in terms of the Second Law of Thermodynamics using the expansion of a gas and the Perfect Gas Law.

$$PV = nRT$$

where P is the pressure in bar, V is the volume in m^3, n is the number of gram molecules of gas in the volume and R is the gas constant: $R = 8.31$ J/mol.K. This measurement requires the use of a constant volume gas thermometer or similar apparatus which is bulky and inconvenient for practical measurements. Since 1968, the International Practical Temperature Scale has been defined using a Platinum Resistance Thermometer calibrated in terms of fixed temperatures. The fixed temperatures are the melting or boiling points of various substances. Lists of these standards are published by international standards organizations. Such a listing is given in Table 3.2.

Table 3.2 International Practical Scale of Temperature for (s) Solid, (l) Liquid, and (v) Vapour

System	Equilibrium			Temperature		Uncertainty
	s	l	v	K	°C	(K)
Hydrogen	x	x	x	13.81	−259.34	0.01
Hydrogen		x	x	17.042	−256.11	0.01
Neon		x	x	27.10	−246.05	0.01
Oxygen	x	x	x	54.36	−218.79	0.01
Oxygen		x	x	90.19	−182.96	0.01
Water	x	x	x	273.16	0.01	exact
Water		x	x	373.15	100.00	0.005
Tin	x	x		505.12	231.97	0.015
Zinc	x	x		692.73	419.58	0.03
Silver	x	x		1234.08	961.93	0.2
Gold	x	x		1337.58	1064.43	0.2

This primary standard is used to calibrate three major types of temperature sensors: resistance thermometers, thermocouples, and thermistors. At higher temperatures, optical effects based on the broadening of spectral lines with temperature are employed.

3.3 TEMPERATURE TRANSDUCERS

Temperature sensors make use of the temperature dependent properties of materials. These include a change in electrical resistance as a function of temperature (resistive temperature detectors—RTDs), the use of voltages generated at junctions between dissimilar materials (thermocouples), and semiconductor temperature detectors (*p-n*-junctions and transistors) [3].

Resistance Thermometers

The resistance of a metal increases approximately linearly with temperature and therefore can be utilized as a *resistive temperature device* (RTD). Figure 3.2 shows that for metals such as platinum the variation is highly linear, while for nickel or tungsten under non-oxidizing environments it is less linear, although it is reproducible. The variation of resistance with temperature can be described by a power series:

$$R_T = R_o(1 + \alpha(T - T_o) + \beta(T - T_o)^2 + ...)$$

where T_o is a reference temperature (20°C). The coefficients of this series for platinum for $T > 0°C$ are $\alpha = 3.664 \times 10^{-3}/°C$ and $\beta = -5.41 \times 10^{-7}/(°C)^2$ [Fig. 3.2].

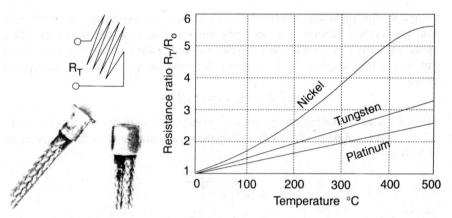

Fig. 3.2 Resistance versus temperature for some common thermometer materials. The inset shows two forms of commercially available transducer— Courtesy: Minco Products Inc., 7300 Commerce Lane, Minneapolis, Minnesota 55432-3177 USA.

Platinum is **linear** over a wide temperature range with the most common sensors having a room temperature resistance of 100 ohms. The wire is carefully wound on a mandrel and care is taken not to strain or mechanically deform it. The type of packaging available is shown in the inset. The stability and reproducibility of well-annealed platinum with time and temperature is such that it is used both as the International Secondary Standard for temperature measurements, and for critical industrial applications. Temperatures within nuclear reactors are generally sensed with platinum RTDs.

The **sensitivity** of platinum resistance thermometers is not particularly high. The calibration can be lost if the wire is bent or damaged. Four terminal (Fig. 2.21) and bridge (Fig. 2.26) measurements are used to avoid electrode effects and to increase the sensitivity of the measurement respectively. Tungsten or nickel RTDs are made in the form of thin films.

Since the transducer is a resistance, current has to be passed through it to measure it. This has the potential to self-heat the sensor by I^2R dissipation. This will cause an error in temperature. The error depends on how well the heat is removed from the sensor, and will be worst for a sensor in static air. For this case the error may be estimated as 1°C per milliwatt of internal dissipation. In any design it is obviously useful to minimize the measuring current consistent with a sufficient value of the signal to noise ratio.

Question (i)   Use a power law series for a platinum RTD to calculate the resistance at 1000°C for a transducer which has a resistance of 100Ω at 20°C. If it was assumed erroneously that the calibration was linear over this temperature range, what would be the error in (a) the resistance and (b) the measured temperature at 500°C?

Thermistors

Complex semiconductor or ceramic materials have a resistivity which changes with temperature according to an Arrhenius relation. Conduction occurs either by the excitation of current carriers (electron-hole pairs) across an energy band gap in a semiconductor, or by hopping between localised sites in the crystal lattice

$$R_T = R_o \exp (E_d/kT)$$

where E_d is an activation energy for the excitation of current carriers ($E_d \approx 0.2$–0.3 eV) and $k = 8.63 \times 10^{-5}$ eV/K is Boltzmann's Constant. T is the temperature in Kelvin (K).

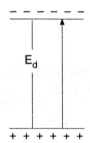

Fig. 3.3   Thermally activated generation of conduction in thermistor materials.

Thermistors are less reproducible than platinum resistance thermometers, and hold their calibration less effectively. Their characteristics are given below.

- The thermistor is a resistance device similar to the platinum RTD, but the value R_T can be chosen to be from ohms to megaohms. This facilitates circuit design.

- The sensitivity is high—there is an exponential variation of R versus $1/T$.

- The thermistor is **non-linear** (Fig. 3.4(a)). Without additional circuitry it is not easy to obtain a simple direct readout. It can be very useful to **control** temperature at a set level.

- Thermistors are available in various sizes: from that of a pinhead to that of a golf ball.

- The response time is rapid in small sizes.

- Self-heating can occur—and a large thermistor takes time to self-heat. Since the resistance is large when the voltage is applied but then decreases as the temperature rises, according to the voltage applied they can be used to create a defined delay before a current reaches a predefined level. This is shown in Fig. 3.4(b).

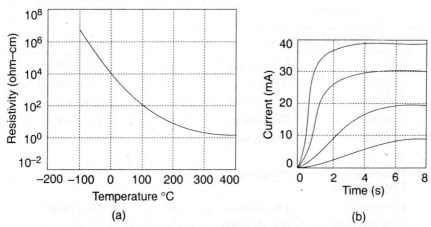

(a)   (b)

Fig. 3.4   (a) A temperature—resistivity $\rho(T)$ plot for a typical thermistor material. The resistance for given dimensions is given by $R = \rho L/A$, where L is the length and A is the area of the material. (b) Current through a thermistor as a result of self-heating.

Question (ii)   The resistivity data for the thermistor material data shown in Fig. 3.4(a) can be represented by $\rho(T) = 3.3 \times 10^{-3} \exp(3850/T)$ Ω–cm, where T is the absolute temperature in Kelvin K.

(a) Calculate the resistance of a block of this material of dimensions 3 mm long and 2 mm^2 cross-sectional area at 100°C, (b) What is the temperature sensitivity in Ω/degree over a 10°C temperature range around 100°C?

Thermocouples

As a result of the thermodynamic Seebeck Effect, two dissimilar materials in contact generate a potential difference between 'cold' and 'hot' junctions held at different temperatures.

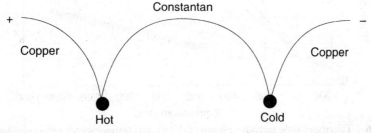

Fig. 3.5   A copper-constantan thermocouple showing hot and cold junctions.

For pairs of metals and alloys, the 'thermocouple' voltage is reproducible and stable over a wide range of temperature. The cold junction is normally maintained at 0°C, either in an ice/water bath or through electronic compensation circuits. Table 3.3 shows some thermocouples which are generally available.

Table 3.3 Commercially Available Thermocouple Pairs Showing ISA Wire Colour Codings

Thermocouple pair		ISA	Colour coding
Iron	Constantan	J	White/Red
Chromel	Alumel	K	Yellow/Red
Copper	Constantan	T	Blue/Red
Chromel	Constantan	E	Purple/Red
Platinum—10% rhodium	Platinum	S	Black/Red
Platinum—13% rhodium	Platinum	R	Black/Red
Platinum—30% rhodium	Platinum—6% rhodium	B	Gray/Red
Tungsten—26% rhenium	Tungsten	G	White/Red

Figure 3.6 shows that the thermocouple voltage is relatively linear with temperature, particularly in the range above room temperature. The output is in the millivolt range, but the device is of low resistance and is reasonably sensitive.

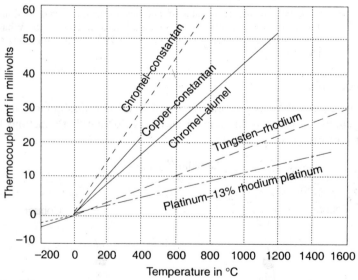

Fig. 3.6 Thermocouple output voltage (mV) versus temperature: reference junction at 0°C.

The Instrument Society of America (ISA) and the National Bureau of Standards publish tables of thermocouple voltages as a function of temperature. Such tables are also available in Handbooks of Tables or in the Omega Handbook [3]. Thermocouple pairs are designated by an ISA number and are sold with a particular colour coding for the outer sheathing. For laboratory and industrial purposes, the most widely used thermocouples are Type K—chromel/alumel and Type T—copper/constantan.

With computer based instrumentation, it is often convenient to use power series to calculate the temperature from a thermocouple voltage measured in volts or millivolts. The coefficients for such calculations for an input voltage in volts are listed in Table 3.4.

$$T = a_0 + a_1v + a_2v^2 + \ldots + a_nv^n$$

Table 3.4 Coefficients for the Calculation of Temperature from Thermocouple emf Measured in mV. Reference at 0°C

Type	Copper/T Constantan -160–400°C	Chromel/E Constantan -100–1000°C	Chromel/K Alumel 0–1250°C	Platinum/R 13%Rh/Pt 0–1450°C
a_0	0.311737	-2.01063	0.0	13.84627
a_1	26.62022	16.89726	25.63659	131.8368
a_2	-0.89656	-0.16202	-0.0971	-6.35701
a_3	0.04451	0.002367	0.001542	0.333189
a_4	-0.0009	-0.000012		-0.00704
Error (±°C)	2	2	2	3

The number of terms used in such series reflects the accuracy with which the temperature is calculated and measured. The number is usually a compromise between accuracy and convenience. The error in the series given is shown in the last row. This applies to the centre section of the range. The errors increase towards the end of the range to which the fit has been made. In using such power series, care must be taken not to exceed the temperature range over which the data fit was made. The original data are taken from the Omega Engineering Handbook of Temperature Measurement [3].

The choice of a thermocouple at high temperatures is determined by the oxidation resistance of the two materials. Chromel-alumel is useful up to about 1250°C. It can also be used below room temperature, but copper-constantan is then a better choice. Figure 3.7 shows the range of application of various thermocouples. Platinum and tungsten based alloys can be used upto a temperature higher than 1500°C.

Electronic cold junction compensation. The requirement to maintain the cold junction at 0°C is a practical inconvenience. Many temperature controllers

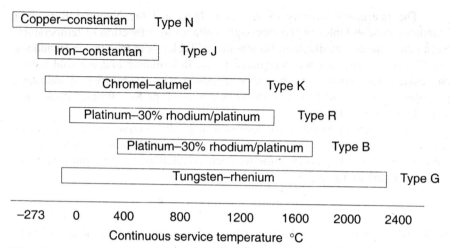

Fig. 3.7 Operating ranges for commercial thermocouples (Omega Temperature Handbook [3].

and temperature measuring devices utilize an electronic circuit to compensate for the approximately 0.9 mV offset between 0 and 20°C. The cold junction can also be kept at 77 K in liquid nitrogen, and an appropriate offset voltage included to determine the temperatures from published tables.

Thermocouple fabrication. Thermocouple junctions must be made carefully, preferably by spot-welding or by fusing the wires using an oxy-acetylene flame. For low temperature measurements a spot of solder on the junction can be of value. When a thermocouple is connected to a metal surface, care should be taken that the thermocouple is not short-circuited by the metal away from the point at which the temperature is being measured.

Care should be taken that connections to external circuits take place in a region where *both connections* are at the same temperature. If a thermocouple has to be taken through the wall of an experimental chamber it is preferable to carry the thermocouple wire through a hollow feedthrough without breaking its continuity.

Question (iii) Set up a power series for a Type *K* chromel-alumel thermocouple operating over a range from 0–1250°C. Plot *T*°C vs *E* (mV). Compare the linearity of the transducer by calculating the emf(mV)/100° (a) from 0 to 100°C, (b) from 500 to 600°C, and (c) from 1100 to 1200°C.

Comparison of Temperature Transducer Characteristics

Table 3.5 compares the properties of the three types of temperature transducer. Assess these properties in terms of the transducer characteristics listed in Table 2.1.

Table 3.5 Comparative Properties for Temperature Transducers

Properties	Thermocouple	Platinum	Thermistor
Repeatability	1°C to 7°C	0.03°C to 0.05°C	0.1°C to 1°C
Stability	1°C to 2°C drift per year	Less than 0.1% drift in five years	0.1°C to 3°C drift in one year
Sensitivity	10 to 50 microvolts/°C	0.2 to 10 ohms/°C	100 ohms/°C
Interchangeability	± 0.75%	± 0.5%	± 0.5%
Temperature range	−200°C to 2200°C	−100°C to 850°C	−100°C to 300°C
Signal output	0 to 60 millivolts	1 to 6 volts	1 to 3 volts
Power (with 100 ohm load)	1.6×10^{-7} watt	4×10^{-2} watt	8.1×10^{-1} watt
Minimum size	0.040 cm diam	1 cm diam. × 1 cm long	0.014 inch diam.
Unique features	Greatest economy; highest range	Greatest accuracy over wide spans; highly stable	Greatest sensitivity
Linearity	Excellent	Excellent	Poor

Semiconductor Temperature Sensors

Semiconductor temperature sensors have recently become available. These use a fundamental property of the silicon transistor. Two identical transistors are operated at a constant ratio r of collector current densities with a sufficiently large collector voltage so that they operate in the saturation region.

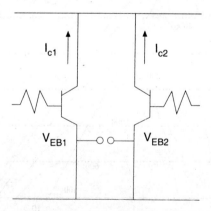

Fig. 3.8 Transistor connections for a semiconductor temperature sensor.

$$r = \frac{I_{c1}}{I_{c2}} \qquad T = \frac{q(V_{EB1} - V_{EB2})}{k \ln r}$$

The collector current under these conditions is given by

$$I_c = \alpha I_{ES} \left(\exp \frac{q V_{EB}}{kT} - 1 \right)$$

where α and I_{ES} are constants, k is Boltzmann's constant (8.625×10^{-5} eV/K), and T is in degrees K.

Simple algebra shows that the difference in the base-emitter voltages between the two transistors is $(kT/q) \ln r$. Since all the constants are known, the output voltage is directly proportional to absolute temperature. The temperature range is from $-55°C$ to $+150°C$. A representative device is the Analog Devices AD590. This is an excellent example of taking advantage of a well defined physical (semiconductor) relationship for measurement purposes.

3.4 THERMAL RADIATION TEMPERATURE MEASUREMENTS

A classical 'black body' emits thermal radiation. The intensity of this radiation is given by the Stefan-Boltzmann Law:

$$E = \sigma(T^4 - T_o^4) \text{ W/m}^2$$

where $\sigma = 5.67 \times 10^{-8}$ Js^{-1}m^{-2}K^{-4} and T_o (K) is the temperature of the surroundings.

The spectral distribution in W/m^2/μ is given by the Planck Radiation Law (Fig. 3.9)

$$E(\lambda) = \frac{2\pi c^2 h}{\lambda^5} \cdot \frac{1}{\exp(hc/k\lambda T) - 1}$$

where c is the velocity of light, h is Planck's Constant and k is Boltzmann's constant.

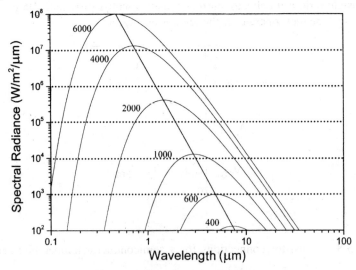

Fig. 3.9 Thermal emission (W/m^2/μ) as a function of wavelength λ for a body at temperature T(K).

A plot of this function shows that the temperature T characterises the spectral distribution of the radiation in two ways. The overall amplitude of the spectrum changes with temperature, and the wavelength at which the maximum emission occurs varies linearly with temperature. This is shown by the line joining the maxima. These features are used in a number of methods of measuring temperature. The units for luminous intensity and flux are discussed in Appendix 1.

A 'black body' thermal distribution is not trivial to observe or measure. In the laboratory it is emitted from a hole cut in the wall of a spherical oven inside which the radiation is in equilibrium with the walls at temperature T. Most real surfaces emit the 'correct' spectral distribution, but radiate less energy by a constant amount over the entire spectrum. This is accounted for by defining the

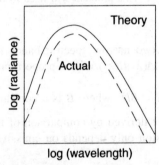

Fig. 3.10 The energy emitted by a 'grey' body has the same spectral distribution as a 'black' body, but is reduced by a constant factor across the entire spectrum.

emissivity ε for an emitting surface. This implies that the body can be treated as a **grey body** and that the energy radiation from a real body is written

$$E = \varepsilon\sigma(T^4 - T_o^4) \ \text{W/m}^2$$

where the T_o term represents the radiation received **back** from the surroundings at temperature T_o. Representative values for the emissivity ε of various materials are given in Table 3.6.

Table 3.6 The Emissivity of Materials under Specific Conditions

Material	Emissivity	Material	Emissivity
Al unoxidized	0.02	C-lampblack	0.95
Al oxidized	0.11	Ni oxidized	0.31–0.46
Platinum	0.11	W (1000°C)	0.15–0.2
Stainless steel	0.55–0.8	Glass	0.80

When radiation falls onto a surface, only a fraction of it will be absorbed. For many computational purposes the amount absorbed can be estimated by defining an **absorbance** α. To a first approximation $\alpha = \varepsilon$.

Accurate measurements of ε are difficult, and changes in ε occur if surfaces oxidize. This constitutes the major difficulty in making temperature measurements by radiation sensing (optical pyrometry). This is partly solved by **comparison** measurements with a known source.

Question (iv)   A thin platinum foil of total surface area 3 cm^2 and emissivity $\varepsilon = 0.11$ has an electrical resistance of 0.3Ω at 293K. Its resistance increases with temperature K approximately as

$$R(T) = R_o\,[1 + 3.7 \times 10^{-3}\,(T - 293)]$$

If the foil is heated electrically in a vacuum by power $P = I^2 R(T)$, heat is lost from the total surface area of the foil and all heat losses other than radiation are neglected, determine the temperature it will attain for a current of 0.5A. This problem will need to be solved numerically.

Ratio Pyrometers

If the light intensity I in two narrow spectral bands is observed as shown in Fig. 3.11, the ratio $R = I(\lambda_1)\Delta(\lambda_1)/I(\lambda_2)\Delta(\lambda_2)$ is proportional to a simple logarithmic function of temperature

$$\ln R = B/T, \text{ where } B \text{ is a constant.}$$

The full expression may be derived by comparison of the Planck Radiation law for λ_1 and λ_2. The expression only depends on emissivity through the variation of ε with λ.

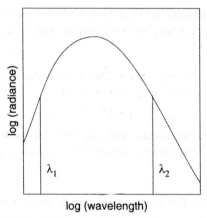

log (wavelength)

Fig. 3.11   Comparative measurements of the spectral intensity of thermal radiation emitted at λ_1 and λ_2. Suitable interference filters are discussed on page 209.

Infra-red Pyrometers

These use sensitive semiconductor infrared detectors. For good calibration the devices often compare the emission from the source to that from an internal oven held at constant temperature T (K). A vibrating shutter or rotating chopper wheel is used to make the comparison. As the shutter moves, it successively exposes an infrared detector to radiation from the source or from the oven. This principle is shown in Fig. 3.12(b).

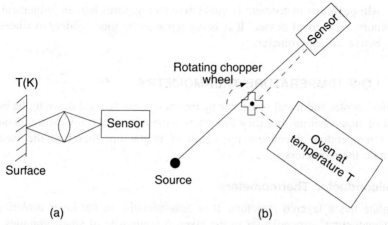

Fig. 3.12 Infrared pyrometer: (a) optics, (b) chopper circuit. The rotating mirror successively presents the source and the standard temperature oven to the sensor.

Many instruments have a voltage output suitable for use in control circuitry, but the lack of knowledge of the emissivity and of its variation with temperature is still a problem. Since lenses can be used to focus the incoming light (Fig. 3.12(a)), a very small area can be viewed. Such instruments are widely used for checking 'hot spots' in integrated circuits.

Disappearing Filament Optical Pyrometer

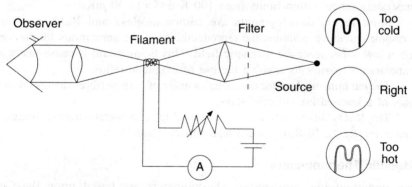

Fig. 3.13 Disappearing filament optical pyrometer. The inset shows the field of view with the hot filament superimposed on the emitting background.

The source is viewed by an optical system, and an image of the source is superimposed on a heated tungsten wire placed at the focal plane. The temperature of the wire is controlled by a current and the two images are viewed through a red filter. When the filament 'disappears' against the background, the spectral emissions are similar and a calibration can be effected.

Although this instrument is qualitative and requires human judgement, it is an important practical device. It is being replaced by more widely available and inexpensive ratio pyrometers.

3.5 LOW TEMPERATURE THERMOMETRY

Thermocouples and metal resistance thermometers can be used down to the boiling point of liquid helium but they lose their sensitivity at the lowest temperatures. Other temperature dependent properties of matter are therefore the basis for alternate thermometers.

Semiconductor Thermometers

Graphite has a layered structure. It is semi-metallic in the layer direction and semiconducting perpendicular to the layer. A composite of small granules, as in a typical carbon resistor, therefore becomes semiconducting with a resistance which increases, often dramatically, as the temperature is decreased. Carbon resistors manufactured by Allen-Bradley, Speer and Matsushita have found favour as resistance thermometers for temperatures down to 1 K, 50 mK and 10 mK respectively.

More stable and reproducible than carbon composites are Au-doped Ge single crystal extrinsic semiconductors. These are available commercially packaged in a small gold-plated sealed metal container fitted with an inert gas. The semiconductor band diagram of a doped semiconductor is shown in Fig. 3.20 on page 73. Control of the dopant and therefore of the energy of the impurity state below the conduction band allows the resistance to be chosen to provide a working temperature range within limits from 100 K down to 30 mK.

More recent developments are carbon-in-glass and RuO_2 resistance thermometers which combine the reproducibility of the germanium thermometer with a low dependence on magnetic field. This is important because many low temperature experiments involve the use of magnetic fields.

The **carbon-in-glass** thermometer is made of carbon fibres deposited in the voids of a leached borosilicate glass.

The **RuO_2** thermometer, a thick-film chip, is a metal-ceramic mixture of conducting RuO_2, Bi_2RuO_2 and a lead-silicate glass.

Magnetic Thermometers

Two important low temperature thermometers are based upon the Curie (Curie-Weiss) paramagnetism of a system of magnetic dipoles. The compound cerium magnesium nitrate has magnetic cerium atoms separated by more than 1 nm by other non-magnetic atoms (magnesium, nitrate molecules and water molecules). At this spacing the magnetic interaction between neighbouring cerium atoms is so small that the magnetization in a small field can be expressed as

$$M = \frac{CV}{T - T_o}$$

where C is the Curie constant/m^3, V is the volume of the sample, and T_o is less than 1 mK. For use below 1 mK, the cerium atoms are diluted to 3% by non-magnetic lanthanum atoms, It is usual to use superconducting electronics (the SQUID) to measure the very small magnetization.

Curie law paramagnetism of Pt nuclei in metallic platinum can be used as a temperature sensor. A very clever technique is used to measure the nuclear magnetization of the platinum in the presence of a possible comparable magnetization due to magnetic impurity atoms in platinum. The instrumentation is based upon nuclear magnetic resonance. Only the platinum nuclei are in resonance and therefore, only their magnetization is detected. Commercial instrumentation using modern spin-echo pulsed nuclear magnetic resonance is also available.

Noise Thermometry

Measurement of thermodynamic fluctuations in the electron gas in resistors can provide a useful measure of temperature. The experimental techniques are discussed fully in Chapter 4.

The root mean square noise voltage from a resistance R is measured by an amplifier of bandwidth Δf. This is related to temperature by (page 101)

$$V_n = \sqrt{4KTR\Delta f}$$

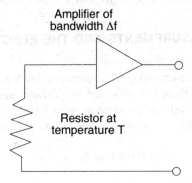

Amplifier of
bandwidth Δf

Resistor at
temperature T

Fig. 3.14 The use of thermally generated Johnson Noise (page 101) as a temperature sensor.

The Melting Curve Thermometer

The standard thermometer in the temperature range 1 mK–30 mK is the ^{3}He melting curve thermometer. The phase diagram of ^{3}He is shown in Fig. 3.15.

^{3}He remains liquid at 0K unless a pressure ~ 30 bar (atmospheres) is applied. The melting line is well defined with a slope given by the Clausius-Clapeyron equation:

$$dP/dT = \Delta s/\Delta V$$

where Δs and ΔV are the differences in the molar entropy and molar volumes of the liquid and solid.

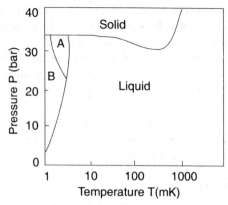

Fig. 3.15 The phase diagram of ^{3}He in the millikelvin range.

Careful work by physicists at Cornell University, Helsinki University of Technology and Bell Laboratories has led to an equation relating pressure to temperature. Thus a measurement of the pressure above a mixture of co-existing solid and liquid gives a direct measure of the temperature. One of the advantages of this thermometer is the existence of two fixed points—the *A* and *B* superfluid ^{3}He transition temperatures at 2.5 mK and 1.9 mK respectively.

3.6 OPTICAL MEASUREMENTS AND THE ELECTROMAGNETIC SPECTRUM

Light forms part of the electromagnetic spectrum. Such electromagnetic radiation can be absorbed, transmitted or reflected by surfaces and hence induce optical, thermal or mechanical effects on systems with which it interacts.

Light of frequency v can also be treated as photons with energy,

$$E \text{ (eV)} = hv = \frac{hc}{\lambda}$$

where E is the photon energy in electron volts (1.6×10^{-19} J), h is Planck's Constant (6.63×10^{-34} J · s), c is the velocity of light (3.0×10^8 m/s), and λ is the wavelength of light.

A convenient relationship between these quantities is

$$E \text{ (eV)} = \frac{1240}{\lambda \text{ (nm)}}$$

For the visible spectrum which extends from 400 to 650 nm, photon energies extend from 3.1 to 2.1 eV, while the infrared spectrum extends to $\lambda = 100$ μm and to photon energies of 0.15 eV.

Most physical systems are **quantized** so that the absorption or emission of a photon causes transitions between discrete levels such that

$$E_2 - E_1 = hv$$

The interactions with the electromagnetic spectrum which are allowed, utilize

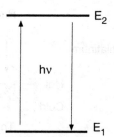

Fig. 3.16 Transitions between states of energy E_1 and E_2 through the absorption of a photon of energy $h\nu$.

specific transitions in the atom, in molecules, and in the atomic nucleus. These energies range from MeV to 10^{-10} eV and lead to various forms of **spectroscopy**. Important optical detectors utilize optical transitions in semiconductors. In the following section, detectors for radiation in the ultraviolet, visible and infrared regions of the spectrum are discussed. The optical instruments in which such detectors are used are reviewed in Chapter 7. The parameters of interest in an optical detector are: (i) wavelength response, (ii) sensitivity, and (iii) speed of response.

Four major classes of detectors can be formed depending on the mechanism of sensing the light.

- **Thermal:** Bolometers, thermopiles, Golay Cell

- **Photoconductive:** Semiconductors such as lead sulphide (PbS), cadmium sulphide (CdS), mercury cadmium telluride (HgCdTe) and gold doped germanium (Ge:Au).

- **Pyroelectric:** Lead titanate ($PbTiO_3$) and lithium tantalate ($LiTaO_3$).

- **Photoemissive:** Photomultipliers, photodiodes, phototransistors.

These detectors are described in the following sections.

Thermal Detectors

Bolometers or thermopiles. These utilize the heating effect of radiation falling on to a blackened surface. The temperature rise of the surface is detected either by the change in the resistance of a platinum strip (bolometer), or by the output from a set of series-connected thermocouples (thermopile) (Fig. 3.17).

Because the effect measured is that of power absorption resulting in heating, the detectors are energy sensitive and there is little wavelength dependence. The sensitivity is poor, but the principle is used in power meters for the calibration of high intensity light sources such as lasers. This type of detector is often used as a standard for absolute source intensity measurements. Practical laboratory calibrations are then carried out using secondary standards.

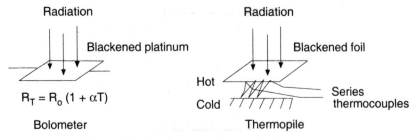

Fig. 3.17   Energy sensitive optical detectors.

Golay cell.   This is an important infra-red detector which is used widely in infra-red spectrometers for chemical analysis. The principle is again that of the absorption of heat, but in this case the temperature rise is measured by a thermo-pneumatic effect (Fig. 3.18).

An increase in pressure in a gas enclosed in a sealed chamber is observed —usually by noting the deviation of the reflection of a beam of light from a polished flexible surface which forms one side of the chamber.

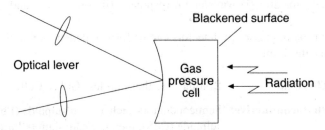

Fig. 3.18   Thermo-pneumatic Golay cell.

These devices are linear, medium to low sensitivity, and again have no wavelength dependence. They are fragile and are only suitable for well controlled laboratory applications.

Photoconductive Detectors

In this type of detectors an increase in conductivity due to the excitation of electrons and holes in semiconductors is observed. This may occur across the band gap (CdS) or by excitation of electrons from impurity states (Ge:Au). Photoconductors can be tailored to have energy transitions appropriate for specific ranges of wavelength. This is particularly the case for compounds such as mercury cadmium telluride. The type of transitions involved are shown in Fig. 3.19.

The **threshold wavelength** is given by $hv = hc/\lambda = E_g$ or E_d if excitation takes place across a band gap, or if E_d is a donor energy respectively. Semiconductors can be chosen or designed with appropriate energy states to respond to a particular wavelength range. However, the correct band gap is not the only criterion for good photoconductive response, and this means that relatively few materials are in practical use. The wavelength sensitivity of important semiconductors and the origin of the photoconductivity is shown in Table 3.7.

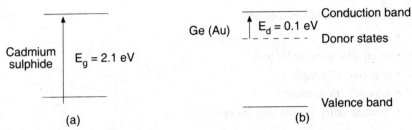

Fig. 3.19 Energy states in semiconductors which lead to photoconductivity: (a) for band to band transitions in cadmium sulphide, (b) excitation from donor or acceptor states in the conduction band.

Table 3.7 Band Gaps and Wavelength Sensitivity of Common Photoconductors

		E(eV)	Wavelength	
Visible:	Cadmium sulphide (CdS)	2.1	550 nm	Band
Infrared:	Lead sulphide (PbS)	1.0	1 μm	Band
Infrared:	Mercury cadmium telluride (HgCdTe)		0.44–12 μm	
Infrared:	Silicon (Si)	0.9	1 μm	Band
Far I.R.	Ge:Au (cooled to 77 K)	0.1	100 μ	Donor

Cadmium sulphide (CdS) is used in many industrial applications. Its spectral response is a maximum near 550 nm, which closely approximates the peak sensitivity of the human eye. A range of dark and light resistance values can be obtained to facilitate circuit design. A representative device could have a dark resistance of 600 kΩ, and a resistance of 9 kΩ when illuminated by 2 ft-candles (35 mW/m^2 at 550 nm). The active area of the device is increased by depositing the CdS in a 'serpentine or meandering' fashion on a ceramic base (Fig. 3.20(b)).

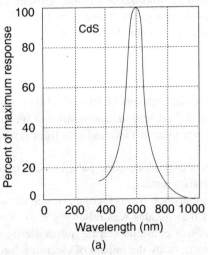

Fig. 3.20 (a) Spectral response, and (b) physical construction of a CdS photocell.

Photoconductive devices are:

- linear
- relatively slow (>10 ms)
- resistive elements
- wavelength sensitive
- mechanically robust and inexpensive.

Lead sulphide (PbS) is important as a conductivity sensor in the near infrared.

Mercury cadmium telluride (HgCdTe) has a band gap which depends on composition and temperature.

Question (v)   Assuming that the resistance change upon illumination of the CdS sensor described above is linear in intensity and the wavelength response is as shown in Fig. 3.20: (a) estimate the illuminated resistance for a light intensity of 10 mW/m^2 at λ = 400 nm. For this illumination (b) devise a bridge circuit using a 9 V battery to give a sensitive output which increases with light intensity from the dark level, (c) calculate the output voltage for 10 mW/m^2 at λ = 400 nm.

Pyroelectric Detectors

Ferroelectric and piezoelectric materials such as *lead titanate* and *lithium tantalate* have an internal electrical polarization (Section 3.7). Changes in temperature with time induced by the absorption of radiation cause a change in the polarization and a pyroelectric current density flows of magnitude:

$$J = p_i \left(\frac{dT}{dt} \right) \text{A/m}^2$$

Because the pyroelectric is a high impedance crystal, current can only be detected when the temperature is changing. The value of p_i for PbTiO$_3$ = 70 $\times$ 10^{-5} m^{-2}K^{-1}. This produces a sensitivity for the detection of infrared radiation of the order of 5000 V/W.

Photoemissive Detectors

Photomultipliers have been the traditional high sensitivity optical detector for the visible and near infra-red for low level light detection, astronomy and spectroscopy. Used with a scintillator which converts nuclear radiation into light flashes, they are used in nuclear physics.

Photomultipliers use **two** effects to produce large pulses of electrons when a photon interacts with a photoemissive surface.

Since the early days of quantum physics, it was recognized that the emission of electrons from alkali metals and their oxides was relatively easy. The **work function** of these materials is low. Such emission can be induced either by the absorption of photons (photo-emission), or by the impact of electrons (secondary electron emission).

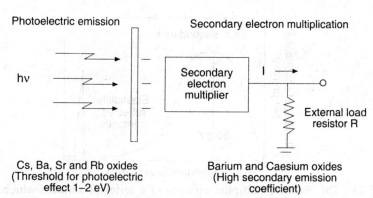

Photoelectric emission   Secondary electron multiplication

hv

Secondary
electron
multiplier

$I \rightarrow$

External load
resistor R

Cs, Ba, Sr and Rb oxides
(Threshold for photoelectric
effect 1–2 eV)

Barium and Caesium oxides
(High secondary emission
coefficient)

Fig. 3.21   Illustration of the phenomena related to the optical and electron beam
stimulated emission of electrons from solids. Important materials are
the alkali metal oxides.

The general construction of a photomultiplier is shown in Fig. 3.22. The
photocathode emits electrons when the incident photon energy exceeds the
photoelectric work function of the photoemissive surface. The photoelectrons are
collected by an electrode structure which creates appropriate electric fields and
are focussed on the first electrode of an electron multiplier.

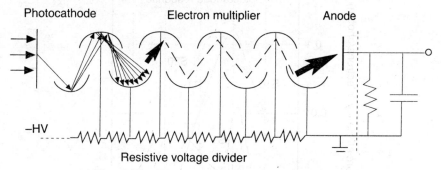

Photocathode   Electron multiplier   Anode

–HV

Resistive voltage divider

Fig. 3.22   The construction of a photomultiplier.

When high energy electrons impinge on a surface, the energy is dissipated
by interaction with electrons in the surface. This forms an excited cloud of
electrons within the material, some of which diffuse to the surface. If the material
has a low work function, there is a strong likelihood that electrons will be
released. The energy spectrum of the secondaries is shown in Fig. 3.23. It includes
a large fraction of electrons with energies in the range of 50 eV, and a peak of
elastically reflected primaries at the incident energy.

If each stage emits n secondaries per primary, and there are G stages (usually
10 or 14), the increase in the number of electrons along the length of the chain is

$$A = n^G$$

For $n = 3$ and $G = 14$, $A \sim 10^8$. A voltage of about 100 V is normally applied
between each stage as a **negative** voltage to the photocathode so that the collection
anode is close to ground potential.

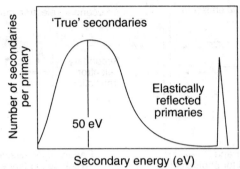

Fig. 3.23 The electron multiplier consists of a series of surfaces which create a cascade of electrons. 3–4 secondary electrons are emitted for each incident primary. The energy spectrum of the emitted 'true' secondaries peaks at around 50 eV.

The **wavelength response** is determined mainly by that of the photocathode. Specialized mixtures of alkali oxides give well characterized responses, known as S10, S4, S11 etc. These are shown in Fig. 3.24.

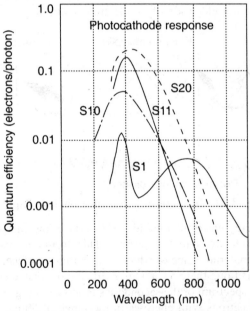

Fig. 3.24 Photocathode wavelength sensitivity.

After a photoelectron is detected the **risetime** of the output voltage is set partly by the electron optics, i.e. by the **transit time** for electrons to propagate through the device (~10 ns), and by the output circuit. For photo-multipliers used for accurate timing strong efforts are made to equalize the length of all electron paths to minimize the **transit time spread**. The **decay time** is given by the RC time constant of the output circuit.

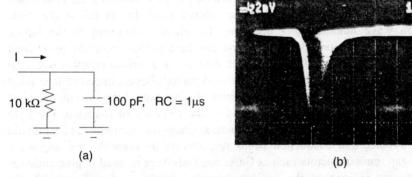

(a)

(b)

Fig. 3.25 (a) RC components which determine the pulse shape at the photo-multiplier anode. (b) Output pulse for a high speed photomultiplier.

The main disadvantage of photomultipliers (apart from size and power supply needs) is that the carrier generation and multiplication is statistical in nature. The collection of individual charges arriving at the photomultiplier anode in a random fashion leads to a large 'shot noise' current (page 102).

Question (vi)   What charge would be delivered to the anode for a single photon detected by a 14 stage photomultiplier having a stage gain of 3.7? Assume that 1 electron is produced by the photon. If the anode of the tube is connected to a parallel circuit consisting of a 1 nF capacitor and a 1 MΩ resistor each having one end connected to ground, draw the amplitude of the output pulse as a function of time.

Photodiodes and phototransistors.   The depletion layer *i* of a semiconductor *p-n* diode or the base region of a *pnp* or *npn* transistor is a region where very low carrier concentrations exist.

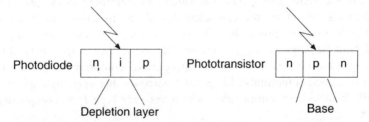

Fig. 3.26   The physical structure of a *np*-junction and a *npn* transistor. When the device functions as a photosensor, the light illuminates the depletion or base regions of the device.

The creation of minority carriers causes major changes in the device current, and this provides a sensitive form of high speed optical detector vital to modern optical communications.

Most photodiodes and phototransistors are based on silicon. Silicon is known as a '**indirect**' gap semiconductor for which the energy minimum in the conduction band and the energy maximum in the valence band do not fall at the same electron wave number. It is necessary for elastic vibrations of the lattice (**photons**) to participate so that photon can be absorbed to create an excited carrier or to recombine to produce light. This is not a serious problem when the junction is used for light absorption and silicon photodiodes are important measuring devices. The indirect band gap is a serious disadvantage if the recombination of electrons and holes is expected to lead to the emission of radiation. For light emitting diodes '**direct**' gap semiconductors, where the maximum and minimum of the valence and conduction bands respectively do coincide, are necessary. Direct gap semiconductors such as GaAs and GaInP are tailored to give emission in different regions of the optical spectrum (green, red, infrared). These semiconductors are used in light emitting diodes (LEDs) and semiconductor lasers.

The voltage or current output for a solar cell or photodiode can be used for radiation detection and power generation. The construction is slightly different than that for a normal diode, in that one of the doped regions is made very thin (the emitter) and the other is made relatively thick (the collector). A metal electrode pattern is prepared on the outer emitter layer to minimize the loss of light capable of being absorbed

At the barrier, the depletion (or intrinsic) layer separates p and n type material. Minority carriers generated within one **diffusion length** of the junction give rise to thermally generated reverse current I_o. The diffusion length is a characteristic of the material

$$L_D = \sqrt{D\tau}$$

where D is the diffusion constant and τ is the minority carrier lifetime. The diffusion length is a measure of how far minority carriers can travel in a semiconductor before they recombine. It is a measure of the effective volume in the vicinity of a junction which can contribute to the photoresponse of a diode. D is set by the characteristics of the material, but τ depends on crystal perfection and processing. While at first sight it may seem that D and τ should both be large, a large τ means that the device response is slow. The design of photodiodes requires an appropriate compromise to be attained. Silicon is a good compromise for general purposes, but very high speed devices are built from indium antimonide—which has very high D to compensate for a small τ.

The various currents in a photoexcited diode are shown in Fig. 3.27. Further information is provided in Appendix 3.

The number of photogenerated carriers is proportional to the incident light intensity, and add I_v to the reverse current. The diode equation then becomes

$$I = I_o \exp \frac{qV}{kT} - I_o - I_v$$

The following two cases can be recognized:

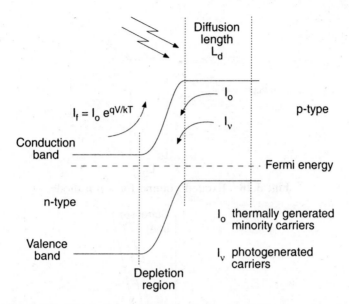

Fig. 3.27   Thermal and photoexcited currents in a *p-n* junction.

Open circuit—photovoltaic mode (no current flow).   When illuminated a bias V_{oc} is developed across the diode given by

$$0 = I_o \exp \frac{V_{oc}}{kT} - I_o - I_v$$

For a good diode $I \gg I_o$, and therefore

$$\frac{I_v}{I_o} \approx \exp \frac{qV_{oc}}{kT} \qquad V_{oc} = \frac{kT}{q} \ln \frac{I_v}{I_o}$$

Reverse biased—photocurrent mode.   Reverse biased current is given by:

$$I = -(I_v + I_o)$$

The light generated current can be estimated in terms of the incident light intensity by recognizing that all carriers generated within one diffusion length $L_D = \sqrt{D\tau}$ contribute to this current, where D is the diffusion constant for minority carriers and τ is the minority carrier lifetime.

$$I = qg_{\text{opt}}AL_D$$

where q is the electronic charge, A is the junction area, and g_{opt} is the rate of production of minority carriers ($\sim$ light intensity).

The *I-V* characteristic for such a diode on which light of different intensity is falling, and the photo generated current I_v is given by Fig. 3.29.

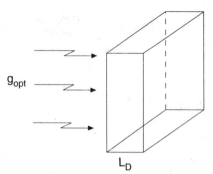

Fig. 3.28 Excited volume for a *p-n* diode.

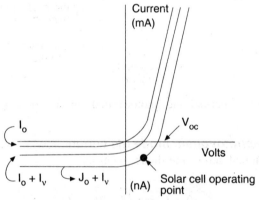

Fig. 3.29 The *I-V* characteristic for a junction diode.

Question (vii) Calculate the open circuit output voltage in a photodiode at 20°C if the photogenerated current is $3 \times I_o$.

Applications of Photodiodes and Phototransistors

Silicon photodiodes and phototransistors have a spectral response that peaks around 880–900 nm. This is in the near infra-red and matches well with the emission from gallium arsenide light emitting diodes and lasers. The spectral response is shown in Fig. 3.30. The dark current is 25 nA. The light current for a light intensity of 20 mW/cm^2 is 1 mA.

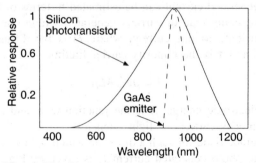

Fig. 3.30 Spectral response of Si and GaAs.

Optical couplers. The LED is driven by the output of the first circuit, while the signal is introduced into the secondary circuit either as a photovoltaic voltage or as a resistance change in a photoconductor.

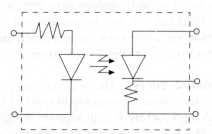

Fig. 3.31 A combination of light emitting diode and photodiode is used to couple digital logic circuits.

Photographic Recording

Photography is one of the oldest methods of optical recording. It consists of the absorption of light by silver halide grains in a photographic emulsion based on gelatine. The original exposure is sufficient to convert only a very small part of the silver halide crystals into metallic silver, but it serves to sensitize the crystallites to respond to chemical processing which 'develops' large silver particles having an optical density D related to the original light exposure.

Optical density. The optical density is a parameter defined to describe the attenuation of radiation in a medium. Suppose we consider a beam of intensity I_o passing through a uniform slab of material of thickness Δx. Due to interactions with the medium, the exit intensity of the beam is I, changed from I_o by ΔI.

$$\Delta I \propto I \qquad \Delta I \propto \Delta x$$

Fig. 3.32 Absorption in slab of thickness Δx.

The **optical density** is defined as $D = \log_{10} I_o/I$. This is related to the **linear attenuation coefficient** μ in the following manner. The attenuation coefficient is defined such that

$$\Delta I = -\mu I \Delta x$$

Hence

$$\int_{I_o}^{I} \frac{\Delta I}{I} = -\int_{o}^{x} \mu dx; \qquad \ln I/I_o = -\mu x \text{ and } I = I_o\, e^{-\mu x}$$

Thus

$$D = \log_{10} \frac{I_o}{I} = \frac{-\mu x}{2.303} \quad (\mu\ \text{cm}^{-1})$$

If the absorption takes place by N absorption centres within a medium, an absorption cross-section σ (in units of m^2) can be associated with each centre such that the linear absorption coefficient $\mu = N\sigma$. This definition of a linear absorption coefficient and a quantity analogous to optical density is important in a number of areas of physics. It will be encountered for the absorption of gamma rays. For gamma ray attenuation the **mass attenuation coefficient** is used, $\mu_m = \mu/\rho$ cm^2/gm, where ρ is the density (page 262).

Photographic exposure curve. The exposure curve for a photographic film shown in Fig. 3.33 has a common form for all types of films when expressed as a graph of D versus log E, where E is in seconds. Such a plot has a straight line section for $0.1 < D < 0.7$. The **speed** of the film is given by the intercept; the **contrast** by the slope γ.

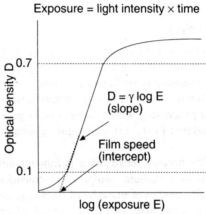

Fig. 3.33 Optical exposure curve for photographic film.

3.7 LINEAR POSITION SENSORS

Large distances can be measured by macroscopic physical phenomena—for example by counting revolutions or part revolutions of a wheel attached to a screwed rod. Ingenuity can be used to create an output which requires the counting of pulses.

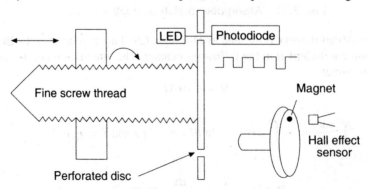

Fig. 3.34 Systems for pulse counting rotary/linear motion.

The linear motion of a screwed rod can be measured by counting the number of holes in a plate attached to rod which interrupts a beam of light passing between an LED and a photodiode. A similar revolution count can be accomplished by a magnet passing near a magnetic Hall effect sensor. The use of computer controlled stepping motors allows **pulse actuation**. A pulse delivered to such a motor rotates its armature through an accurately defined angle.

Strain Gauges

Metal or semiconducting films are deposited on a flexible backing which is attached to the body under investigation. The resistance of the elements changes under strain through a phenomenon known as piezo-resistance.

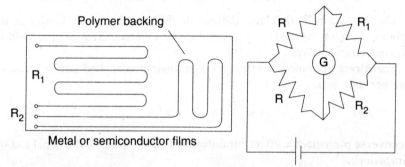

Fig. 3.35 Piezoresistive strain sensors.

The change of resistance with strain is not large and temperature effects can be of importance. This can be partly eliminated by bridge measurements (Fig. 2.26).

Piezoelectric Transducers

Piezoelectric materials are a class of crystals in which the centre of charge in the unit cell of the crystal does not coincide with the centre of mass. When an electric field is applied to a piezoelectric material, a mechanical distortion takes place resulting in a change of dimensions. If a mechanical stress is applied, charge separation occurs and an electric field is induced.

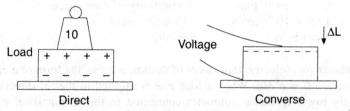

Fig. 3.36 The direct and converse piezoelectric effects.

Quartz is the best known piezoelectric single crystal. For practical applications, a ceramic—lead zirconate titanate ($PbZrO_3$:$PbTiO_3$—often known as PZT) is widely used. The polycrystalline ceramic with randomly oriented grains is caused to become piezoelectric by 'poling'. After fabrication of the ceramic by high temperature sintering, a large electric field is applied to the ceramic at relatively high temperature to orient the piezoelectric domains within the grains. This process of 'poling' is the electrical equivalent of magnetizing a permanent magnet. The direction in which the field is applied is known as the poling axis. This defines a direction (known as the '3' direction in the crystal) to which mechanical effects can be referred. In a ceramic, all directions in the plane perpendicular to the poling axis are equivalent. These are known as the '1' or '2' directions.

The piezoelectric equations of state relate four electric and elastic variables. The effects are linked by tensor equations

E—Field	$D = \varepsilon E$—Displacement	S—Strain	T—Stress
(V/m)	(C/m^2)	($\Delta L/L$)	(N/m^2)

The displacement D is also defined as the surface charge density at the electrodes $\sigma = Q/A$, where Q is generated either by an external electric field or by piezoelectric activity.

The **direct** piezoelectric effect (charge separation/applied stress) creates a charge per unit area.

$$D = \frac{Q}{A} = dT + \varepsilon^T E$$

The **converse** piezoelectric effect (strain/applied field) creates a fractional change in dimensions.

$$S = s^E T + dE$$

The coefficients ε^T and s^E are defined for constant stress T and field E respectively. A piezoelectric is often used under conditions of small E or no stress where one of the terms in the equations can be neglected. The piezoelectric parameters are tensors and the nomenclature d_{33} means that something done in the '3' direction results in an effect in the '3' direction. d_{31} means an action in the '3' direction and a result in the '1' direction. Representative values of the parameters for PZT are

$d_{33} = 285 \times 10^{-12}$ C/N	Charge constant
$d_{31} = -115 \times 10^{-12}$ C/N	Charge constant
$\varepsilon^T = 1350 \times \varepsilon_o$	Permittivity
$\varepsilon_o = 8.85 \times 10^{-12}$ F/m	Permittivity of free space
$s^E = 12.5 \times 10^{-12}$ m^2/N	1/Young's modulus
$\rho_d = 7600$ kg/m^3	Density

A planar piezoelectric transducer of thickness L has the form of a capacitor of capacitance $C = \varepsilon^T A/d$. When a load F/A is applied in the '3' direction with a relatively low impedance voltmeter connected to the terminals ($E = 0$), the piezoelectric effect causes a charge separation Q/A on the electrodes. This charge produces a voltage $V = Q/C$.

$$D = \frac{Q}{A} = d_{33}\frac{F}{A}, \quad \text{so that } Q = d_{33} F$$

$$V = \frac{Q}{C} = \frac{d_{33}L}{\varepsilon^T \varepsilon_o A} \quad (V)$$

If a voltage V is applied across a planar transducer of thickness L, the stress induced is $\Delta L/L = d_{33}$. V/L, so that $\Delta L = d_{33}$ V.

Applications of piezoelectrics. Piezoelectric discs and rings can be driven into mechanical resonance by an ac electric field. In general, for free-standing resonators, the resonance modes can be deduced by assuming that the dimension setting the frequency corresponds to one half the vibrational wavelength at the resonant frequency (Fig. 3.37).

The thickness mode resonance of a plate of thickness d corresponds to an ultrasonic wave having wavelength $\lambda = 2d$, propagating with velocity $c = \sqrt{1/s^E \rho_d}$ m/s in the plate ($f = c/\lambda$). Similar calculations can be made for the width or length of a slab transducer.

Cantilever or bimorph beams with PZT fabricated on silicon and operating in the d_{31} mode have made the fine motions required in atomic force microscopy (page 318) possible.

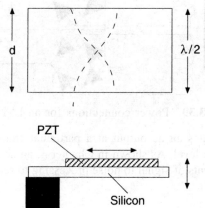

Fig. 3.37 Elastic strains or displacements leading to mechanical vibrations.

The high impedance of ceramic PZT ($10^9 - 10^{12}$ Ω) or quartz requires that high input impedance preamplifiers must be used for measurement as force or vibrational sensors.

Linear Variable Differential Transformers (LVDT)

This is an important sensor for linear motion. LVDTs create inductive ac voltages in two coils connected so that the induced voltages are opposed. As shown in Fig. 3.38(b) at the centre position of the armature the coil voltages are equal and the output voltage is zero. Movement of the armature changes the coupling to each coil and a net output voltage is recorded. The output is in dc after rectification, with the frequency of operation being normally between 60 Hz to 20 kHz. The sensitivity is high at millivolts/25 μm.

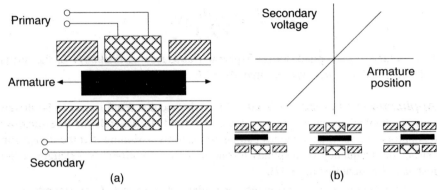

Fig. 3.38 (a) Construction and (b) operation of a linear variable differential transformer.

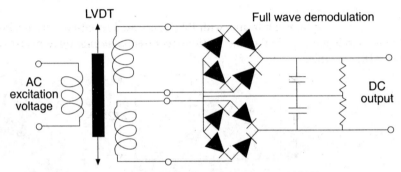

Fig. 3.39 Power connections for an LVDT.

This device creates an ac output at a particular frequency which is then rectified to create a dc level. As shown in Chapter 4, an ac output fits well with many of the requirements of signal to noise processing to reduce spurious signals at other frequencies.

Capacitance Transducers

If a parallel plate capacitance is maintained at constant potential, a change in capacitor spacing will be detected as a change in capacitor charge, i.e. as a flow of current in an external circuit. This system has good high frequency response and is highly sensitive (see Problems 3.3 and 3.10).

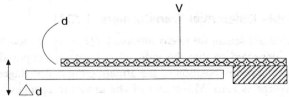

Fig. 3.40 A capacitance sensor utilizing the charge flow to the plates when the plate separation is changed. The voltage V is a dc bias applied from an external supply.

3.8 SUMMARY AND CONCLUSIONS

This review of transducer design and characteristics shows:

1. Transducers may be designed which have characteristics suitable for a particular circuit application.
2. They may be resistive, or current or voltage sources.
3. Ingenuity in physics can lead to novel devices and materials of high practical importance.

REFERENCES AND FURTHER INFORMATION

1. **General Purpose Transducers, Devices and Instrumentation**

   ElectroSonic Inc., 1100 Gordon Baker Road, Willowdale, Ontario M2H 3B3, Canada.

   Edmund Scientific Catalog, Edmund Scientific Company, 101 E. Gloucester Pike, Barrington, New Jersey 08007-1380, USA.

   Efstonscience Inc., 3350 Dufferin St, Toronto, Ontario M6A 3A4, Canada.

   Microswitch Specifier's Guide for Solid State Sensors, Microswitch, Honeywell Ltd, (Current, Hall Transducers, current sensors, airflow), 740 Ellesmere Road, Scarborough, Ontario M1P 2V9, Canada.

   Building Scientific Apparatus: A Practical Guide to Design and Construction, 2nd ed., J.H. Moore, C.C. Davis and M.A. Coplan, Addison-Wesley Inc., Redwood City, California, USA.

2. **Data Tables and Materials Characteristics**

   Handbook of Chemistry and Physics (Various editions), The Chemical Rubber Co, 18901 Cranwood Parkway, Cleveland, Ohio 44128 (USA) (Thermocouple data, X-ray crystallographic data, properties of inorganic compounds).

   American Institute of Physics Handbook, American Institute of Physics (Optical data, spectroscopy, various thermal properties, emissivity).

   American Ceramic Society—*Annual Source Books 1986-1995.*

   These are valuable sources of data on a wide range of ceramic and dielectric materials.

3. **Temperature Measurements (and other sensors)**

   Omega Temperature Measurement Handbook and Encyclopaedia, Omega Engineering Inc., P.O. Box 2284, Stanford CT 06906, USA.

   Omega also have an excellent set of handbooks related to pressure and strain, pH and conductivity, flow and level gauges, and data acquisition.

   Thermistor Sensor Handbook, Thermometrics, 808 US Highway 1, Edison, New Jersey 08817, USA.

4. **Low Temperature Sensors**

   G.K. White, *Experimental Techniques in Low Temperature Physics,* 3rd ed., Oxford, 1979, UK.

F. Pobell, *Matter and Methods at Low Temperatures,* Springer-Verlag, Germany, 1992.

Lake Shore Measurement and Control Technologies, Lake Shore Cryotronics Inc., 64 E. Walnut St., Westerville, Ohio 43081-2339, M USA Canberra Packard Canada Ltd., 6470 Van Deemter Court, Mississauga, Ontario L5T 1S1, Canada.

5. **Optoelectronics and Light Sensors**

EG&G Optoelectronics Data Book, '91 Edition, EG&G Vactec, 10900 Page Boulevard, St. Louis, 1991, Missouri, 63132, USA.

RCA Photomultiplier Handbook—PMT-62, RCA, New Products Division, Tube Operations Marketing, Lancaster, Pennsylvania 17604, USA.

6. **Piezoelectrics and other Ceramic and Dielectric materials**

Piezoelectric Ceramics, Philips Appln. Books, Edited by J. van Raderat, Technical Publications Department, Philips Electron Devices, 116, Vanderhoof Ave., Toronto 17, Canada.

Piezoelectric Ceramics— Product Catalogue and Application Notes, Sensor Technology Ltd (BM Hi-Tech Division), P.O. Box 97, 20 Stewart Road, Collingwood, Ontario L9Y 3Z4, Canada

Pyroelectric Sensors: Application Notes, GEC-Marconi Materials Technology Limited, Caswell, Towcester, Northamptonshire NN12 8EQ, UK.

Convenient packaged *high voltage power operational amplifier circuits for exciting ceramic piezoelectric resonators* (PA85 or PA87) are now available from:

Apex Microtechnology Corporation, 5980 North Shannon Road, Tucson, Arizona 85741, USA (Application Hotline 800–546–2739).

7. **Linear Motion**

Kulite Strain Gauge Manual, Kulite Semiconductor Products Inc., 1 Willow Tree Road, Leonia, New Jersey 07605, USA. Durham Instruments, P.O. Box 426 Pickering, Ontario L1V 2R7, Canada.

Linear Variable Displacement Transducers, Schaevitz Inc., US Route 130 and Union Avenue, Pennsauken, New Jersay 08110, USA. (In Canada, Durham Instruments—see above)

Sensotec, *Pressure Transducers, Load Cells, Torque Transducers, Accelerometers, LVDTs,* 1200 Chesapeake Ave, Columbus, Ohio 43212, USA.

8. **Integrated Sensors**

Optoelectronics, Hall Effect Magnetic Field Sensors, Stepping Motor Controls, Allegro MicroSystems Inc., 115 Northeast Cutoff, Box 15036, Worcester, Massachusetts 01615, USA.

ANSWERS TO QUESTIONS

(i) 276 compared to 263 Ω, 31°C low

(ii) 1500 Ω, 42 Ω/°C

(iii) 3.96 mV, 4.23 mV, 3.74 mV

(iv) 270°C

(v) 570 kΩ,

Since the output change is only from 600 kΩ to 572 kΩ, it would be most effective to use a bridge circuit using 600 kΩ resistors.

(vi) 1.45×10^{-11}C, 14.5 mV

(vii) 27.6 mV

DESIGN PROBLEMS

3.1 (a) A 14 stage photomultiplier tube has a stage gain of 3.5 secondary electrons per incident primary. What is the overall amplification of the tube?

(b) If such a tube has a gain of 10^8, a dark current of 1.2×10^{-10} A and the photocathode is of area 20 cm^2 with a quantum yield (electron/photon) of 3.5×10^{-2}, what is the minimum detectable light intensity in watts/ cm^2 if the wavelength of the incident light is 550 nm. Assume that the minimum signal current is equal to the dark current.

3.2 (a) Design a 5A fuse using copper wire. The standard fuse cartridge is 3 cm in length.

(b) Estimate the time such a fuse takes to "blow" if the normal current is 1 A and an overload of 10 A is experienced.

It is suggested that to a first approximation, the wire is considered to cool by radiation and is treated as a 'grey body' with an emissivity of 0.6. Neglect heat loss by conduction to the air or to the contact caps.

Copper: resistivity $\rho(T) = \rho_{20} (1 + \alpha (T - 20))$
$\rho(20) = 1.67 \times 10^{-6}$ Ω-cm
$\alpha = .0068$/°C
Melting point of copper = 1083°C
Specific heat = 0.38 J/g
Density = 8.92 g/cm^3

3.3 Design a non-contacting capacitance technique to measure the absolute value of the thickness mode vibration of a planar ultrasonic transducer operating at a frequency of ω_o = 27 kHz. The latter may be considered to be a disc of area 3 cm^2 with a thickness which varies as

$$t = t_o + t' \sin \omega_o t$$

Use a capacitance sensor with an area of 1 cm^2 set at distance $x_o = 1 \times 10^{-4}$ m (100 μm) from the top surface of the disc, $t' \ll x_o$. The surfaces of the transducer are silvered to provide conducting electrodes. Provide an output signal across a 1 MΩ resistor which could be observed on an oscilloscope. Estimate the sensitivity in V/m if a 300 V power supply is available.

Briefly compare the features of this arrangement with other potential measurement methods for linear displacement.

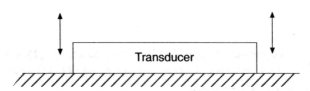

3.4 It is required to design a laser beam trimmer for printed circuit thick film resistors. It is suggested that this may be done by focussing a pulsed laser onto the materials so as to evaporate the material in the illuminated area. To a first approximation this occurs at a temperature roughly twice that required for melting. What pulse power is required in the laser if the following parameters apply. Give a dimensioned drawing of the optics.

> Laser: yttrium aluminium garnet λ = 900 nm
> Pulse length 100 μs
> Clearance above sample 0.1 m
> Beam diameter 4 mm
> Optics: limited only by diffraction-use the Rayleigh Criterion (see page 200)
> Material: gold in a borosilicate glass matrix
> Melting temperature 800°C
> Specific heat 800 J/kg-K
> Absorption coefficient for laser beam in material
> $\mu = 5 \times 10^5$ m^{-1} (see page 81)
> Density = 2400 kg/m^3

Note that heat conduction during the pulse will be small. The Rayleigh Criterion for the angular size of the first maximum in the image of a circular aperature is $\Theta = 1.22\lambda/d$, where d is the diameter of the aperture (page 200).

3.5 It is required to measure the temperature change near the surface of paper when it is illuminated in a photocopier during the copying cycle. The period of the cycle is approximately 1 second. Absolute temperature values are not as important as the change during the cycle. What kind of sensor would you suggest, and what limits would you have to consider on its size, location etc?

3.6 It is required to design a vacuum chamber in which to heat a cylindrical platinum crucible of diameter 4.5 cm and height 6 cm to a temperature of 1400°C by radio-frequency induction heating. The proposed system utilizes a cylindrical glass tube of internal diameter 7.5 cm. The thickness of the glass is 3 mm.

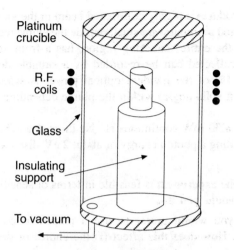

Platinum crucible

R.F. coils

Glass

Insulating support

To vacuum

In the worst possible case, one might assume that all radiation falling on the glass tube is absorbed. This would be the case if an opaque film forms on the glass. Under these circumstances, the coefficient of absorption for the glass equals the emissivity. Can you use ordinary glass of softening point 450°C for the wall, or is a vitreous quartz tube of melting point >1200°C required?

Emissivity   Pt 0.3
                    Glass or quartz 0.35

In the best case of a clean tube, a major part of the radiation will be transmitted by the glass. How might you take this into account? Estimate the reduction in wall temperature which might result. Would you now use quartz or glass?

3.7 Spectral emission lines from two elements are recorded on a photographic plate with four successive exposure times. The measured line densities are the following

Exposure (units)	Density ratio
1	22/78
2	36/64
3	36/64
4	50/50

Explain these results in terms of the exposure characteristics of the film.

3.8 It is required to measure the density of particulates emitted in the plume of smoke from the stack of an electrical generating station.

It is proposed to measure the particulates remotely by observing the scattering of light from a laser beam directed through the plume from a position 500 m distant on the ground. The laser and telescope system are mounted together.

The laser produces a highly collimated beam in the vicinity of the plume and scattering and absorption are small outside the plume region. The telescope system views the entire scattering region, has a front aperture of 10 cm² and all light collected can be recorded by a suitable detector. It may be assumed that 1% of the incident optical power is scattered isotropically (i.e., over a 4π solid angle) within the plume depending on the particulate concentration.

If you have a 30 mW continuous He-Ne laser or a 1.5 J per 30 ns pulsed ruby laser emitting a photon energy of about 2 eV, discuss as **quantitatively** as possible:

(a) Whether the experiment is feasible in terms of sensitivity. What optical detector would you use?

(b) Whether you would prefer to use a pulsed or continuous mode of operation. How does this affect (i) the choice of detector and (ii) the information you can obtain?

(c) What are the main factors which determine the detectability of the signal?

3.9 Electron induced thermoluminescence after a solid which is capable of emitting light is bombarded by an electron beam at 77 K. The beam is switched off and the solid is heated at a steady rate of increase in temperature. Free electrons created by the action of the high energy electron beam are trapped at impurity levels within the solid at low temperature, and are released with the emission of light when the temperature becomes sufficiently high.

Design an experimental apparatus suitable for investigation of such a phenomenon for electron beam energies of 0–30 kV.

The temperature range will be from 77 to 400 K with heating rates of the order of 10°/s, and excitation and viewing of the sample should be possible without moving the sample.

Give a block diagram of the overall components of the apparatus, a sketch of a suitable sample holder and electron beam system. Briefly outline the method of operation of any major items of equipment you use, and describe the transducers used to measure temperature, pressure, light intensity etc.

3.10 A capacitance transducer may be designed to sense the displacement of metal surfaces using one electrode kept at constant potential V:

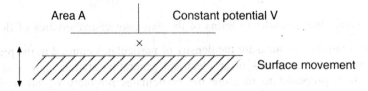

Movement of the surface causes the capacitance between the grounded surface and the plate to change, and hence to cause a change in charge on the top plate which can be measured.

(a) Work out an expression for the sensitivity of such a device in terms of the charge induced for a change in gap 'x' due to surface movement (cm^{-1}).

(b) Assume that a charge amplifier is available for connection to the device to produce a voltage output of 10^{-3} V/pC (1 pC $= 10^{-12}$ C). Design the transducer so that, for a plate area of 0.5 cm^2, an output of 1 mV is available for a surface movement of 10^{-8}m (10 nm). The device operates in air at atmospheric pressure ($\varepsilon_r = 1$ and $\varepsilon_o = 8.85 \times 10^{-12}$ F/m). The breakdown field of the air gap is 3×10^6 V/m, and the field applied should not exceed 80% of this value.

(c) Comment on some of the design features necessary for the successful operation of such a device.

3.11 A ratio pyrometer is designed using an infrared detector and two narrow band filters. The temperature $T < 1500$ K and the wavelengths used are in the visible region of the spectrum.

(a) If the bandwidths of the two filters defining the wavelength regions λ_1 and λ_2 are $\Delta\lambda_1$ and $\Delta\lambda_2$ respectively, work out the ratio between the ratio of the measured intensities in each wavelength region and the temperature T.

(b) Why may this ratio and the resulting temperature be weakly dependent on emissivity?

WORKED EXAMPLE

Problem Definition

It is proposed to measure the reflectivity (R) of a high T_c superconductor as a function of temperature (T) in the range 77 K to 120 K over a narrow band of wavelengths in the visible region of the spectrum. Design an experimental arrangement to undertake this measurement.

Measure temperature and reflected light intensity from an appropriate source. The sample is a disc 1 mm thick and 1 cm^2 in area. In order to prevent the condensation of water vapour from the atmosphere on the surface of the sample, it must be enclosed within a vacuum chamber 5×10^{-6} mbar. Temperature control should be based on the use of liquid nitrogen as a coolant and a heater.

Sketch an appropriate apparatus for the measurement and draw a block diagram showing all sources, detectors, vacuum systems and measurement transducers needed to provide a plot on an x-y recorder of R vs T. Briefly describe the main features of each component.

What are the potential sources of error in the measurement of R and T for the sample. What signal processing would you suggest to minimize these uncertainties?

The *correct solution* of such a design problem is that the apparatus should *work*. The full solution to this problem requires information from a number of chapters of this book. The problem is summarized here to show the stages through which the information is brought together. These include making a careful list of the requirements of the experiment, a statement of the main principles in a block diagram and details of specific units.

Design parameters required

Reflectivity in visible region; Source–He/Ne Laser; Detector Photodiode

Temperature; 77 K to 120 K—Thermocouple: copper/constantan

Vacuum; 5×10^{-6} mbar—Diffusion/backing pump system

Vary temperature continuously (but slowly) and plot output on recorder.

The experimental arrangement is discussed (a) as an overall block diagram, (b) with special attention to the design of the cryostat and vacuum system.

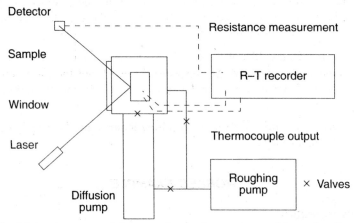

Fig. 3.41   Main elements of the experimental apparatus without details of each component.

Cryostat design.   A source of heat, a source of cold and a way of stopping condensation all over the sample is required to allow the apparatus to operate continuously (Fig. 3.42).

The liquid nitrogen reservoir is filled and used to cool the copper block as a source of *cold*. The latter is isolated by the vacuum space and by the double-walled thin wall stainless tubes. *Heat* is added by a heater wound on the constriction on the copper rod between the bottom of the reservoir and the sample. Measurements up to room temperature can be made by allowing the liquid nitrogen to evaporate and then allowing the block to warm naturally, or to add heat through the heater. The reflectivity measurement should have no major time dependence so that measurements during heating should lead to small error. The temperature is measured by a thermocouple with voltages in the range –5 mV to +5 mV.

Principle of Operation

The reflectance is measured by assuming that the laser has a constant output and recording the output of the photodiode after the beam has been reflected from

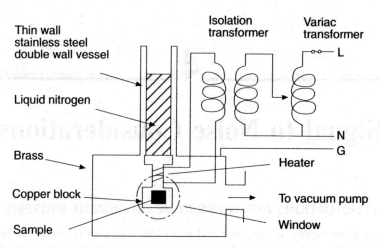

Fig. 3.42 Design features of the apparatus (This sketch would eventually form the basis of an engineering drawing which would enable the apparatus to be constructed.)

the sample surface. For linearity, use the photodiode in its reverse-biased mode. (page 79).

The light beam has to be admitted through a window. For a He/Ne laser source at 653.2 nm this can be glass. If an UV or IR source is used, use quartz or LiF (see page 198).

The temperature is measured as a voltage output from the thermocouple. This is non-linear in this range, but the temperature can be determined by reference to tables or from a computer power series (see page 61). Use an isolation transformer to provide some protection to the operator if the heater short circuits (see page 332).

The vacuum system uses an oil based diffusion pump, backed with a mechanical rotary pump (page 176). Valves are added to isolate the chamber and to allow air to be admitted for sample changing. A Penning Gauge is used to measure the system pressure (see page 180).

Potential Sources of Error

Reflectance. Stray light in the room may give a spurious reading on the detector. Use a chopper wheel and lock in detector to only measure the reflected component (see page 113).

The laser intensity may change. Use two photocells to measure signals proportional to the incident and reflected beams and use an analog divider to provide a ratio signal (see page 145).

Temperature. The thermocouple is attached to the block and may not be the real temperature of the sample. Attach a fine thermocouple wire to the sample itself. Use a surrounding radiation shield of aluminum foil to reduce the radiated heat falling on the sample from the surroundings (page 65).

4

Signal to Noise Considerations

4.1 FLUCTUATIONS AND NOISE IN MEASUREMENT SYSTEMS

When a phenomenon is observed and recorded, information about the process involved is obtained from the value of a quantity, e.g. light intensity or number of counts per second. It is desirable that this value reflect the stable or 'true' behaviour of the phenomenon. However, the recorded quantity usually shows **variations** with time as may be observed from a plot of the data shown in Fig. 4.1.

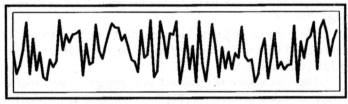

Magnitude x

Time

Fig. 4.1   Fluctuations in a signal voltage.

The origin of these fluctuations may be:

- (a) **inherent** in the process, e.g. the flickering of a candle flame, or
- (b) **measurement-related**, e.g. due to instabilities in the amplification of a measuring instrument.

In instrumentation, these fluctuations are termed 'noise'. If the noise obeys Poisson statistics, the width of the distribution of measured voltages is given by the square root of the mean value (page 14).

Signal to Noise Ratio

The relative magnitude of noise voltage is defined by

$$\frac{S}{N} = \frac{\text{average value of signal} \quad \text{(volts)}}{\text{rms value of fluctuations} \quad \text{(rms volts)}}$$

although it can also be defined in terms of a **power ratio**.

$$\frac{S}{N} \text{ (dB)} = 10 \log \frac{\text{(signal power)}}{\text{noise power}}$$

The purpose of many measurement techniques is to increase the signal to noise ratio so that the **detectability** of the signal is increased. Simple amplification **cannot** do this since the noise is amplified by the same factor as the signal. Means have to be sought *to distinguish the signal from the noise.*

Noise Figure

Noise may be contributed by each stage of a measurement system: the initial phenomenon, the detecting instrument, the amplifier chain and the display. The noise voltage contributed by each stage is defined by

$$\text{Noise Figure } F = \frac{S/N \text{ (input)}}{S/N \text{ (output)}}$$

$F = 1$ if no noise is contributed by the stage, $F > 1$ for all real systems; $F = 1.3$ for many semiconductor devices.

Question (i)   The voltage output from a transducer has a steady value of 0.95 V with a fluctuating component of 0.35 V rms. If the noise figure of the transducer is 1.3, what is the signal to noise ratio in the measured quantity?

4.2   NOISE IN THE FREQUENCY DOMAIN

The *x-t* graph in Fig. 4.1 shows the time variation of a quantity. The average (signal) value is constant (or slowly varying) with time (dc) while the noise has a complex waveform. These differences are most conveniently described by the frequency spectrum.

Fourier Analysis

The Fourier Transform converts an **amplitude** variation as a function of time [$f(t)$] to a **complex amplitude and phase** spectrum as a function of frequency [$F(\omega)$].

$$F(\omega) = \int_{-}^{+} f(t) \, e^{-j\omega t} \, dt \quad (\omega = 2\pi f)$$

The amplitude $|F(\omega)|$ has units of volts per Hz. $\omega = 2\pi f$ is the angular frequency. The Fourier power spectrum is calibrated by noting that, in principle, the power spectrum S_p is given by the square of the voltage amplitude v_n generated across a standard 1 ohm resistance ($S_p = v_n^2/(1)$).

Examples of such calculations for particular waveforms are given in Appendix 2. Important technical examples are shown in Fig. 4.2.

For a completely random waveform, an infinite range of frequencies exists. For a pulse waveform, if the pulse is infinitely short ($T \to 0$), $F(\omega)$ extends to infinite frequency. Instrumental considerations invariably preclude both of these possibilities.

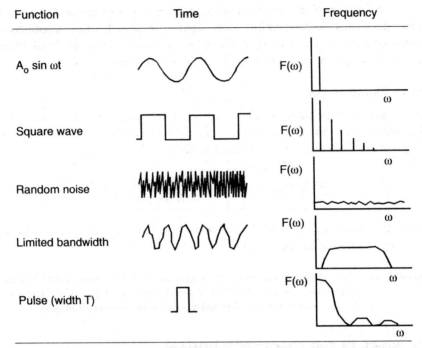

Fig. 4.2   The Fourier transform of various waveforms of technical importance.

Question (ii)   (a) Figure 4.2 and Appendix 2 show that the frequency components of a symmetrical square wave include only <u>odd</u> harmonics of the fundamental frequency. Sketch the waveform and by successively adding frequency components justify this fact.

(b) If you consider the **shape** of the waveform, what information is carried in the frequency spectrum (i) at low frequencies, (ii) at high frequencies?

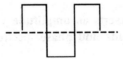

Frequency Bandwidth

The importance of the frequency domain is that it can describe the frequency bandwidth in which the signal information is carried, and distinguish this from the frequency bandwidth over which noise exists. For example, the waveform shown in Fig. 4.1 can be analyzed in the frequency domain as having two components: one defining the signal and other defining the noise. This is shown in Fig. 4.3.

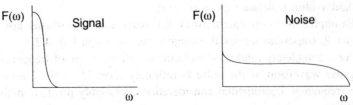

Fig. 4.3   Frequency spectra for signal and noise waveforms.

Because the signal is likely to be well defined in time, the frequency spectrum which contains the signal information is relatively limited in width. The noise spectrum originates from a wide variety of sources and is much broader in frequency bandwidth.

The frequency limiting effects of various stages in the measurement system must also be described. Instruments (amplifiers) are built with a specific frequency bandwidth Δf. Circuit components form low pass and high pass filters (page 38). All frequency components of a signal falling within the measurement bandwidth will be detected, all those outside will not be observed. Signal to noise ratios must therefore be calculated *taking into account the frequency bandwidth of the measurement system.*

The frequency bandwidth for a measurement system may be represented by a frequency dependent transfer function $T(\omega)$. In the ideal case (a), the frequency response is. limited at a specific frequency (Fig. 4.4(a)).

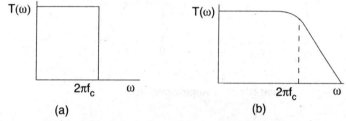

(a)   (b)

Fig. 4.4   (a) An ideal response with a sharp cut-off; (b) Non-ideal response of a *RC* filter.

For analog circuits such a response cannot be achieved, although it is possible using digital techniques. The transfer function for real analog systems falls off more slowly (Fig. 4.4(b)). For such a **low-pass** *R-C* circuit, the cut-off frequency f_c is defined as the frequency at which $T(\omega)$ falls by 3dB or to $1/\sqrt{2}$ of its initial value. The analysis for the transfer function is given on page 39.

The frequency-axis for graphs of the type given in Fig. 4.4 is usually plotted on a logarithmic scale. This can be misleading with respect to the range of frequencies over which the response is frequency dependent. A linear plot of the data shown in Fig. 4.4(b) will clearly demonstrate that the rate of fall-off of circuit response with frequency for analog *R-C* circuits is relatively slow.

Equivalent Bandwidth of a Pulse of Duration T

The waveform of a pulse of width T seconds and amplitude V_0 is of importance in digital logic. It is also of general importance in instrumentation as the transfer function $T(t)$ for measurement systems which act (measure or count) for a fixed time. Thus the action of closing a switch for a fixed period can be represented by a rectangular pulse in time. The **equivalent bandwidth** for this waveform will be required on a number of occasions.

The Fourier Transform of a pulse having amplitude V_o over a time from $-T/2$ to $+T/2$ has the mathematical form of

$$F(\omega) = V_o T \frac{\sin(\omega T/2)}{(\omega T/2)}$$

This function has a primary peak extending to an angular frequency of $2\pi/T$, with smaller maxima at higher frequencies.

While in principle it would be possible to work with the actual form of the transform in circuit calculations, a reasonable approximation can be achieved by defining an **equivalent bandwidth** f_o. This is defined by drawing a rectangle of the same height as $F(\omega)$, but extending up to an angular frequency $2\pi f = \pi/T$ such that the total area under the first lobe of $F(\omega)$ equals the area under the rectangle.

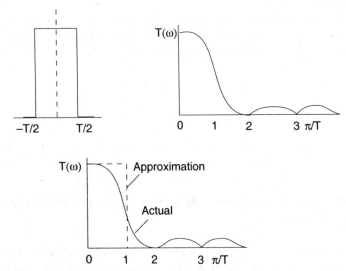

Fig. 4.5   Successive approximations for the Fourier Transform of a pulse.

An analysis on this basis gives

$$f_o = \frac{1}{2T}$$

The bandwidth of a pulse 1 millisecond in duration is 500 Hz. If a gate is opened for 1 ms, the frequency response has a bandwidth of this value. This relationship is particularly important for instruments which have a gating action as part of their operation.

Question (iii)   A dual slope digital voltmeter (page 111) has a gate time of 400 ms reflecting the measurement cycle. What is the frequency bandwidth of this instrument?

4.3 SOURCES OF NOISE

Inherent Fluctuations

Some experimental systems are inherently unstable and lead to measurements which fluctuate in time. Examples are optical fluctuations in starlight passing through the earth's atmosphere, radiowaves in the ionosphere, or bad electrical connections at contacts to a semiconductor. In general, this inherent noise has to be treated statistically or by noting the variation over periods of time. It is always important to treat instabilities with interest—observations of fluctuations in starlight led to the discovery of pulsars and quasars, while consideration of unstable microwave emissions from metal junctions at low temperatures led Brian Josephson, then a graduate student at Cambridge University, to win a Nobel Prize for the Josephson Effect!

Some sources of noise are inherent to particular physical systems. These include Johnson Noise in resistors, and Shot Noise in systems where charge collection occurs.

Thermal or Johnson Noise

> Occurs due to the thermodynamic fluctuations of the electron gas in a conductor.

The density of the electron gas in a resistance R fluctuates spatially leading to variations in potential difference v_n between the contacts. The variations arise because of the thermal motion of the electron gas, and the possible frequencies are, in principle, unlimited. This is termed **white noise**. The magnitude of Johnson noise may be estimated by considering the energy which is available to generate the fluctuations, and the manner whereby the power is communicated to an external circuit.

Thermal energy available at temperature T is: $\sim kT$
(k is Boltzmann's Constant 1.38×10^{-23} J/K)

This energy is spread across the entire frequency spectrum. If the measurement system has a limited bandwidth Δf, only a fraction of the energy will be measured.

Noise power available to a measurement system of frequency bandwidth Δf is given by: $\sim kT \cdot \Delta f$

The resistor may be treated as a perfect 'noiseless' resistance R in series with an internal voltage source v_n. The maximum power that can be transferred to an external load resistance R_E occurs when $R_E = R$. The terminal voltage is then $v_o = v_n/2$.

Noise power $= kT \cdot \Delta f = \dfrac{v_o^2}{R_E} = \dfrac{v_n^2}{4R}$

rms noise voltage $v_n = \sqrt{4kTR\Delta f}$

The frequency spectrum is flat, in principle to infinite frequency. In practice, it is limited by circuit connections which act as high or low pass filters. Johnson noise exists in all resistive circuits; it cannot be eliminated, but its effects can be reduced.

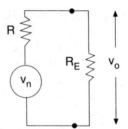

Fig. 4.6 Circuit diagram for Johnson noise measurements.

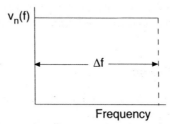

Fig. 4.7 Frequency spectrum of Johnson noise—flat over bandwidth Δf.

Shot Noise

> Occurs due to the collection of electrons at an electrode, or by diversion over a barrier.

This type of noise is characteristic of any system in which charge collection occurs statistically. This takes place in the current flowing to the anode of a photomultiplier, in minority carriers flowing across a junction diode, or across the base of a junction transistor to the collector. It is a direct consequence of Poisson statistics.

Consider N number of charges q collected in time T

$$\text{Mean current } I_o = q\,N/T$$

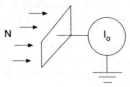

Fig. 4.8 Charge collection at an electrode.

Since this is a random process, there will be a statistical uncertainty of $\pm \sqrt{N}$ in the number of charges collected in this time,

$$I = I_o \pm I_{rms}$$

$$\frac{I_{rms}}{I_o} = \frac{\sqrt{N}}{N} = \frac{1}{\sqrt{N}} = \frac{1}{\left(I_o\dfrac{T}{q}\right)^{1/2}}$$

The root mean square value of the shot noise current is

$$I_{rms} = \left(\frac{qI_o}{T}\right)^{1/2}$$

This is a result of a counting process over a time interval T. We can therefore replace it with the equivalent bandwidth of a pulse of this time duration ($\Delta f = 1/2T$).

rms shot noise current $\boxed{I_{rms} = \sqrt{2qI_o\Delta f}}$

These fluctuations can occur at any frequency, and therefore the noise is 'white'.

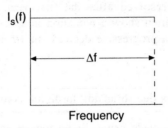

Fig 4.9 Frequency spectrum of shot noise—flat over bandwidth Δf.

1/f **Noise**

> 'device' or 'excess' noise
> semiconductor generation/recombination noise

Many systems show more noise **power** at low frequencies than is predicted by the thermodynamic formulae. This excess noise is often determined by unknown effects at surfaces, contacts and barriers and improves as devices are better designed or manufactured for a longer period of time. 1/f noise is also associated with the generation and recombination of minority carriers in semiconductors.

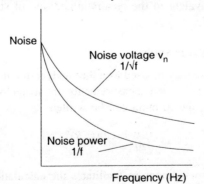

Fig. 4.10 The frequency dependence of noise **power** and noise voltage of 1/f noise.

> Noise power varies as 1/(frequency).
> Noise voltage varies as 1/$\sqrt{}$(frequency)

An empirical law known as the 'Hooge Law' describes many results in semiconductor materials.

$$\text{Noise power} = \left(\frac{\alpha}{N_f}\right) \cdot \frac{\Delta f}{f}$$

where N_f is the density of electrons or holes, Δf is the measuring bandwidth, f is the frequency and $\alpha = 2 \times 10^{-3}$ W/cm^3 is an empirical constant.

This type of noise is not well understood, but it is very prevalent and consideration of its effects is an essential part of instrumentation design. The phenomenon extends into everyday life. It is often recommended that one should not buy an automobile from the first year of introducing a model because initial faults and defects are resolved after the first year. Specification sheets for semiconductor devices often show a reduction in specified noise levels as years proceed. This is due to a progressive decrease in $1/f$ noise.

Quantization Noise

Occurs due to analog to digital conversion.

Many measurements are made using digital meters having a certain number of digits in the display. Such an instrument can be no more accurate than ±1 in the last digit.

Multiple Noise Sources

In experimental situations where several sources contribute to the total noise power

$$NP(\text{total}) = NP(1) + NP(2) + NP(3) + \dots$$

In terms of rms voltage contributions, this implies that

$$v_n^2 \text{ (total)} = v_n^2(1) + v_n^2(2) + v_n^2(3) + \dots$$

This expression is equivalent to the quadratic addition of standard deviations in error analysis (page 17).

Johnson Noise Contour Plots

Johnson or thermal noise may be used as a thermodynamic method of temperature measurement (page 69). If such measurements are made with a rms voltmeter having a bandwidth Δf, the formula can be written

$$\frac{v_{\text{rms}}}{\sqrt{\Delta f}} = (4kTR)^{1/2}$$

A graphical plot or nomogram facilitates the calculation of Johnson noise voltage for a resistance R at a given temperature for any bandwidth (Fig. 4.11). For example, a horizontal line drawn from 1 MΩ on the resistance axis will intersect with a vertical line extending up from 300 K at a point close to 100 nV/$\sqrt{\text{Hz}}$. If the band width of the measurement system is known, the noise voltage can be calculated.

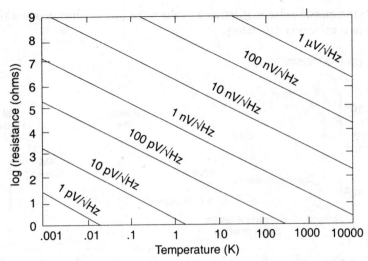

Fig. 4.11 Nomogram to determine Johnson noise voltage from resistance R at absolute temperature T.

Question (iv) An average photogenerated current of 100 µA in a reverse-biased photodiode is measured using a series resistance of 10 kΩ and an amplifier with a bandwidth of 100 kHz: (a) Compare the relative contributions of Johnson and Shot noise voltage, (b) What is the signal to noise ratio at the amplifier output? Assume that the noise figure is 1 and the temperature is 293 K.

4.4 SIGNAL TO NOISE AND EXPERIMENTAL DESIGN

The design of experiments and instrumentation must optimize the inherent signal to noise ratio and consider how the final signal to noise ratio may be enhanced by suitable signal processing. The factors to be considered are:

A good initial experiment having high *S/N*.

Minimum noise introduced by sensors and electronic signal processing.

The frequency bandwidth for: (a) the signal information, (b) the noise.

Specialized methods for *S/N* enhancement.

Experimental Design

The requirement for the highest signal to noise ratio in the basic experiment would seem self-evident. However, in these days of easily available sophisticated signal processing and computing methods it is easy to forget. One requires:

(a) an experimental arrangement giving the highest possible signal—this is a measure of the ingenuity of the experimenter.

(b) the highest possible reduction of spurious or environmental noise. This is usually a case of careful electronic design.

Environmental noise is illustrated by the noise power generated at a typical location in a university laboratory.

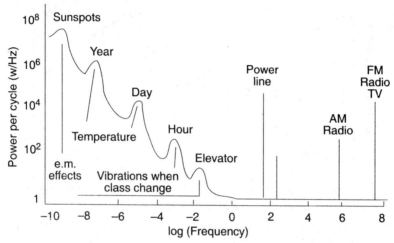

Fig. 4.12 Contributions to the noise power spectrum by noise sources in a university laboratory.

This summation includes the effects of temperature, vibrations and the electromagnetic (11 year) sun spot cycle. It is of interest that the general envelope is consistent with a $1/f$ pattern. Various harmonics of the 60 Hz power line frequency, TV and radio, and local sources of noise are of obvious importance.

Much electromagnetic noise can be eliminated or reduced by careful attention to shielding and ground connections. All equipment should be enclosed in metal with well-made coaxial connecting cables. On occasions, some sources of noise cannot be eliminated and other more serious measures must be taken. A Nobel Prize was awarded to workers at the University of Birmingham for experiments carried out exclusively between 1 a.m. and 5 a.m.—the period when the Birmingham electric trams were not operating.

Multi-Stage Measuring Systems

In general, a measurement system consists of a measuring device followed by an amplifier chain. Each of these units will introduce noise, but it is of interest to consider which part of the chain must be designed with maximum care.

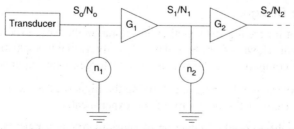

Fig. 4.13 A multistage amplifier system with amplifier noise contributions included.

Consider each amplifier as having gain G_m, with a noise voltage generator at its input. The inherent signal to noise voltage from the experiment is S_o/N_o. The various noise voltage components add in quadrature.

Signal amplitude, $S_2 = G_1 G_2 S_o$

First stage noise, $N_1 = G_1\sqrt{(N_o^2 + n_1^2)}$

Second stage noise, $N_2 = G_2\sqrt{(N_1^2 + n_2^2)}$

Hence $\quad\quad\quad\quad\quad\quad N_2 = G_2\sqrt{\left[G_1^2 \,(N_o^2 + n_1^2) + n_2^2\right]}$

Rearranging $\quad\quad\quad\quad N_2 = G_1 G_2 N_o\sqrt{1 + \dfrac{n_1^2}{N_o^2} + \dfrac{n_2^2}{G_1^2 N_o^2}}$

If the first stage gain G_1 is large, all higher terms have G_1 in the denominator and are small. The effect of noise in subsequent stages can therefore be neglected. In designing instrumentation systems, the most effort should go into the first stage. Higher stages are much less critical.

Low noise, high gain pre-amplifiers are essential to good instrumentation systems.

Noise temperature. If G_1 is large and the first stage noise n_1 is low compared to N_o, the above result can be rearranged to calculate the noise figure for a multistage system.

$$F = \frac{S/N \;(\text{Input})}{S/N \;(\text{Output})} = 1 + \frac{n_1}{N_o}$$

Since N_o often arises substantially from thermal noise in a transducer at temperature T and $N_o \sim kT$, the noise figure is often written

$$F = 1 + \frac{T_a}{T}$$

where T_a is the **equivalent noise temperature** of the system. For the best pre-amplifiers T_a = 3–10 K.

First Stage Preamplifier Design

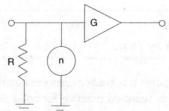

Fig. 4.14 First stage preamplifier.

An amplifier may be represented as

- an input resistance R
- a noise source n
- a 'perfect' gain G

R may include the parallel combination of an external measuring resistance R_E and the actual internal input impedance of the amplifier.

Amplifier components. Junction transistors are minority carrier devices which involve the collection of charges at a reverse biased *p-n* junction. They are therefore subject to shot noise.

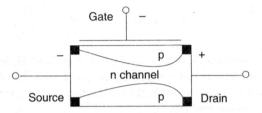

Fig. 4.15 The resistive nature of a field effect device.

Field effect transistors (FETs) are majority carrier devices which involve no collection of electrons at a junction, and therefore noise is much reduced compared with conventional junction transistors. Comparative noise figures for FET and bipolar transistors are:

FET 1–2 dB at 2 GHz

NPN 2–5 dB at 2 GHz

FET input preamplifiers are standard practice for most instrumentation. In designing circuits, metal film resistors are less 'noisy' than standard carbon resistors.

Cooling of the input stage. The input resistance R is a source of Johnson noise, while even shot noise tends to diminish at low temperatures. If the temperature is reduced from T to T_o, Johnson noise is reduced by a factor of $(T/T_o)^{1/2}$.

Temperature K	300	77	4.2
Noise reduced by factor	1	2	8.5

In low temperature experiments it is convenient to mount the first stage preamplifier onto a cooled stage. This is standard practice for solid state gamma ray detectors used in nuclear physics.

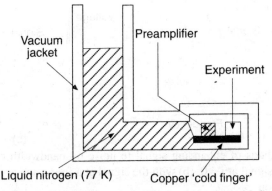

Fig. 4.16 Preamplifier mounted on a copper cold finger.

4.5 FREQUENCY AND BANDWIDTH CONSIDERATIONS

Signal Frequency Bandwidth

In designing a measurement system for a particular experiment, the frequency bandwidth within which both the signal and the noise information is contained should be known. For example, a system to monitor the solar intensity looks at slow variations in an essentially constant quantity (longest duration, day long; shortest duration, 1 second as clouds cover the sun). On the other hand, faithful transmission of a 10 ns digital logic pulse may require a bandwidth of more than 100 MHz. Some common signal frequency bandwidths are:

Experiment	Bandwidth used
Solar intensity	0.001–10 Hz
Telephone speech	0.2–5 kHz
Audio stereo system	10 Hz–25 kHz
Digital logic	20 Hz–100 MHz

A different amplifier should be used for the solar experiment than for digital logic.

> The bandwidth of the measurement system must be chosen to fit the signal bandwidth.

Noise Bandwidth

Noise voltages are a combination of Johnson, Shot and $1/f$ noise. The resulting frequency response is in Fig. 4.17.

If the signal is at dc ($f = 0$), and a **low pass filter** is used to restrict the bandwidth (a), the signal to noise is improved by removal of all high frequency noise. However, noise at low frequencies is still large. If the signal frequency is at $f = f_o$, and a **band pass filter** is used (b), the noise is reduced to its minimum possible value.

> The signal frequency should minimize external noise contributions.

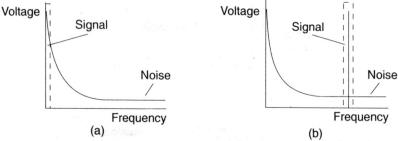

Fig. 4.17 Methods of enhancing signal to noise by bandwidth control: (a) with the signal at $f = 0$, (b) with the signal at a frequency above the $1/f$ noise region.

4.6 BANDWIDTH CONTROL

The frequency bandwidth can be controlled by **passive** or **active** methods. Passive methods are based on the circuit action of resistance, capacitance and inductance. Active methods use specific instrumentation techniques.

Low Pass Techniques

Passive low pass R-C circuits

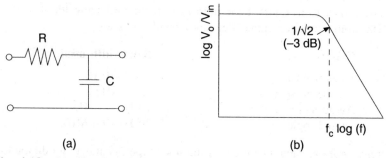

Fig. 4.18 (a) Circuit and (b) Frequency response for a low pass *RC* filter.

The frequency response of an *RC* circuit was derived on page 38. It may be noted that a large capacitance connected across the output of a 'noisy' instrument measuring a steady quantity will "smooth" the output.

Operational amplifier filters.   The same response but with a low (controlled) output impedance can be set up using an operational amplifier (Sec. 5.1).

$$f_c = \frac{1}{(2\pi RC)}$$

Active filters have the advantage that several stages can be easily put in series without interaction between the stages.

Digital voltmeters.   **Analog** voltage and current measurements are commonly

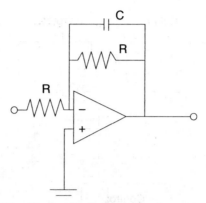

Fig. 4.19 An *RC* filter in the feedback loop of an operational amplifier.

made using **digital** multimeters. Such meters require a certain sampling time *T* to make a measurement and therefore act as an **active** low pass filter of bandwidth $\Delta f = 1/2T$ (see page 100).

As shown in Fig. 4.20, signal current i_s charges a capacitor *C* for time *T* to voltage *V*, where

$$V = \frac{Q}{C} = \frac{i_s T}{C}$$

The meter displays the results, shorts *C*, and repeats.

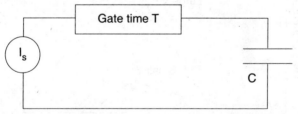

Fig. 4.20 The basic principle of an integrating voltmeter.

During time *T*, the mean value of i_s contributes to the final charge *Q*; the noise fluctuations average to zero.

Dual slope integrating digital voltmeter. In the dual slope technique, the input voltage is first converted to a positive charge by integration over a set period, and then the quantity of charge is determined by counting the number of pulses of negative charge required to discharge the integration capacitor back to zero. The number displayed is this number of pulses.

The first stage is a buffer amplifier with a gain of 1, while the second operational amplifier serves to integrate the input signal. The control circuitry zeros the output during the **autozero** period. During the **integration** period the signal charges the capacitor for a fixed number of clock pulses. At the end of this period, the voltage level on the capacitor is proportional to the average of the input signal during the integration period. During the **reference** period the capacitor is

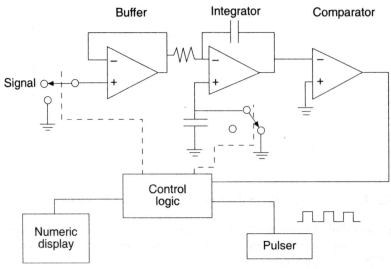

Fig. 4.21 A block diagram for the operation of a typical dual slope integrating voltmeter. (With permission—Keithley Instruments Inc.)

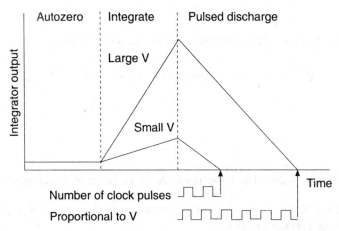

Fig. 4.22 The measurement sequence for a dual slope integrating voltmeter.

discharged by pulses of charge. The number of pulses required is the final display. In a representative voltmeter—the Keithley Instruments Model 177, the integration period T is 100 ms. The total sampling time is 400 ms so that the effective frequency response is 1.25 Hz.

Band Pass Techniques

Since $1/f$ noise introduces a large amount of noise at low frequencies it is useful to ensure that the signal frequency f_s is sufficiently high that $1/f$ noise is negligible, and then to limit the bandwidth to close to f_s. It is first considered how to do this for signals which are slowly varying with time.

Methods for Frequency Modulation

Optical 'chopper' wheels. An optical signal of slowly varying intensity is shone through a 'chopper' wheel. This consists of a disc with N holes around its periphery which is driven by a synchronous motor of frequency f. The beam is interrupted at a frequency of Nf and the output signal is modulated at this frequency. The modulation not only differentiates against noise within the signal, but also discriminates against 'background' light. Only 'signal' light which has passed through the wheel is modulated; 'background' light which may enter the optical instrument remains at dc.

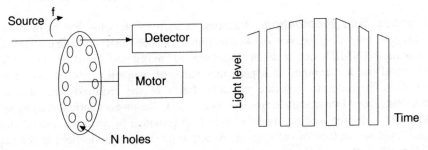

Fig. 4.23 A chopper wheel for the modulation of light beams. The signal light level is interrupted as the wheel rotates to give a series of 'pulses' at frequency Nf.

Electronic 'chopper' circuits. A dc voltage is converted to ac by an electrically driven switch or relay. Such 'chopper' amplifiers are available as integrated circuits.

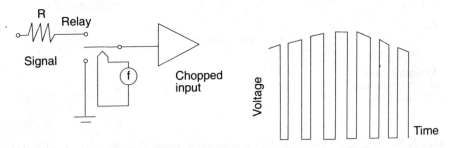

Fig. 4.24 Electronic choppers for current or voltage signals. An input resistance R is connected in series with a relay switch. The waveform presented to the amplifier is similar to that for the optical modulator.

Current or voltage modulation. In two or four probe resistance, or in magnetic Hall Effect measurements, the applied voltage or current can be alternating at frequency f. Spurious frequency independent components of the response due to thermal effects, for example, can be eliminated. Responses at harmonics ($2f$ or $3f$) can be measured.

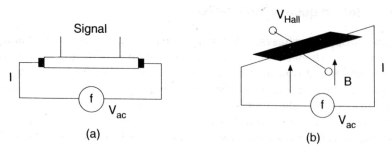

Fig. 4.25 AC driven circuits: (a) Four probe *V-I* (b) Hall effect.

Magnetic field modulation. Phenomena such as the Hall Effect in semi-conductors, magnetoresistance, or electron spin resonance (ESR) or nuclear magnetic resonance (NMR) occur as a function of magnetic field *B*. In ESR, the signal has a maximum at a particular value of magnetic field, as shown in Fig. 4.26. In order to generate magnetic fields of the required intensity and uniformity, large iron-cored electromagnets are used. The magnet current is increased smoothly with time, and the modulation field is provided by ac driven modulation coils fixed to the face of the magnet. At any time, the magnetic field in the gap is $B = B_{dc} + B_o \sin 2\pi f_s t$.

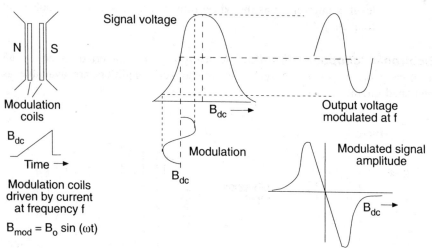

Fig. 4.26 Magnetic field modulation utilizing a steady field B_{dc} and a modulation field $B_o(f_s)$.

The output waveform is derived by the construction shown. To a first approximation if the modulation amplitude B_o is small compared to both B_{dc} and to the range of field over which the signal occurs, the output is an ac signal at the modulation frequency f_s with an amplitude proportional to the differential of the original waveform. The fact that this output signal has a zero at the magnetic field at which the original signal is a maximum, assists in locating the 'resonance'.

Bandpass Filters

The modulated signal is amplified only within a narrow band of frequencies about f_s. Circuits which have a bandpass response are as follows.

Tuned LCR circuits. A parallel LCR circuit has a high impedance when $1/\omega C = \omega L$ at a resonance frequency $\omega_o^2 LC = 1$ where $\omega_o = 2\pi f_o$. This circuit is used in the tuning of a simple amplitude modulated (AM) radio receiver.

The width of resonance peak is denoted by its full width at half maximum (FWHM) Δf and the **quality factor** Q is defined as $Q = f_o/\Delta f$. Circuit calculations show that for an LCR circuit Q is given by the ratio of the inductive impedance

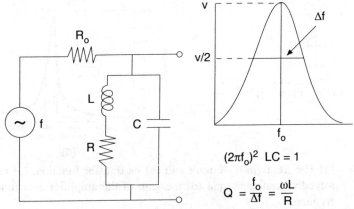

$$(2\pi f_o)^2\ LC = 1$$

$$Q = \frac{f_o}{\Delta f} = \frac{\omega L}{R}$$

Fig. 4.27 The parallel circuit and output voltage across a tuned LCR circuit. The relative frequency bandwidth $Q = f_o/\Delta f = \omega L/R$ is large at radio frequencies such as 455 kHz. Values in the frequency range of 1–10 kHz range are much smaller (<5).

(ωL) to the circuit resistance R. Most coils are wound with copper wire and the inductance L obtainable is limited for an allowed R. A typical coil for use in the 1–10 kHz range will have $L = 10$ mH and $R = 160$ ohms with a Q of the order of 4.

Question (v) Design an LCR filter for a resonance frequency of 50 kHz. A coil with an inductance of $L = 390\ \mu H$ and resistance $R = 35\ \Omega$ is available. The circuit is supplied from a 2 V rms source with a 600 Ω output impedance. (a) Determine the capacitance C and (b) the Q of the circuit. Sketch the response curve on the frequency axis. What is the voltage across the coil at $f = 10$ kHz?

R-C twin-tee filters. Inductors are large and inefficient. The restriction imposed by the LCR circuit can be avoided by the twin-tee R-C circuit. The circuit and its response is shown in Fig. 4.28.

If the components have exactly the ratios shown in Fig. 4.28(a), the transfer function at the resonance frequency f_o, where $\omega_o = 1/RC$, is identically zero. The transfer function for an actual circuit is shown in Fig. 4.28(b). Such passive RC

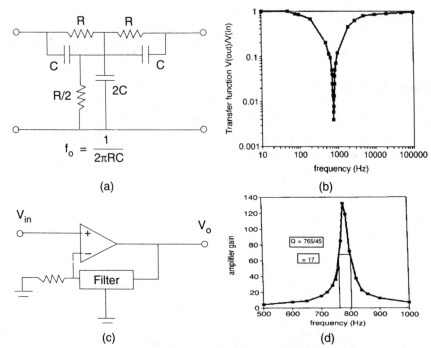

Fig. 4.28   (a) The *RC* twin-tee circuit and (b) its transfer function; (c) its use as a feedback element and (d) the gain of the amplifier as a function of frequency.

circuits can be used as 'notch' filters to eliminate, say, 60 Hz from extraneous mains voltages in a measurement. A bandpass filter can be constructed by employing the circuit as the feedback element in a non-inverting operational amplifier circuit (page 143). Figures 4.28(c) and (d) show such a circuit and its response. The *Q* value is 17.

4.7 SIGNAL TO NOISE ENHANCEMENT

The bandwidth restriction methods discussed in the previous sections provide some improvement in signal to noise, but with relatively poor quality factor *Q*. There is also no guarantee that the signal frequency and frequency response of the amplifier will remain accurately matched. In this section **correlation techniques** are discussed which are capable of changing the signal to noise ratio by orders of magnitude.

Phase Sensitive (Lock-in) Amplification

Consider a signal waveform of frequency f_s at the input of the circuit shown.

A switch allows the signal to pass to some form of averaging circuit either directly (+) or via an inverting amplifier (−). The switch is driven by an ac voltage at frequency *f* and its position is denoted by the polarity shown. The averaging circuit produces a time average of the waveform existing at the output of the switch.

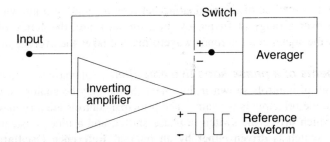

Fig. 4.29 The basic circuit of a phase sensitive amplifier.

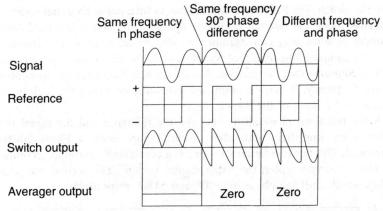

Fig. 4.30 The output of a phase sensitive detector: (i) when signal and reference have similar frequency and phase, (ii) when the phase of the signal frequency is shifted by 90°, and (iii) when there is no correlation between the signal and reference.

The operation of the circuit may be understood by considering its output under the three circumstances shown in Figs. 4.30(i), (ii) and (iii) respectively. The waveforms existing in this circuit at different parts of the ac cycle under various circumstances are given below:

- if the switch frequency and the signal frequency are identical and the two waveforms are in phase, the negative going parts of the signal waveform are inverted and the output of the averager is positive and finite.

- if the switch frequency and the signal frequency are identical but the two waveforms are 90° out of phase, the output of the switch is a complex ac waveform equally spaced about the origin. The output of the averager is zero.

- if the switch frequency and phase bear no relation to the signal frequency and phase, the waveform output is even more complex, but over a period of time the averager output is zero.

If the switch and signal frequencies are correlated there is a steady output, otherwise all signals average to zero.

This is a method of discriminating between a **signal** at a known frequency, and **noise** over a range of frequency. In a formal sense, the action of a lock-in multiplies the signal and reference waveforms and takes the average of the result.

Components of a phase sensitive amplifier.   All phase sensitive amplifiers have the set of controls shown in Fig. 4.31. Some may be analog, some may be digital, but the principle is the same. The principal controls are a frequency source or input which drives the switch, a phase shifter, and a time constant control.

The switch is driven either by an internal **Reference Oscillator** or by a signal extracted from the experiment. Normally the same connector is used to provide a voltage to drive the experiment, or to input a voltage from the experiment to drive the switch. The **Reference** switch is set to **Internal** or **External** respectively. In **Internal** mode the internal oscillator operates the switch and modulates the experiment in some way (for example, by driving an ac current at frequency f_s through the sample). The reference waveform may be sinusoidal, square or pulsed.

The **Signal** voltage is generated with a major frequency component at the reference frequency. It may be amplified externally or within the lock-in, and is presented to the input of the switch.

Since unknown phase shifts between the reference and the signal may be introduced by amplifiers and other circuit components, a **Phase Shifter** is incorporated. This has two functions—one is a continuous control to accommodate fixed phase changes within the experimental system. The second provides two switches which shift the phase by ±90° and ±180° respectively.

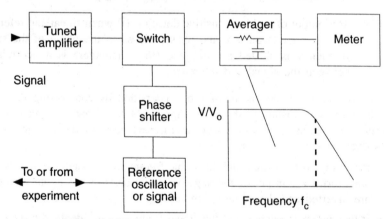

Fig. 4.31   Components of a phase sensitive amplifier.

The **Time Constant** control sets the time constant for the averaging circuit. This may be calibrated as an *RC* time constant, a frequency bandwidth for a single or double pole filter with time constant τ ($1/4\tau$ or $1/8\tau$ respectively), or as an *RC* frequency bandwidth $1/2\pi RC$. The **Output** of the averaging circuit is displayed on a meter.

In operating a lock-in, the following procedure is of value. Connect the signal and reference cables. With a time constant of the order of 100–300 ms, set the 90° and 180° phase and input sensitivity switches to give a meter reading which is on-scale. Adjust the continuous phase control until the output falls to zero. When this condition is established, switch the phase by 90° so that the meter reads a maximum. Operate the 180° switch and check that the respective readings are symmetrical about zero. Finally increase the time constant until an appropriate level of signal to noise enhancement is achieved.

A representative phase sensitive amplifier is shown below.

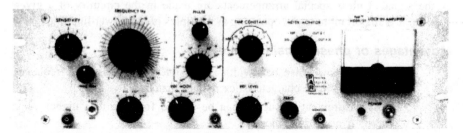

Fig. 4.32 A phase sensitive amplifier (with permission—EG&G Instruments).

Bandwidth of the measuring circuit. The averaging circuit is a low pass RC filter with a cut-off frequency of $f_c = 1/(2\pi RC)$. However, in the lock-in the frequencies which enter the filter result from both sum and difference frequencies between the signal and the frequency of the switch. Thus the effective bandwidth about the lock-in operating frequency is equivalent to $\pm f_c$.

If the filter is a simple RC circuit, the bandwidth for the averager is

$$\boxed{\Delta f = 2f_c = 1/(\pi RC)}$$

For more complex single and double pole filters the bandwidth Δf is $1/4RC$ and $1/8RC$ respectively.

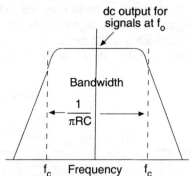

Fig. 4.33 The effective bandwidth of a lock-in amplifier is $\pm 1/(2\pi RC)$.

If the measuring frequency $f_o = 10^4$ and $RC = 1s$, the lock-in bandwidth $\Delta f = 2f_c = 0.3$ Hz if the averager is a simple low pass RC circuit. This implies that the Q of the measurement is $Q = f_o/\Delta f = 30,000$. This is a substantial improvement over the value of $Q = 4$ for the LCR circuit (Fig. 4.27). For these settings, the system is primarily sensitive to frequencies from 9999.7 to 10000.3 Hz. This implies a major improvement in signal to noise ratio.

A **disadvantage** of a phase sensitive amplifier is that the system takes roughly a time of 5 RC to respond to a change in signal level. This is because the averager acts by charging a capacitor. RC must be chosen to be small compared to the period of time variation of the signal. In practice compromises are required between the time available for a signal to stabilize and the degree of signal enhancement which is desired. The switch also responds to harmonics of the switch frequency in the signal. Unless special arrangements are made in the circuitry of a given lock-in, it is important to avoid operating at multiples of the power line frequency.

Advantages of phase sensitive detection

1. A very narrow noise bandwidth is possible. The measuring frequency and the amplifier characteristics remain matched.
2. The measurement frequency can be shifted from dc to a higher value where $1/f$ noise is reduced.
3. Modulation of the signal at frequency f_o allows experiments to be made against a steady or unmodulated background.

For example, if the light beam in a spectrometer is modulated, the measurement circuit will only respond to light passing through the sample and will not respond to stray ambient light.

Disadvantages of phase sensitive detection

1. The measurement is slow: it is essentially a technique for slowly varying phenomena.
2. It cannot deal with rapid transient phenomena.

A lock-in amplifier can be used to measure a weak optical signal. The set-up for this is, that the light signal is passed through an optical chopper to a deflector which acts as signal for lock-in amplifier. The reference signal is obtained by using a subsidiary source such as LED before the chopper and a simple photo-diode, to give the reference signal as shown in Fig. 4.34.

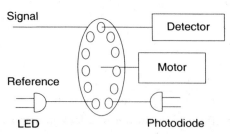

Fig. 4.34   A set-up for modulating optical signals for measurement by a lock-in amplifier.

Question (vi)   An optical measurement results in a noisy signal which varies with time over a period of 3 s. The signal is modulated by passing the light incident on the sample through a wheel with 20 holes rotating at an angular speed of 600 revolutions per minute. What lock-in time constant would you select and what would be the Q of the measurement.

Signal Averaging

If measurement of a signal voltage leads to a digital number (counts) N, the uncertainty in this number is $\sqrt{N}$. As noted earlier if the measurement is repeated M times, the total number of counts is MN and the relative uncertainty is decreased by a factor of $\sqrt{M}$.

For the sum of M repeated measurements, the noise adds to zero, the signal builds up, and the signal to noise ratio improves as $\sqrt{M}$. This mechanism is illustrated in Fig. 4.35. Panel 6 shows the summation of five repeated traces of a sine wave plus random noise.

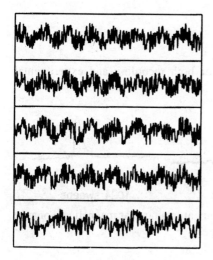

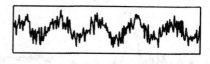

Fig. 4.35   Signal averaging of five traces of a sine wave plus random noise. Trace 6 is the summation of the five waveforms. The signal to noise ratio is improved by a factor of $\sqrt{5}$.

Multichannel Averaging

This technique is used for experiments that can be repeated in a reproducible manner. An analog signal voltage is successively recorded as a function of some parameter. For example, in magnetic resonance, an absorption signal is recorded by a phase sensitive amplifier as the magnetic field B is changed from B_1 to B_2. The sweep of B is repeated and the successive values of (signal + noise) are added into memory channels linked to specific values of B. At the end of the M repeated experiments, the average value of the signal in any channel is given by (total channel counts/M). As shown in Fig. 4.36 a combination of analog plus digital techniques are employed, with the instrument often being based on a multichannel analyzer with 2048 or 4096 channels. Such instruments include delay circuitry to allow time for associated equipment such as electromagnets to recycle in a reproducible manner. Signal averaging is now a standard facility in most digital oscilloscopes. A measurement example is shown in Fig. 4.37.

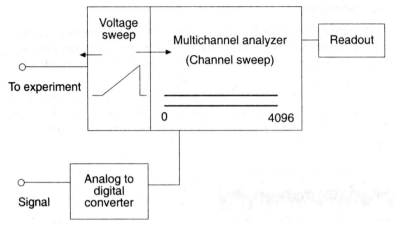

Fig. 4.36   A multichannel analyzer used for both recording and experimental control.

Fig. 4.37   An example of *S/N* noise improvement in a nuclear magnetic resonance experiment. (Courtesy—S.L. Segel)

Boxcar Integration

Repetitive transient signals are common in biological experiments or, say, in the response of a semiconductor to a flash of light from a pulsed laser. Signal averaging for such signals is carried out by boxcar integration as shown in Fig. 4.38.

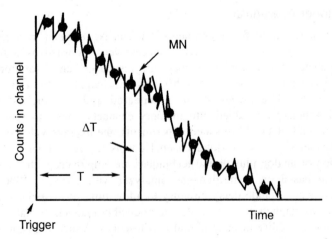

Fig. 4.38   Boxcar integration of repeated transient waveforms.

The waveform is triggered at time $t = 0$. A gate is set to open for a fixed period ΔT after a time interval T. The signal in time interval ΔT is measured either as an analog voltage or by conversion to a digital count. This measurement is stored in a channel of a memory. The process is repeated for M transients and the data is accumulated. If the original signal corresponds to N counts, the signal to noise ratio is $\dfrac{N}{\sqrt{N}} = \sqrt{N}$. The accumulated signal is MN with a standard deviation of $\sqrt{MN}$ so that the new signal to noise ratio is increased by $\sqrt{M}$. The gate is now shifted to open at a new T and the process is repeated. At the end of the total measurement cycle, the set of data points are displayed to show the form of the waveform.

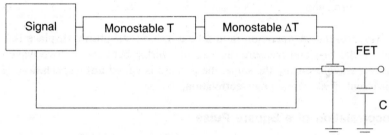

Fig. 4.39 A simple boxcar integrator implemented by an FET switch.

At the end of the experiment, the signal to noise has improved by $\sqrt{M}$, but it has taken M times as long to obtain this information as for a single pulse. If there are P gate settings along the waveform, the total time for a measurement is MP.

This type of function can be implemented by a simple 'sample and hold' circuit using a FET switch. A diagram of such a circuit is shown in Fig. 4.39.

Question (vii) A waveform consists of a signal of approximate level 10 μV with random fluctuations of 50 μV rms. If the waveform is signal averaged, what will be the magnitude (in volts) of the dc and ac components of the stored waveform at the output of the averager at the end of (a) 100 averages, (b) 1000 averages, (c) how many averages would it require to achieve a signal to noise ratio of 5:1?

4.8 DIGITAL CORRELATION AND AUTOCORRELATION METHODS

High speed analog to digital (A to D) converters allow signal to noise enhancement and signal processing to be carried out using digital or computer methods. As shown in Fig. 4.40, a waveform is sampled at a regular interval Δt to create a series of numbers. Digital techniques are based on the concept of **correlation**. This has become technically important because of the development of integrated circuits which can carry out a Fast Fourier transform. This is discussed in Chapter 5. In order to retain all the analog information, sampling theory (the Nyquist Theorem) (page 156) states that the signal must be sampled at a rate that is 2 x the highest frequency component in the signal. For example, if the analog bandwidth of the signal is 0–300 Hz, the sampling rate must be at least 600 samples per sec.

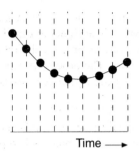

Time ⟶

Fig. 4.40 A segment of waveform is sampled at regular intervals Δt and the analog value is digitized and stored in a set of memory channels in a computer.

The correlation operation is to take a second waveform, displace this with respect to the first and measure the area of overlap between the two waveforms. If the two waveforms are the same, the process is called **autocorrelation**. If they are different it is termed cross-correlation.

Autocorrelation of a Square Pulse

A square pulse is stored in a segment of memory, each memory channel corresponding to successive instants in time ΔT. A copy of this is stored elsewhere. The second trace is now successively shifted in memory location by an amount corresponding to time ΔT and the area of overlap between the two traces is computed. A plot of the area of overlap versus the relative shift of the two waveforms is termed the autocorrelation function (Fig. 4.41).

The mathematical description of this procedure corresponds to multiplying the two waveforms and integrating:

$$cc_{ab} = \frac{1}{2t} \int_{-\infty}^{+\infty} a(t)\, b(t + \delta t)\, dt$$

In digital form

$$cc_{ab} = \sum_n a(t)\, b(t + n\Delta t)$$

where Δt is the sampling interval.

The waveforms are moved relative to each other by increasing the value of the integer n, each shift being by the interval ΔT between the data points.

The waveform of the autocorrelation function is a characteristic of the original waveshape. An important result is that the value of the autocorrelation function is unity (1) when the two waveshapes are superimposed.

As an extension of the pulse case, the autocorrelation function for a segment of a square wave is a series of increasing triangles which build up to a maximum when the two waveforms completely overlap (Fig. 4.42).

Mathematically, the correlation operation is equivalent to multiplying the Fourier Transform of each of the two waves, and taking the Inverse Fourier Transform of the result. This can be shown in the following manner.

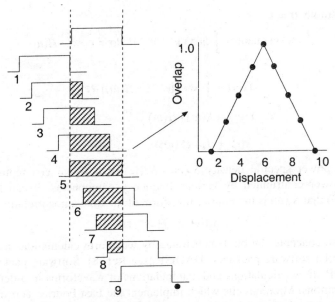

Fig. 4.41 The construction of the autocorrelation function of a waveform containing a single square pulse.

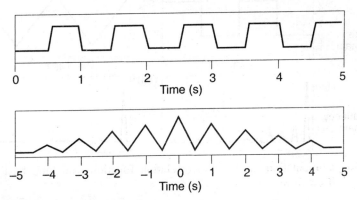

Fig. 4.42 Autocorrelation function of a segment of a square wave.

$$y(t) = \int_{-\infty}^{+\infty} x(\tau) \cdot h(t - \tau) \, d\tau$$

Take the Fourier transform of both sides

$$\int y(t) \, e^{-j\omega t} \, dt = \int \left[\int x(\tau) \, h(t - \tau) \, d\tau \right] e^{-j\omega t} \, d\tau$$

$$Y(\omega) = \int x(t) \left[\int h(t - \tau) \, e^{-j\omega t} \, dt \right] d\tau$$

Substitute $\sigma = t - \tau$

Bracket quantity $\int h(\sigma)\, e^{-j\omega(\sigma + \tau)}\, d\sigma = e^{-j\omega\tau} \cdot H(\omega)$

Therefore

$$Y(\omega) = \int x(\tau)\, [e^{-j\omega\tau} \cdot H(\omega)]\, d\tau$$

$$Y(\omega) = X(\omega) \cdot H(\omega)$$

$$y(t) = I_{FT}\,(Y(\omega))$$

The power spectral density $S_g(\omega)$ (W/Hz) of a signal $g(t)$ represents the relative power contributed by various frequency components. It can be shown (see Ref. 9) that $S_g(\omega)$ is the Fourier transform of the time autocorrelation function

$$S_g(\omega) = (\mathcal{F}[cc^*(\tau)]$$

These concepts can be demonstrated by waveform calculations made using a commercial software package—DADiSP [see Ref. 8]. Software packages such as DADiSP allow calculations and manipulations of waveforms in order to create these operations. Microcircuits which implement the Fast Fourier Transform have brought autocorrelation into widespread use in such digital equipment (Fig. 4.43).

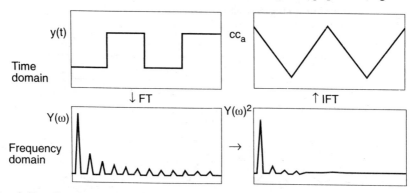

Fig. 4.43 Computation of the autocorrelation function using the Fast Fourier Transform.

Applications of Autocorrelation

Correlation implies that 'memory' exists over some period of time, i.e. the value of a point on a waveform depends on its value some time earlier. This is of importance in a number of areas.

Signal to noise enhancement.
The autocorrelation function for a segment of a 'noisy' sine wave (Fig. 4.44(a) is shown in Fig. 4.44(b)).

When any two waveforms are completely superimposed ($\Delta t = 0$), the value of the autocorrelation function is 1. This is true for completely random noise as well as for waves with a periodicity. However, as the waveforms are moved with respect to each other the degree of overlap for a random wave diminishes rapidly and is zero when all relationships between the parts of the waveform is lost. For a completely random wave this will occur for any $\Delta t > 0$. The autocorrelation function is therefore a delta function at the origin (Fig. 4.44(b)). By comparison, the autocorrelation function for a periodic wave will have its characteristic shape.

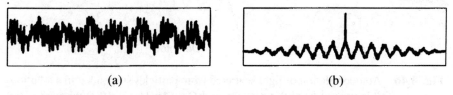

(a)   (b)

Fig. 4.44 Autocorrelation function (b) of a 'noisy' sine wave (a).

In a system having a finite bandwidth (such as in a circuit of resistance R and input capacitance C (page 75), a change in signal voltage from $V(t)$ to $V(t + \Delta t)$ requires a change in the charge on the capacitor. This takes a time related to the RC time constant of the measuring circuit $\tau = RC$, and the form of the autocorrelation function is $e^{-t/\tau}$. This is shown in Fig. 4.45. Calculation of the autocorrelation function can therefore be used to determine the frequency bandwidth of the system.

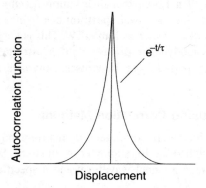

Displacement

Fig. 4.45 The autocorrelation function of a random noise waveform measured on a system having finite bandwidth is broadened by a factor $e^{-t/\tau}$, where τ is the time constant of the measuring system.

Autocorrelation function for physical systems. The autocorrelation function of a signal can contain information about the system in which the signal was generated. For example, in a system of diffusing particles such as fine particles in a water solution, the particles interact with each other and with the molecules of the water. This gives rise to 'Brownian motion' which implies that the motion of a given particle is described by the dynamics of the diffusion in water. The positions of a particle in successive intervals of time are therefore correlated or interlinked. This shows up in an experiment in which a laser beam of intensity $\langle I^2 \rangle$ illuminates the solution and light **scattered** in a direction perpendicular to the initial beam direction is recorded as a function of time (Fig. 4.46).

The autocorrelation function of the signal is related to the diffusion constant D for particles in the solution by

$$cc^*_{aa} = \exp(-2DQ^2T) \langle I^2 \rangle$$

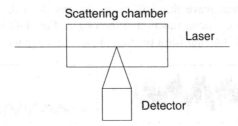

Fig. 4.46 Autocorrelation of light scattered from particles suspended in a solution. R is the radius of the particle, and $Q = (4\pi/\lambda n)\sin(\theta/2)$ where λ is the wavelength of light, n the refractive index of the medium ($n = 1.33$ for water), and θ is the scattering angle. $\langle I^2 \rangle$ is the laser intensity.

Hydrodynamics relates D to the particle radius R by $D = kT/6\pi\eta R$, where η is the viscosity ($\eta = 0.01$ P for water at 293 K). From the relationship between the spectral power density and the autocorrelation coefficient (page 127), and from direct calculation [see Ref. 4-10] the spectral power density for the scattered light scattered has the form of a Lorentzian distribution $S_g(\omega) = 2\Gamma/(\Gamma^2+\omega^2)$, where $\Gamma = 2DQ^2$ and $\omega = 2\pi f$. Such a power spectrum can be determined conveniently using instrumentation software such as LabVIEW. This is the basis of an important technique known as homodyne or **diffuse light scattering spectroscopy** used for the measurement of particle size in viruses, polymers and macromolecules [Ref. 10].

Signal Processing using Correlation Methods

The advent of powerful small computers in instrumentation systems allows correlation techniques to be used for a number of functions. Some applications use **cross-correlation** between a signal waveform and a specific reference pattern.

Signal to noise enhancement. Cross-correlation can be undertaken between a signal waveform and an arbitrary theoretical waveshape. Noise has no correlation while the signal waveform will reflect the major common features between the real and artificial signals. The shape of the reference waveform is adjusted to make the cross-correlation a maximum. The cross-correlation with a triangular waveform smooths a noisy waveform (Fig. 4.47).

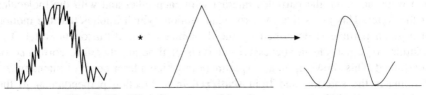

Fig. 4.47 Waveform smoothing by cross-correlation.

Spectrum line location. A common requirement is to accurately define the wavelength of a spectral line. Most spectrum lines are not infinitely sharp but are broadened, reflecting difficulties in measurement or factors related to atomic excitation. The cross-correlation shown converts the waveshape to a well-defined crossing of the zero axis (Fig. 4.48).

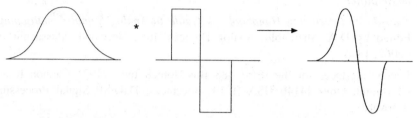

Fig. 4.48 Waveform shaping using cross-correlation.

Pattern recognition. The cross-correlation function shown in Fig. 4.49 exhibits a large peak when the small pattern is duplicated in the larger pattern.

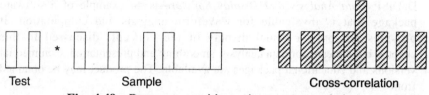

Test   Sample   Cross-correlation

Fig. 4.49 Pattern recognition using cross correlation.

Such a cross-correlation technique can be employed if it is required to identify, say, an X-ray or optical spectrum by comparison with a library of standard spectra. For a complex spectrum, the cross-correlation function will have a single peak at $\Delta x = 0$ if the spectrum is identical with the standard. If it is not, peaks will be absent even at $\Delta x = 0$ (Fig. 4.50).

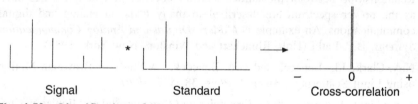

Signal   Standard   Cross-correlation

Fig. 4.50 Identification of spectra or complex patterns using cross-correlation.

This technique is used in the matching of fingerprints with data banks. Note that the correlation can again be undertaken either by the basic time shift process noted above, or by multiplying the Fourier Transforms of the waveforms.

REFERENCES AND FURTHER INFORMATION

1. *Data Reduction and Error Analysis for the Physical Sciences,* 2nd ed., P.R. Bevington and D.K Robinson, McGraw Hill, New York, 1992.

2. Ithaco Inc. produce an excellent set of *Application Notes for signal to noise processing using lock-in amplifiers,* Ithaco Inc., 735 West Clinton St., P.O. Box 6437, Ithaca, New York, USA.

3. Some of the best treatments of amplifier considerations for low noise processing are given in instrument manuals: See for example: *Ithaco Model 3962 Lock-In Amplifier.*

4. *Transducer Interfacing Handbook—A Guide to Analog Signal Conditioning*, Edited by D.A. Sheingold, Analog Devices, Inc., Norwood, Massachusetts 02062, USA.

5. Iotech Interfaces for the IEEE 488 Bus, Iotech Inc., 25971 Cannon Road, Cleveland, Ohio 44146 USA (IEEE interfaces, DaDiSP Signal Processing Software).

6. Rapid Systems, Inc., 433N, 34th St., Seattle, Washington 98103, USA (Fast Fourier Transform Software).

7. *Nuclear Instruments and Systems, EE&G Ortec Catalog*, EE&G Ortec Inc., 100 Midland Road, Oak Ridge, Tennessee 37830-9912, USA.

8. DaDiSP *Data Analysis and Display Software* is an example of a software package that is invaluable for waveform analysis and computation. It provides not only a visual display of the concepts discussed in this chapter, but a means for data analysis, smoothing and presentation. Commercial versions and educational packages are available. The product may be obtained from

    DaDiSP Development Corporation
    One Kendall Square
    Cambridge, MA 02139 USA
    1-800-424-3131
    dspdev @ world.std.com

9. Relationships between the autocorrelation coefficient and quantities such as the power spectrum are described in many texts on analog and digital communications. An example is *Modern Digital and Analog Communication Systems*, B.P. Lathi (Holt, Rhinehart and Winston, New York, 1983).

10. N.A. Clark, J.L. Lunacek and G.B. Benedek, "A Study of Brownian Motion using Light Scattering", *Amer. J. Phys.* **38**, 575, 1970.

    Laser Light Scattering—Basic Principles and Practice, 2nd ed., Benjamin Chu, Academic Press Inc., Boston, 1991.

ANSWERS TO QUESTIONS

(i) 3.5

(ii) (a) The manner in which odd harmonics build up a square wave is shown.

The odd harmonics add or subtract to the fundamental in a manner which constructs the square waveform over the period of the fundamental. The even harmonics add or subtract in ways which distort the fundamental and eventually lead to zero amplitude.

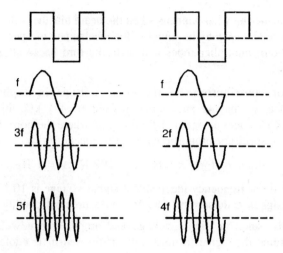

(b) (1) periodicity, (2) edges and corners

(iii) 1.25 Hz

(iv) 4 μV, 17.9 μV, 5.44 × 10⁴

(v) 26 nF, 3.49, FWHM = 14.32 kHz, 0.16 V

(vi) 0.53 Hz, 377

(vii) 1 mV, 0.5 mV; 10 mV, 0.158 mV; 625

DESIGN PROBLEMS

4.1 In the fabrication of a *p-n* junction by ion implantation, a beam of Li^+ ions of total current 10^{-8} A bombards a *p*-type Si wafer. The resistance of the wafer across its thickness is a few ohms. It is proposed to measure the beam current incident on the wafer by measuring the voltage developed across a 1 MΩ resistance connected in series with the wafer. The input impedance of the voltmeter is 10 MΩ, its frequency bandwidth is 70 kHz and the noise figure is 1.4.

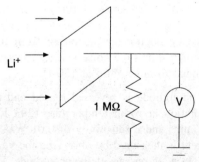

(a) Calculate the signal/noise ratio at the voltmeter output at 300 K.

(b) It is possible that the specimen current recorded may not be a good measure of the incident ion current? Why is this? Illustrate your answer by considering what happens when the beam hits the wafer surface. Some useful analogies may be found in the mechanism of operation of electron (or photo)-multiplier tubes and in Rutherford backscattering in nuclear physics.

4.2 The major contributions to the noise voltage in a photoconductive infrared detector having an internal resistance of 1 kΩ are thermal noise $v_n = (4kTR\Delta f)^{1/2}$ and device or $1/f$ noise, $v_{1/f}$, where the noise voltage $v_{1/f}$ has a frequency spectrum proportional to $\Delta f/(\text{frequency})^{0.5}$.

$$v_{1/f} = v_n \text{ for } f = 1 \text{ Hz}, \; T = 290 \text{ K}, \; \Delta f = 1 \text{ Hz}.$$

(a) If the dc or frequency independent signal voltage is 10^{-8} V calculate the total signal to noise ratio for these measurement conditions.

(b) For the same dc signal voltage and amplifier bandwidth $\Delta f = 1$ Hz, determine the signal to noise ratio if the sensor is cooled to 77 K and the light intensity is modulated by an optical chopper wheel at a frequency of 150 Hz. Assume that the internal impedance of 1 kΩ and the amplitude of $v_{1/f}$ as a function of frequency does not change with temperature.

4.3 A photoconductive material can be used as a photodetector in a circuit of the type shown.

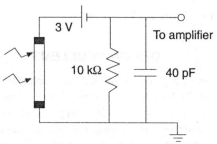

The load resistance is 10 kΩ, while the capacitance associated with the amplifier cables and input is 40 μF.

The detector has a dark resistance of 100 kΩ, an illuminated resistance of 50 kΩ and an *excess* $1/f$ noise contribution of 5 $\times$ thermal noise at 10 Hz. If the light source is an electric arc operating from the ac mains (60 Hz), (a) draw a graph which shows the noise voltage spectrum for the system, and (b) estimate the signal to noise ratio at 150 Hz.

4.4 A semiconductor photodiode is reversed biased and is employed to detect visible optical radiation at room temperature (293 K). The device has an active area of 0.5 mm^2 and a sensitivity of 0.01 W/A.

The device current is measured by observing the voltage developed across a 1 MΩ resistance placed in series with the photodiode. This voltage is measured using a digital voltmeter having an input impedance of 10 MΩ in parallel with 30 pF. The voltmeter has a full scale reading of 10 mV with a 4 digit display (i.e. the display reads 9.999 mV maximum).

(a) If the dark current of the device is 5×10^{-9} A, what will the display read in the dark, and which digits will be fluctuating? $1/f$ noise generated by the device is negligible.

(b) If the output voltage increased by 0.05 mV on illumination, what is the light intensity in W/m^2.

(c) How could the signal to noise ratio in this experiment be enhanced using an additional capacitor. The average light intensity and dark current is constant in time.

4.5 A flight is used to measure the absorption coefficients of liquids contained in a 10 cm thick cell, through which the light passes. The transmitted signal is monitored by a photodiode in series with a 5 kΩ resistor and a 1.5 V battery, and the voltage across the resistor is displayed using an amplifier and oscilloscope with a 300 kHz bandwidth. At a particular wavelength, the absorption coefficient of the liquid is 0.2 cm^{-1}. If the signal input to the detection amplifier is 0.1 mV with no liquid in the cell, estimate whether a signal waveform would be visible above the noise on the oscilloscope when the liquid is present.

4.6 A transient electrical signal lasting 10 ms and of a few microvolts in amplitude is triggered repetitively from a muscle in a biophysics experiment. The repetition rate is 20 Hz. What experimental methods might be employed to determine the form of the signal as a function of time with a good signal to noise ratio, assuming that the wave shape is also repetitive. If the statistical variation V_{p-p} of the waveform about its average value V_A at any time is approximately 30%, how long would it take to accumulate sufficient data to determine the shape of the waveform to 1%?

4.7 The collector of a scanning electron microscope receives a mean current of 4.0×10^{-11}A when operated in the back scattered electron mode at 300 K. The collector has a capacitance of 35 pF to ground, the input resistance of the preamplifier is $10^7 \, \Omega$.

(a) What is the signal/noise voltage ratio at the amplifier input?

(b) If both preamplifier and amplifier stages introduce a noise voltage of 10 μV/stage, and it is required to amplify the signal by a factor of 1000 in a two-stage amplifier (preamplifier followed by an amplifier), what would be the best stage gains for the preamplifier and the amplifier respectively. Justify your answer.:

$$10 : 100; \quad 33 : 33; \quad 100 : 10$$

(c) If the beam is kept stationary, show how lock-in detection could be incorporated to improve the signal/noise ratio. What lock-in frequency would you use and why, and what improvement in *S/N* would be expected for a lock-in time constant of 0.5 s.

4.8 A scanning electron microscope operates in conductive mode at room temperature.

An electron beam carrying a current of 1 μA scans a specimen and up to 15 % of the current is collected at a rear electrode. The bias battery of 9 V is included to enhance the collection process. The large capacitance C eliminates this dc bias and passes the varying signal voltage to the amplifier.

The output signal voltage V_s is developed across a load resistance of 10 kΩ and the sample resistance is of the order of 500 kΩ. It is expected that there will be noise contributions from both Johnson and Shot noise at 300 K.

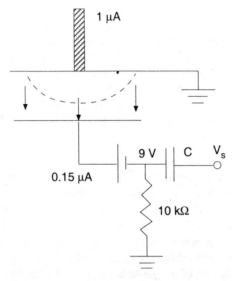

(a) If the bandwidth of the measuring system is 500 kHz, calculate the signal to noise ratio.

(b) Why would it be of value to pulse or switch on and off the electron current incident on the sample. If the stray capacitance connected across the 10 K load resistance is 30 pF, at what time intervals would you switch the beam?

4.9 (a) Explain the concept of correlation, and give a practical application of autocorrelation and crosscorrelation in noise reduction techniques.

(b) A rectangular pulse of 1 ms duration is symmetrical about $t = 0$. Using a spreadsheet plot its power spectral density $F(\omega)^2$ for frequencies up to 6 kHz. By numerical integration demonstrate the basis for the equivalent bandwidth of the pulse $\Delta f = 1/2T$.

(c) How would you determine the autocorrelation function of the original waveform from $F(\omega)$?

4.10 In a low level light intensity measurement (which may be regarded as photon counting), a 14 stage, 10 cm diameter photomultiplier is used to measure a light intensity of 2×10^{-16} W/m². The photon energy is 2.2 eV. The steady dark current of the photomultiplier is 20 pA, the secondary electron multiplication factor per stage is 4.3, and the photocathode quantum efficiency is close to 1 for a photon energy of 2.2 eV. The output current is measured using a 1 MΩ load resistor with a stray capacitance of 30 pF associated with the output circuit.

Discuss the sources of noise in this experiment and calculate the signal to noise ratio. How would you improve the *S/N* ratio to obtain the best <u>mean</u> value of the light intensity?

4.11 Distinguish objects identified by type using reflective stickers imprinted with a *bar code* similar to those found on grocery items (e.g. ‖ ‖ | ‖ | ‖ ‖ | ‖). Suggest a measurement circuit for reading the bar code at constant speed and a computer technique for making the identification. Estimate the signal voltage in terms of the component values of your proposed circuit. The computer has a known internal clock frequency.

4.12 Discuss the mode of operation of a phase sensitive detector. Why is such an amplifier with, say an averaging time constant of 1 s, a more effective band pass filter at 10 kHz than a tuned LCR filter using a coil of $L = 25$ mH and $R = 150 \ \Omega$?

4.13 A weak optical absorption band in a solid is to be measured by determining the absorption of light as a function of wavelength. A spectrometer is available whose wavelength drive is fitted with a controllable stepping motor. Discuss how a measurement of the transmission of the sample as a function of wavelength may be set up based on this experiment. What kind of signal to noise system would be useful to incorporate in order to improve the accuracy of measurement of the form of the absorption band as a function of wavelength.

4.14 A cadmium sulphide photoconductor cell of dark resistance 100 kΩ is used to measure a low level light intensity using the circuit shown. The oscilloscope amplifier and input lead have an input resistance of 1 MΩ and an input capacitance of about 75 pF respectively. The photoconductivity within the

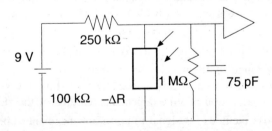

CdS has a linear sensitivity of 10 kΩ/mW/cm^2 at low light levels.

(a) Calculate the light level for which a signal to noise ratio of 1.5 is expected. The major frequency component of the signal is 3 Hz and the magnitude of l/*f* noise voltage at this frequency is 5 × the thermal noise voltage. Assume that l/*f* noise voltage falls off as 1/√*f*.

(b) If the CdS has a response time of about 0.5 ms (i.e. a light induced change in resistance can occur from 10% to 90% in 0.5 ms), how might you use a lock-in amplifier (and any associated equipment) to improve the *S/N* ratio. Define an appropriate frequency, lock-in time constant etc. and calculate the resulting *S/N* ratio for the light intensity used in (a).

4.15 You wish to measure the recombination lifetime of photo-generated carriers in a sample of semiconductor. Suggest a method for doing this, including one technique for improving the signal to noise ratio. Any circuit shown should work.

4.16 An electroluminescent display produces light when a voltage is applied to it. The light emission has a characteristic time dependence of 20 μs.

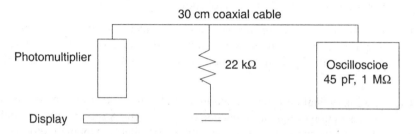

It is proposed to measure the light output both as a function of the frequency of the applied voltage and in response to a fast voltage *step*. A photomultiplier is selected for this purpose and the experimental arrangement is as shown. Power supplies are not indicated. It can be assumed that all light emitted from the display is captured by the photocathode. The current from the photomultiplier is recorded using a 22 kΩ load resistance. The output is connected to an oscilloscope of input impedance 45 pF in parallel with 1 MΩ by a coaxial cable 30 cm in length. The cable has a capacitance of 88 pF/m.

The *dark* current from the photomultiplier is 2.4×10^{-9} A, the gain of the multiplier chain is 10^7 and the quantum efficiency (photoelectrons/photon) for light of photon energy 1.6 eV is 0.17 (1 eV – 1.6×10^{-19} J).

(a) What is the minimum light intensity in mW that you might expect to record? Use a criterion of a *S/N* ratio of one.

(b) What range of frequencies are you able to measure with a minimum level of distortion in this system. Discuss how the system will affect your observation of time dependent features of the step response.

(c) Propose methods for signal to noise enhancement to be used for (i) the ac measurements and (ii) the step response.

Be as quantitative in your choice of parameters as possible.

4.17 It is proposed to make a device to monitor the output of an X-ray tube in terms of power W (watts) emerging from an exit slit. A resistive device having a resistance of 5000 Ω at 300 K is constructed and measured in the circuit shown. The multimeter has a input impedance of 1 MΩ and

30 pF. 30 cm of coaxial cable having a capacitance of 88 pF/m is used to connect the resistor to the voltmeter. The circuit is operated at a temperature of 300 K and a signal to noise ratio of 2 : 1 is sought for all measurements.

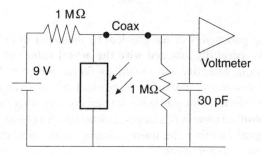

(a) What is the output voltage of this circuit?

(b) A thin film of thermistor material which is prepared as a detector has a resistance—temperature relation over a small temperature range (ΔT) K of

$$R(T) = 5000 \times [1 - .04(\Delta T)]$$

As noted above, the value of $R(300)$ is 5000 Ω. The film thickness and area are such that all X-rays falling on it are absorbed within its bulk so that a power input W increases the temperature of the film. On the other hand, careful measurements show that the rate of heat loss from the device if its temperature rises by ΔT K is given by $H_{loss} = 10^{-3} \times \Delta T$ W.

If equilibrium voltage values are recorded, what is the smallest output power that you can measure using this arrangement?

4.18 Photoemitted electrons from a metallic surface are to be collected by a well insulated positively biased electrode which has a capacitance to ground of 15 pF. It is proposed to use the voltage developed across a 100 kΩ resistor to measure the photoelectron current. A preamplifier is available which has a gain of ×100, an input impedance of 1 MΩ and 30 pF, a noise figure of 1.3 and a frequency bandwidth of 500 kHz. $1/f$ noise is negligible and the measurement is made at room temperature (300 K).

(a) What is the frequency response of this circuit?

(b) If the signal voltage at the output of the amplifier must be greater than 2 mV, what is the smallest rms signal current that can theoretically be measured at the input?

(c) What is the rms signal to noise ratio at the amplifier output at this current level?

(d) Show how a lock-in amplifier and associated equipment could be used to increase the detectability of the signal current by a factor of 50. Give quantitative values for lock-in parameters such as frequency and time constant.

WORKED EXAMPLES

PROBLEM 4.1

General Activities

Clearly define the parameters of the problem by making a full circuit diagram of the measurement—**carefully labelled with the actual values of the components used in the design**. Simplify the circuit as far as possible. Work out which components will define the measuring bandwidth of the circuit. Decide on the types of noise involved—Johnson noise for resistors, or Shot noise if collection phenomena are involved (beams of charges, junction diodes or transistors). Calculate separately the **signal** and then the **noise** voltages. Such calculations can be done very effectively in a spread sheet.

(a) *Signal to Noise Ratio*

Type of source   High impedance current source: The collection current is set by the Li^+ ion flux.

Type of noise   The collection of ions over time is a statistical process: Shot noise (see page 102). Thermal noise will be generated in the load resistor (see page 101).

Frequency bandwidth   Set in the problem to be $\Delta f = 70$ kHz

Circuit

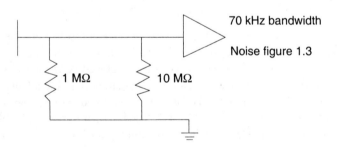

The voltmeter can be treated as a *perfect* voltmeter which takes no current (infinite input impedance) with an input resistance of 10 MΩ to ground. This is in parallel with the external load resistance.

Effective load impedance is 1 M // 10 M $= \dfrac{R_1 R_2}{R_1 + R_2} = \dfrac{10}{11} = 0.909$ MΩ

Signal voltage $V_s = 10^{-8} \times 9.09 \times 10^5 = 9.09 \times 10^{-3}$ V

Noise Voltage = Thermal Noise + Shot Noise
(voltage adds in quadrature—power adds linearly)

Thermal noise voltage $= \sqrt{4kTR\Delta f}$ (page 102)

$$= (4 \times 1.38 \times 10^{-23} \times 300 \times 9.09 \times 10^5 \times 7 \times 10^4)^{0.5}$$

$$v_t = 4.25 \times 10^{-5} \text{ V}$$

Shot noise current, $i_n = \sqrt{2qI\Delta f}$ (page 103)

Shot noise voltage, $v_n = \sqrt{2qI\,\Delta f} \cdot R$

$$= (2 \times 1.6 \times 10^{-19} \times 10^{-8} \times 7 \times 10^4)^{0.5} \times 9.09 \times 10^5$$

$$= 1.15 \times 10^{-12} \times 9.09 \times 10^5$$

$$v_n = 1.36 \times 10^{-6} \text{ V}$$

Total noise voltage, $v_t = (v_t^2 + v_n^2)^{0.5} = 4.25 \times 10^{-5}$ V (page 104)

Amplifier noise figure, $F = 1.3 = \dfrac{\text{Noise at output}}{\text{Noise at input}}$ (page 97)

Signal/Noise at output $= 9.09 \times 10^{-3}/(1.3 \times 4.25 \times 10^{-5}) = 215$

(b) The specimen current may not be a good measure of the incident ion current.

   (i) Elastic scattering of ions may occur with scattered energy close to that of the incident ions. The measured current will be less than the actual incident current.

  (ii) Secondary emission may occur in which several negative electrons may be lost from the collector for each incident positive ion. The measured current will be greater than the actual incident current.

PROBLEM 4.2

The circuit can be rearranged to have the form shown below:

Signal

$$V_{s1} = 3 \cdot \frac{10}{10 + 100}, \quad V_{s2} = 3 \cdot \frac{10}{10 + 50}$$

Signal voltage $= V_{s2} - V_{s1} = .50 - .272 = 0.228$ V

The noise voltage needs to be calculated in the illuminated state.

By Thevenin's theorem (page 37) the circuit can be resolved in the following manner:

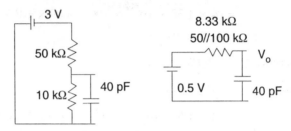

The frequency bandwidth (page 98) of the circuit leading to the output voltage v_o is

$$\Delta f = \frac{1}{2\pi R C} = \frac{1}{6.28 \times 8.33 \times 10^3 \times 4 \times 10^{-11}} = 4.77 \times 10^5 \text{ Hz}$$

Thermal noise voltage, $v_n = \sqrt{4kTR\Delta f}$ (page 102)

$$= (4 \times 1.38 \times 10^{-23} \times 300 \times 8.33 \times 10^3 \times 4.77 \times 10^5)^{0.5}$$

$$= 8.12 \times 10^{-6} \text{ V}$$

Excess noise at 10 Hz = 5 × thermal noise = 4.06×10^{-5} V   (page 103)

Excess noise **power** varies as $1/f$; excess noise **voltage** varies as $1/\sqrt{f}$.

Excess noise at 150 Hz: $\dfrac{V_f(150)}{V_f(10)} = \dfrac{(10^{0.5})}{(150^{0.5})}$ or $V_f(150) = 1.05 \times 10^{-5}$ V

Note: This is an approximation since the $1/f$ noise can only be measured over the entire bandwidth—which in this case is large. The reduction in the overall magnitude should be reasonably correct.

Total noise voltage at 150 Hz is added in quadrature   (page 104)

$$V_t = \sqrt{v_n^2 + V_f^2} = 1.33 \times 10^{-5} \text{ V}$$

Signal/Noise (150) Hz = $0.23/1.33 \times 10^{-5} = 1.71 \times 10^4$

Noise voltage spectrum

Note the advantages of making measurements at higher frequencies.

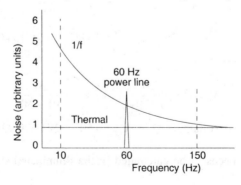

5

Instrumentation Electronics

The universal use of microcircuits for instrumentation purposes has simplified the design of measurement systems. In this chapter basic concepts of operational amplifiers, analog to digital conversion, and other electronic functions used in the 'tool kit' of the instrumentation engineer are summarized. Important concepts related to data sampling and the Fast Fourier Transform are reviewed.

5.1 OPERATIONAL AMPLIFIERS

An operational amplifier is a **high gain**, **high input impedance** amplifier used in a **feedback** mode. Such amplifiers are available in integrated circuit form and are the backbone of modern analog instrumentation. Their major advantage is that the amplifier properties depend only on the external feedback components and are independent of the amplifier chip itself. The basic principles of operation are summarized, and some of the more important configurations used in instrumentation are listed.

Basic Principles—An inverting amplifier

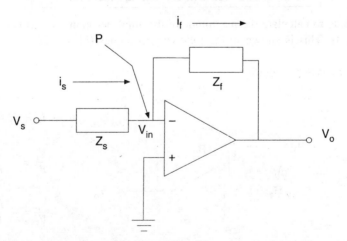

Fig. 5.1 An inverting operational amplifier. The output voltage $V_{out} = -V_{in}$, and the input current to the amplifier $i_a \rightarrow 0$.

For the amplifier $V_o = - A \, V_{in}$

At the input: $i_s + i_f + i_a = 0$

If A is large, $i_a = 0$ $V_{in} = 0$ and $i_s = -i_f$

The point P is known as the summing point. The action of the high gain amplifier causes the voltage at this point to be close to zero at all times (virtual ground). Because of the high input impedance of the amplifier, its input current is negligible and hence any source current i_s must be matched by an equal current i_f through the feedback loop.

For the input circuit: $V_s - i_s Z_s - V_{in} = 0$

For the feedback loop: $V_{in} - i_f Z_f - (-V_o) = 0$

Eliminating i_s, i_f and V_{in} gives

$$\frac{V_o}{V_s} = A_f = \frac{A}{1 - (A-1)\dfrac{Z_s}{Z_f}}$$

If $A \gg 1$, A cancels from the equation and the gain of the feedback amplifier is,

$$A' = - \frac{Z_f}{Z_s}$$

This is purely a property of the circuit components making up Z_s and Z_f. The output impedance is $(Z_o \times A'/A)$, where Z_o is the output impedance without feedback. For most practical operational amplifiers the output impedance with feedback is of the order of 5 to 100 ohms.

The properties for an operational amplifier that,

> the summing point P is always at virtual ground
> the input and feedback currents are identical

make it easy to calculate the magnitude of the amplifier gain for various feedback components. This is shown in the following examples (Figs. 5.2(a)–(h)).

(a) Inverting amplifier

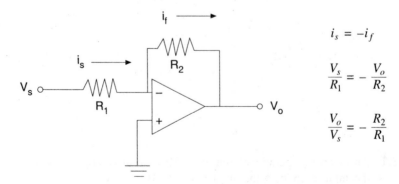

$$i_s = -i_f$$

$$\frac{V_s}{R_1} = - \frac{V_o}{R_2}$$

$$\frac{V_o}{V_s} = - \frac{R_2}{R_1}$$

(b) Non-inverting amplifier

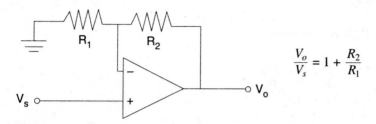

$$\frac{V_o}{V_s} = 1 + \frac{R_2}{R_1}$$

(c) Summing amplifier

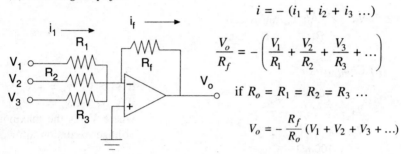

$$i = -(i_1 + i_2 + i_3 \ldots)$$

$$\frac{V_o}{R_f} = -\left(\frac{V_1}{R_1} + \frac{V_2}{R_2} + \frac{V_3}{R_3} + \ldots\right)$$

if $R_o = R_1 = R_2 = R_3 \ldots$

$$V_o = -\frac{R_f}{R_o}(V_1 + V_2 + V_3 + \ldots)$$

(d) High input impedance/low output impedance buffer amplifier

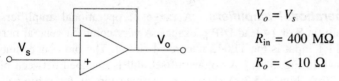

$V_o = V_s$

$R_{in} = 400 \text{ M}\Omega$

$R_o = < 10 \ \Omega$

(e) Integrator (low pass filter)

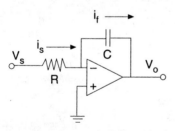

$i_s = -i_f$

$$\frac{V_s}{R} = -C\frac{dV_o}{dt}$$

$$V_o = -\frac{1}{RC}\int_o^t V_s \, dt + C$$

(f) Differentiator (high pass filter)

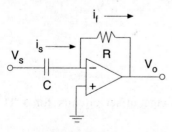

$i_s = -i_f$

$$C\frac{dV_s}{dt} = -\frac{V_o}{R}$$

$$V_o = -RC\frac{dV_s}{dt}$$

If $V_s = V_o \sin \omega t$, $V_o = V_s \omega RC \cos \omega t$ and higher frequency harmonics are amplified preferentially.

(g) Difference amplifier

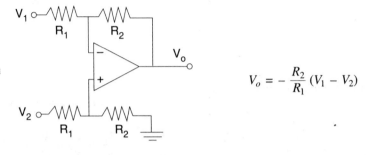

$$V_o = -\frac{R_2}{R_1}(V_1 - V_2)$$

(h) Comparator

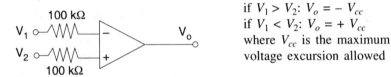

if $V_1 > V_2$: $V_o = -V_{cc}$
if $V_1 < V_2$: $V_o = +V_{cc}$
where V_{cc} is the maximum
voltage excursion allowed

Fig. 5.2(a)–(h)   Basic operational amplifier circuits.

Practical operational amplifiers.   A range of operational amplifiers are available, usually in 8 or 14 lead DIP packages. A very common general purpose unit using an FET input is the TL072 integrated circuit. The pin connections for this are as shown in Fig. 5.3(a). A voltage offset null is obtained between pins 1 and 5 (Fig. 5.3(b)). Figure 5.3(c) shows an alternate circuit for setting the dc output level of an amplifier with a gain of ×10 using a summing amplifier.

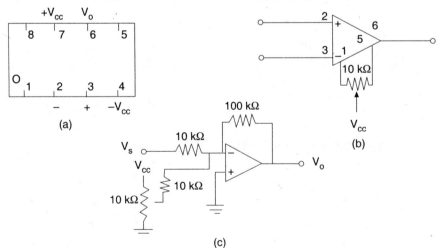

Fig. 5.3   Pin connections and alternate voltage offset circuits for a TL072 operational amplifier.

Question (i) (a) Design an inverting amplifier with a gain of 200 and an input impedance of 1 kΩ.

(b) A piezoelectric transducer is a voltage source of 10 V rms with an internal impedance of 10 MΩ. It is connected to a digital oscilloscope with an input impedance of 10 MΩ (i) directly and (ii) using a high impedance buffer amplifier. What voltage will be measured by the oscilloscope in the two cases?

(c) Design an integrating circuit with a time constant of 1 s. If the input resistance is 1 MΩ, what will be the output voltage as a function of time if a voltage of 1.5 V is applied for 3 s?

5.2 ANALOG SIGNAL PROCESSING

It is often desired to multiply, divide or take the inverse of analog voltage signals which are proportional to a measured physical quantity. Specialized integrated circuits allow such signal processing as an integral part of a measurement system. Figure 5.4(a) shows the type of applications and the accuracy achievable for circuits using **Burr-Brown multi-function converters** and logarithmic amplifiers as noted in the manufacturers literature. Figure 5.4(b) shows the connections for a multiplier.

Burr-Brown's multifunction converter model 4302 is a low cost solution in many analog conversion needs. Much more than just a multiplier/divider, the 4302 performs many analog circuit functions with a high degree of accuracy at low total cost.

Functions	Accuracy
Multiply	± 0.25%
Divide	± 0.25%
Square	± 0.03%
Square root	± 0.07%
Exponentiate	± 0.15% ($m = 5$)
Roots	± 0.2% ($m = 0.2$)
Sine θ	± 0.5%
Cosine θ	± 0.8%
Tan^{-1} (Y/X)	± 0.6%
$\sqrt{X^2 + Y^2}$	± 0.07%

Typical accuracies expressed as % of output full scale (+10 VDC) at 25°C.

(a)

In multiplier applications the 4302 provides high accuracy. The 4302 accepts inputs up to +10 VDC and provides an accuracy of ±0.25% of full scale.

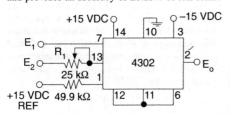

Transfer function	$E_o = + E_1 E_2/10$
Accuracy	
Typical at +25°C	± 25 mV
Maximum at +25°C	± 50 mV
Output Offset (at 25°C)	± 10 mV
vs Temperature	± 0.2 mV/°C
Noise (10 Hz to 1 kHz)	100 µVrms
Bandwidth (E_1, E_2)	
Small signal (−3 dB)	500 kHz
Full output	60 kHz

(b)

Fig. 5.4 (a) General functions obtainable from a Burr Brown 4302 multifunction converter. (b) Connections for a multiplier. Copyright 1995 Burr-Brown Corporation. Reprinted, in whole or in part, with the permission of the Burr-Brown Corporation.

(a) Exponential Functions

Model 4302 may be used as an exponentiator over a range of exponents from 0.2 to 5. The exponents 0.5 and 2 square rooting and squaring respectively, are often used functions and are treated below. Other values of exponents (m) may be useful in terms of linearization of nonlinear functions or simply for producing the mathematical conversions. Charateristics of m between 0.2 and 5 are presented below.

Transfer Function	$E_o = 10$ $(E_i/10)^m$
Total Conversion Error; $m = 0.2$ (Typical 0.5 VDC – 10 VDC)	± 2mVDC
Total Conversion Error; $m = 5$ (Typical 1 VDC – 10 VDC)	± 15 mVDC
Input voltage range	0–10 VDC

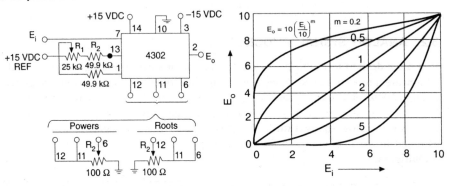

(b) Logarithmic Amplifiers

- ACCEPTS INPUT VOLTAGES OR CURRENTS OF EITHER POLARITY
- WIDE INPUT DYNAMIC RANGE
  6 decades of Current
  4 decades of Voltage
- VERSATILE
  Log, Antilog and Log Ratio Capability

Packaged in a ceramic double wide DIP, the 4127 is the first hybrid logarithmic amplifier that accepts sigrals of either polarity from current of voltage sources. A special purpose monolithic chip, developed specially for logarithmic conversions, functions accurately for up to six decades of input current and four decades of input voltage. In addition, a current inverter and a precise internal references allows pin programming

of the 4127 as a logarithmic, log ratio, or antilog amplifier.

To further increase its versatility and reduce system cost, the 4127 has an uncommitted operation amplifier in its pakage that can be used as a buffer, inverter, filter or gain element. The 4127 is available with initial accuracies (log conformity) of 0.5% and 1%, and operates over an ambient temperature range of –10°C to +70°C.

With its versatility and high performance, the 4127 has many application in signal compression, transducer linearization, and phototube buffering. Manufacturers of medical equipment, analytical instruments and process control instrumentation will find the 4127 a low cost solution to many signal processing problems.

(a)   (b)

Fig. 5.5 (a) Exponential or power function ($E_o = E_i^m$) implemented using a 4302 multi-function converter, (b) a logarithmic amplifier using a Burr-Brown 4127. Copyright 1995 Burr-Brown Corporation. Reprinted, in whole or in part, with the permission of Burr-Brown Corporation.

5.3 HIGH SPEED ANALOG TO DIGITAL CONVERSION

Analog to digital (A to D) conversion using a **dual slope** integrating voltmeter technique was outlined in Section 4.6. The long sampling time makes this method unsuitable for high speed conversions. The **successive approximation** A to D converter is most often employed for instrumentation purposes.

Digital or binary numbers.   A binary number is a sequence of **bits** which are 0 or 1, the bit representing a power of 2.

Numeric	32	16	8	4	2	1
Binary	1	0	1	0	1	1

MSB   LSB
(most significant bit) (least significant bit)

A **byte** is an 8-bit **word**. A **baud** is a transmission rate of 1 **byte/s**. The numeric value of the 6-bit binary number above is

$$(32 + 8 + 2 + 1) = 43$$

Successive approximation D/A converter.   Using a standard reference source, a pulser creates a series of pulses of fixed duration of an amplitude which decreases by factor of 2 for each successive pulse. If the standard voltage is V, successive pulse heights will be

$$\frac{V}{2}, \frac{V}{4}, \frac{V}{8}, \cdots \frac{V}{2^m}$$

where m is the number of bits in the digital number. The amplitude of the successive pulses are added successively into a memory based on a decision made by a comparator circuit. The comparison is made of the signal voltage with the sum of the pulse heights added into the memory. The sequence of events described in Fig. 5.6 is as follows:

1. A pulse of amplitude $V/2$ is added to the sum in memory.

2. If the memory voltage is *less* than the signal voltage, the added amplitude is permanently included into the memory. Simultaneously with this, a binary 1 is written into a memory location corresponding to the most significant bit.

3. If the memory voltage is **greater** than the signal voltage, the additional amplitude is ignored and the voltage in the memory is unchanged. A digital 0 is written at the MSB location.

4. The sequence is repeated for the next pulse of half the previous amplitude, the summation building up until it is a good approximation to the voltage to be measured.

The trial addition occurs during the test (*T*) period, the logic bit is written during the post (*P*) period. The sequence is shown for a 6-bit A to D converter.

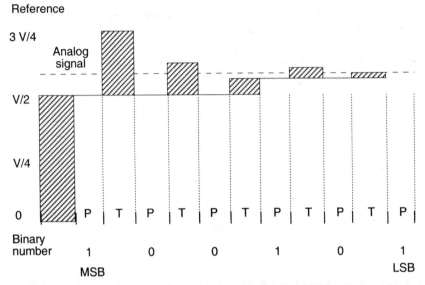

Fig. 5.6　6-bit successive approximation analog to digital converter.

This sequence occupies the **conversion time**. This is generally 0.1–0.22 μs/bit with a total conversion time for an 8- or 12-bit word being 1–2 μs. The frequency bandwidth is therefore $\Delta f = 1/2\Delta T = 500$ kHz or 200 kHz respectively.

Higher speed ADC's which make a conversion within 10–100 ns using simultaneous sets of comparators are known as **flash ADCs**.

Question (ii)　The reference voltage of an ADC is 1 V. What is the smallest voltage step that you can record using (a) an 8-bit and (b) a 12-bit converter.
If the conversion time is 0.2 μs/bit what is the highest frequency (c) sine wave and (d) square wave you can record? Give a reasoned estimate for case (d).

5.4　DIGITAL TO ANALOG CONVERSION

The reverse conversion of a binary number to an analog voltage output is necessary where a computer is required to drive an experiment. The sequence is shown in Fig. 5.7. A current is generated from a standard voltage source which is proportional to the weight of the bit in the binary number. These currents or voltages are then summed using an operational amplifier. The circuit is often known as a ladder network. For example, the binary number representing 28 (11100) appears as a set of low or high voltage levels from a set of logic gates, where the 'high' level is *V*. Each bit is connected to a summing amplifier with a resistance of the magnitude illustrated in Fig. 5.7.

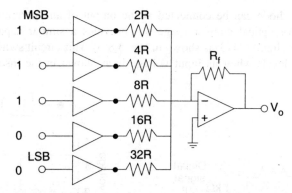

Fig. 5.7 A 5-bit ladder digital to analog converter.

Each bit (m) contributes an output voltage of $V_m = \dfrac{R_f}{2^m R} V_s$,

where V_s is the standard voltage. For a 5-bit converter and assuming that $V_s = 1$ V, the voltage contributions are

	MSB				LSB	
Bit	1	2	3	4	5	
V_o	0.5	0.25	0.125	0.0625	0.0312	V

Thus a binary number of (11100) gives rise to a voltage output

$$(.5 + .25 + .125)V = 0.875 \text{ V}$$

Such circuitry is normally obtained within an integrated circuit chip. The number of bits gives the resolution with which the voltage can be set. If $V_s = 1$ V, this is 4 millivolts for an 8-bit converter, and 0.2 millivolts for a 12-bit converter.

Question (iii) A thermocouple voltage is approximately 4 mV for a temperature rise of 100°C. If it is desired to reproduce the output voltage of a thermocouple for 1°C steps, what resolution DAC is required if the standard voltage is 1 V?

5.5 DIGITAL LOGIC LEVELS

In the conversion from analog to digital signal processing it is necessary to create signal levels which are compatible with digital circuits. While a number of types of digital logic have been developed, the important levels have remained consistent with those of 'transistor-transistor logic' (TTL). These are:

Digital 'high' 3.3 V   Digital 'low' < 0.3 V

Several methods can be used to create these levels from analog signals. A limiting Zener diode can be connected at the output of an operational amplifier (Fig 5.8(a)). An optical coupler (page 81) provides a similar capacity. A TTL Schmitt trigger circuit (7413) is shown in Fig. 5.8(b). This circuit switches between the TTL logic levels when its input voltage is as shown in the inset.

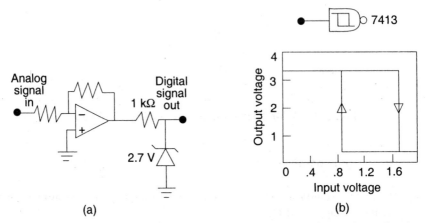

Fig. 5.8 (a) Comparator/Zener diode circuit to limit analog voltage levels to digital TTL. (b) A 7413 Schmitt trigger for 5 V operation.

5.6 DIGITAL INSTRUMENTATION

Digitizing oscilloscopes provide versatile data display and signal processing capabilities. One of the most prominent performance parameters is the sample rate. This sets the time interval at which a scope measures the instantaneous value of the signal. Higher sample rates (Sa/s) give higher bandwidths and better definition of pulse edges, but are more costly. Specific techniques are employed to extend the bandwidth, but utilize a cheaper and lower sampling speed A to D converter.

This leads to the definition of two kinds of frequency bandwidth for a digital oscilloscope.

Repetitive bandwidth.   This is the result of multiple sweeps over the same signal waveform. The scope is triggered in a manner similar to that required to create a stationary pattern in a conventional analog scope. At the limit of its performance the number of samples it takes may be very small. A small random time interval is included between the trigger time and the time when a sweep of the channels commences. Successive traces build up a representation of the signal as shown in Fig. 5.9.

Real time bandwidth.   This is the highest frequency a digital scope can capture in a single sweep. It is determined almost completely by the digitizing time of the input A to D converters. The single real time or single sweep bandwidth can be a factor of 4 or more, less than the repetitive bandwidth. Expensive digitizing

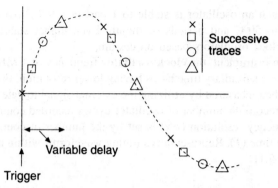

Fig. 5.9 Successive measurements (shown by different symbols) of a waveform build up after a random trigger delay to create a waveform in a digital oscilloscope.

scopes use sampling rates of 8 GSa/s, but many laboratory versions use less costly A to D converters and repetitive methods.

A digital oscilloscope normally incorporates multichannel averaging for signal to noise enhancement. It has excellent capability both for data transfer to and control by an associated computer. Instrument probes (page 42) are mandatory for such high speed instrumentation. Care has to be taken to avoid problems due to aliasing (page 156) and triggering (page 157).

5.7 FREQUENCY MEASUREMENTS

Frequency is determined by comparison against a standard clock oscillator of frequency 1 or 10 MHz. High stability quartz crystal oscillators utilize the precise vibrational frequency of a piezoelectric quartz crystal. A representative oscillator is shown in Fig. 5.10.

The temperature and time stability of such oscillators are maintained by compensating circuit components or by the use of constant temperature ovens. The

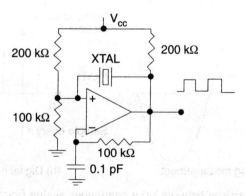

Fig. 5.10 A simple crystal controlled oscillator.

frequency of such an oscillator is stable to 1 part in 5×10^6 over a temperature range from 0 to 50°C and periods of month. Even higher stability is achieved using atomic clock oscillators based on cesium.

Within an instrument, the **clock oscillator** frequency of 1 MHz is divided by circuits to provide subsidiary time bases having lower resolution. Frequency, period and times are then measured by utilizing the external signal to gate the appropriate time base and record the number of oscillator cycles recorded counts within a gate time. The frequency resolution ($\sqrt{N}$) is set by the number of counts (N) recorded in a given gate time (T). Representative timing waveforms within an oscillator are shown in Fig. 5.11.

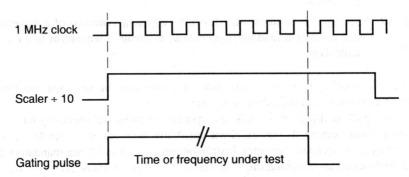

Fig. 5.11 Timing and gating waveforms in a counter: (a) 1 MHz quartz clock oscillator, (b) scaling by a factor of 10, (c) gated signal from an external source.

5.8 THE FAST FOURIER TRANSFORM

The analog Fourier Transform outlined in Appendix 2 involves an integration process from – infinity to + infinity. In digital signal processing, a waveform in the time domain is digitized at sampling intervals of ΔT for a finite period of time T.

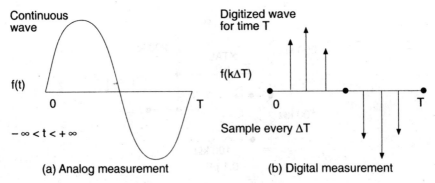

Fig. 5.12 Comparison between (a) a continuous analog function and (b) discrete measurements made in the digital domain (b).

Thus a continuous **analog** waveform after the digitizer becomes a finite series of **discrete values** separated by time ΔT. The analog time function $f(t) = f(k\ \Delta T)$, where k is a channel number $0 < k < T/\Delta T$. The total number of steps $N = T/\Delta T$ where N is normally chosen to be a power of 2.

The **Discrete Fourier Transform** provides a rapid digital technique for calculation of Fourier Transforms in terms of a set of discrete values in the frequency domain. The algorithm uses a major simplifying assumption that

> Any signal recorded within the gate time T is periodic in time T

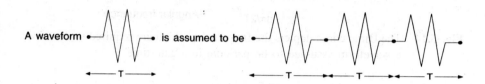

A waveform •—   is assumed to be

The angular frequency components required to describe the waveform within this gate time T are periodic in T as shown in Fig. 5.13.

$$\omega_n = 2\pi \left[\frac{1}{T}, \frac{2}{T}, \frac{3}{T}, \frac{4}{T}, \dots \frac{n}{T}\right]$$

Fig. 5.13 The waveform within gate time T is regarded as having originated from a continuous waveform repetitive in T. The frequency components ω_n, are periodic in time T.

The frequency spectrum is then periodic in the gate time T, and the only angular frequencies $\omega = 2\pi f$ that have to be considered in evaluating the discrete Fourier Transform of a signal within this gate time are plotted on the **Dirac Comb** shown in Fig. 5.14.

The analog expression for $f(t)$ may then be replaced by its digital equivalent

$$F\left(\frac{2\pi n}{T}\right) = \sum_{k=1}^{n} f(k\Delta T) \exp -j\left(\left[\frac{2\pi n}{T} (k\Delta T)\right]\right)$$

Since $N = T/\Delta T$

$$F\left(\frac{2\pi n}{T}\right) = \sum_{k=1}^{N} f(k\Delta T) \exp -j\left[2\pi nN\right]k$$

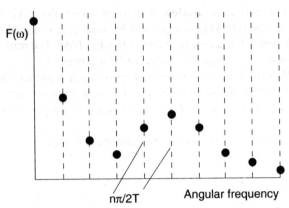

Fig. 5.14 The Dirac comb depicts the frequencies which are required to describe a waveform assumed to be periodic in a gate time *T*.

If the exponential terms $e^{-j\omega T}$ are written as (cos ωt + j sin ωt), all multiplications become summations, the integral is carried out over a finite relatively small number of terms, and the whole process can be accomplished rapidly in a manner which can be implemented conveniently in a microprocessor circuit.

The **Fast Fourier Transform** utilizes some further simplifications and algorithms to make the calculations even more rapid without substantial loss of accuracy. One such algorithm is the Cooley-Tukey algorithm where the number of operations is reduced to $N \log_2 N = 1024 \log_2 1024$ instead of $(1024)^2$ for $N = 2^{10}$. This reduces the number of computations by a factor of 100. The algorithm uses a radix 2 and the number of frequency steps equals the number of time samples $N = T/\Delta T$.

5.9 SAMPLING TIME AND ALIASING

In measuring Fast Fourier Transforms (FFTs) in practice, careful consideration must be given both to the value of the gate time *T* and to the sampling time ΔT or sampling rate (Sa/s) with respect to the frequencies and waveform being measured. This is analogous to the care required to trigger an analog oscilloscope to maintain a steady trace.

Gate time T. For a given digitizing time ΔT the frequency intervals at which computations are performed is set by the sampling or gate time *T* since $N = T/\Delta T$. The larger *T*, the closer the separation of frequency components in the transform. It is desirable to choose *T* as long as possible.

Sampling time ΔT or sampling rate (Sa/s). This has to be chosen with particular care. As discussed below the Nyquist Theorem states that sampling must be carried out done at a rate greater than twice the highest frequency component present in the waveform. This may also be stated alternatively that the maximum frequency that can be digitized adequately is given by one half the sampling rate. The origin of this condition is seen as follows:

Sampling Rate—the Nyquist Theorem

Consider sampling a sine wave with the following different sampling rates:

(a) If sampling occurs at a rate much faster than the frequency of the waveform the digitized information is unambiguous.

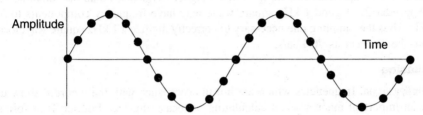

Fig. 5.15(a) A wave is sampled at intervals much smaller than half the period of the wave—at a frequency much greater than twice the signal frequency.

(b) If the signal waveform is sampled at exactly twice the frequency of the sine wave, depending on the triggering, a situation could be encountered in which it is predicted that either no wave is present (●) or that the wave is a square or triangular wave (o).

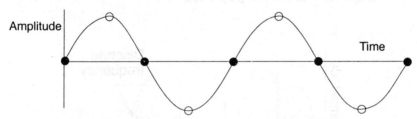

Fig. 5.15(b) Inconclusive sampling at a frequency exactly equal to twice the signal frequency.

(c) If sampling is carried out at a slower rate (longer time intervals), the digitized waveform is still related to the actual waveform, but it will be interpreted as a sine wave of lower frequency.

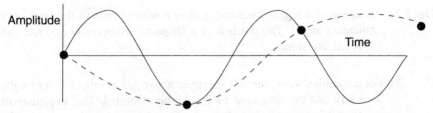

Fig. 5.15(c) A wave is sampled at intervals larger than half the period of the wave—at a frequency lower than twice the signal frequency.

The diagrams in Fig. 5.15 are a statement of the **Nyquist Theorem**:

> a signal must be sampled at a rate **greater than twice the highest frequency** present in its spectrum.

This is easy to resolve if the signal is a sine wave characterized by only a single frequency. However, a complex waveform such as a square wave has frequency components extending to much higher frequencies than the fundamental (Appendix 2). A good 1 kHz square wave may have frequency components to 35 kHz. Thus the sampling rate necessary to correctly digitize a 1 kHz square waveform may be as great as 70 kSa/s.

Aliasing

Higher signal frequencies which are not in accordance with the theorem show up in an important manner when calculating or measuring Fast Fourier Transforms. In general, if the sampling rate is set at f_s (Sa/s) the Fourier Transform will be calculated and plotted over a range of frequency from 0 to $f_s/2$. In calculating the FFT, the effect of frequency components at greater than half the sampling frequency f_s are 'folded back' around $f_s/2$ and cause distortion of the lower part of the spectrum. This is called **aliasing**. Two examples are shown below:

(a) For a sine wave, the frequency component is 'reflected' in the frequency boundary and shows up at a frequency as much **below** $f_s/2$ as the actual frequency is **above** $f_s/2$ (Fig. 5.16).

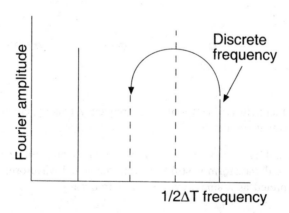

Fig. 5.16   Aliasing of a single frequency f shown with respect to the frequency boundary at $f_s/2$. The position of a frequency component $f_1 < f_s/2$ and $f_2 > f_s/2$ are shown.

(b) For a complex wave, the full spectrum above $f_s/2$ is reflected across the boundary and the measured FFT is highly distorted. The amplitude of the frequency components above the boundary add to those below the boundary (Fig. 5.17).

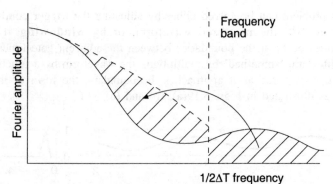

Fig. 5.17 Aliasing of the frequency spectrum for a complex wave about $f_s/2$.

Question (iv) The Fast Fourier Transform of 4 kHz square wave made using a 30 kSa/s sampling rate is as shown. Explain and interpret the position and relative magnitude of all the peaks seen on this transform.

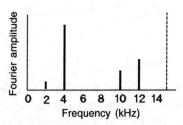

Windowing

The synchronization of the phase and frequency of the measured waveform to the gate time must also be considered. Figure 5.18 shows two waveforms of the same frequency, but with different phases with respect to the gate time. The Fast Fourier Transform assumes that the waveform repeats in the gate time.

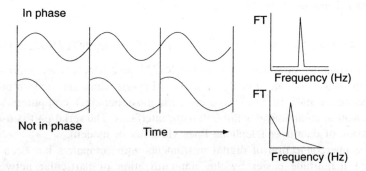

Fig. 5.18 Two waveforms which are phased differently with respect to the gate time: (a) The FFT sees a continuous sine wave. The FFT is well defined. (b) Poor triggering with respect to the gate time T leads to apparent discontinuities. The computed FFT has a broad spectrum.

This problem can be solved either by adjusting the trigger conditions and gate time to 'fit' the measured waveform, or by **windowing**. The major discontinuities occur at the boundaries between the adjacent gate times. If these discontinuities are 'smoothed' by multiplying the waveform by a function which has a decreasing value as it approaches the boundary, the form of the FFT is improved as illustrated in Figs. 5.19(a), (b) and (c).

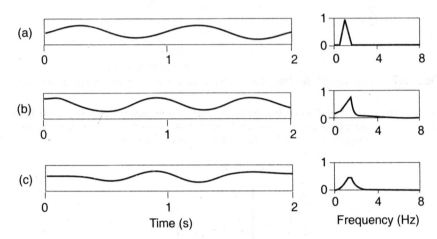

Fig. 5.19 Smoothing of apparent discontinuities of a function using a Hanning filter: (a) FFT of a signal having a frequency and phase which matches the gate time. (b) FFT of a signal which is poorly matched to the gate time. (c) FFT of signal (b) after multiplication by a Hanning function.

A number of windowing functions have been devised. Important functions are a **rectangular** window which has a value of zero until a fixed time interval away from the start and end of the gate time, or a **Hanning** window which has a smoother decrease.

5.10   TALKING AND LISTENING—THE iEEE 488 INTERFACE BUS

Common computer interfaces include the RS232 port which is an interface for **serial** transmission of data, and the parallel (LPT) ports which are used for **parallel** transmission of data to printers used with most personal computers. Some instrumentation communicates through these interfaces. The serial port is used for transmission of data over telephone lines via a fax or modem.

The interconnection of digital instruments with computers has been made orders of magnitude easier by the standardization of particular networking arrangements or **interface** buses. Most instrumentation can now be purchased fitted with a particular bus—the IEEE 488. This was developed by Hewlett Packard as the General Purpose Interface Bus (GPIB) and the interface has become the industry standard for the parallel transmission of data bits on a clock pulse.

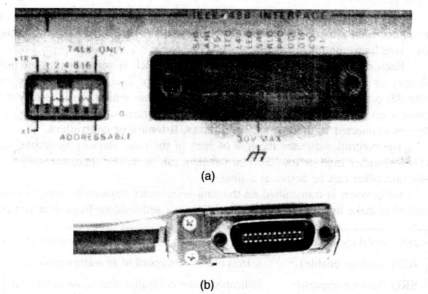

(a)

(b)

Fig. 5.20 Characteristic features of the General Purpose Interface Bus (GPIB):
(a) The switch setting for the device number and the GPIB connection
plug. (b) The GPIB cable and connector.

The GPIB parallel interface allows the interconnection of equipment including
computers, computer peripherals (printers, disk drives etc.), and instrumentation
units by a single special passive 16 line cable. In order to do this some compromises
were made. These include:

- a limit of 15 devices
- maximum cable length (20 m)
- relatively slow data transfer rate
- message length of 10–20 bits
- requirements for low cost and ease of use

Instrumentation fitted with an IEEE-488 bus can be recognized by (a) a
special plug on the back of an instrument, (b) a switch to set a device number, and
(c) the use of the special cable. These are shown in Fig. 5.20. The structure of the
GPIB instrumentation bus is shown in Fig. 5.21.

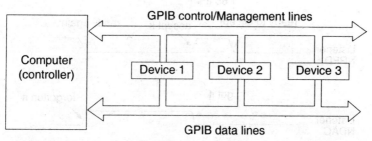

Fig. 5.21 The structure of the GPIB instrumentation bus.

Eight of the lines associated with the bus are reserved for the transfer of data, three lines are known as **handshake** lines (DAV, NRFD, NDAC), while the other five lines control bus activity.

Each device has its own '**device number**' which is set either by internal circuitry or by switch settings on the instrument. For example, disk drives are often #8, printers #4. Digital voltmeters have a number which can be set by the operator using a switch found either within the instrument or on the back casing. Devices connected to the bus may be **talkers**, **listeners** or **controllers**.

The controller dictates the role of each of the other devices by setting the ATN line either high or low. Several listeners can be active simultaneously, but only one talker can be active at a time.

Information is transmitted on the data lines under sequential control of the three handshake lines. The other control lines are self evident from their names.

IFC (interface clear)	Place the interface in a quiescent condition.
REN (remote enable)	selects remote or control of an addressed device.
SRQ (service request)	indicates to the controller that some device on the bus requires attention.
EOI (end or identify)	is used by a device to indicate the end of a data transfer sequence.

The handshake sequence. The handshake sequence overcomes problems due to synchronization between instruments, propagation delays etc. It involves three control lines

talker	DAV	(Data AVailable)
listener	NRFD	(Not Ready for Data)
	NDAC	(Not Data ACcepted)

controller: set bus ready: address specific device number
talker – expects to talk
listener – expects to listen

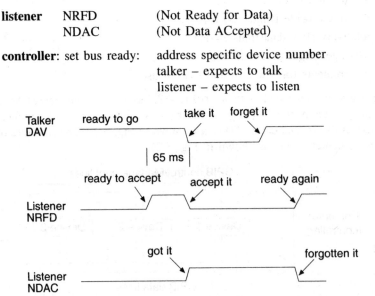

Fig. 5.22 Actions taking place on the GPIB bus during a data transfer.

The bus sets up the initial conditions when the ATN line goes low, and then performs the sequence shown below. All instruments must be set up within 65 ms, otherwise the software will indicate '**timeout**'. At the end of the sequence the bus has transferred the data and is ready to accept further data.

The GPIB bus is sold as a hardware card which fits into the back of a personal computer plus a software program which allows instruments to communicate through the bus. The software can be programmed in a wide range of programming languages.

5.11 INSTRUMENTATION SOFTWARE—LABVIEW™

Instrumentation software such as National Instruments 'LabView™' allows both convenient communication with a wide range of instruments via the IEEE488, parallel and RS232 ports, and a comprehensive portfolio of arithmetic, graphical, filter and analytic functions to be performed on acquired signals. This includes Fourier Transforms (Sec. 5.8), power spectra (page 126), and statistics (Chapter 1).

LabVIEW™ utilizes the concept of a virtual instrument (VI) with a **front panel** containing all controls and indicators, and a 'back panel' or wiring diagram in which the functions are assembled. An introduction to LabVIEW™ programming particularly with respect to the IEEE488 bus is provided in Appendix 5. A particularly valuable service for complex instruments is the ability to download an appropriate VI for driving the instrument from the National Instruments home page on the World Wide Web at www:http.natinst.com.

5.12 NUCLEAR INSTRUMENTATION

Much of the signal processing employed in nuclear instrumentation is carried out using standard modules or NIM bins (Nuclear Instrument Modules) defined by the US Atomic Energy Commission and the National Bureau of Standards. CAMAC or FASTBUS are interfaced systems which allow easier computer access to modules.

Definitions

Logic signals have a fixed shape and amplitude, and convey information by their presence, absence or relation to time.

Logic signals may be positive or negative (shown as + or – on panel symbols). Positive logic is a positive pulse of amplitude +4 to +12 (nominally 10) V and nominal pulse width 0.5 µs. High speed logic and timing pulses use negative pulses of –10 V with a rise time of 2 µs.

Linear signals convey information in the voltage amplitude—such as a voltage proportional to a particle energy. Such signals may be 'shaped' to reduce pulse overlap, improve the signal to noise ratio and to allow accurate timing while still retaining amplitude information. The pulse shapes employed are unipolar and bipolar. Only the bipolar signal swings both positive and negative.

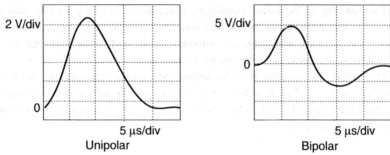

2 V/div

0

5 μs/div
Unipolar

5 V/div

0

5 μs/div
Bipolar

Fig. 5.23   Unipolar and bipolar pulse shapes used in general purpose nuclear instrumentation. Circuit time constant is 3 μs (After EE&G/ORTECH).

Linear Instrumentation Modules

Preamplifiers.   These are normally connected close to a detector. They may be cooled to reduce electronic noise. The input has a fast risetime with an output time constant of about 500 μs. No other pulse shaping is included.

Amplifiers.   These are of variable gain and/or attenuation to avoid overloading by high amplitude pulses. Provision is included for pulse shaping using differentiating and integrating controls. The outputs may be bipolar or unipolar.

Linear delay amplifier.   A means of delaying a pulse by a fixed period of time with no relative change in pulse shape or amplitude. This is of value to allow time for logic circuits to operate.

Linear gate.   A logic pulse opens a circuit for a fixed period of time to allow

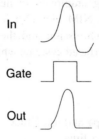

In

Gate

Out

a linear pulse to pass. This is used to limit pulse lengths to minimize overlap between sequential pulses and to remove negative going parts of the pulse.

Biased amplifier.   This allows the expansion of a certain portion of the amplitude

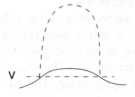

V

of a linear signal above a bias voltage *V*. This effectively reduces the pulse width.

Pulse stretcher.   This often follows the above. It takes the maximum value of a linear pulse and prolongs it at that value for a set time.

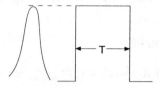

Analog to digital converter (ADC).   This takes a linear pulse and converts the amplitude into digital form suitable for analysis by **logic** units.

Time to pulse height converter.   Produces a pulse of amplitude related to the difference in time between two events.

Single channel analyzer (SCA).   This examines linear pulses of varying amplitude, emits a logic pulse if the amplitude is: (a) greater than a threshold *E* (integral mode), and (b) lies in a 'window' *E* to *E* + *ΔE* (differential mode).

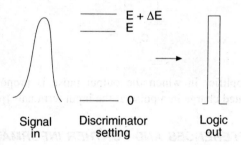

Fig. 5.24   Schematic representation of a SCA.

Controls for *E* and *E* + *ΔE* may be referred to as LLD and ULD, the lower and upper level **discriminators.**

Pulse Shapes and Logic Sequence for Nuclear Instrumentation

Timing SCA.   It produces a logic signal at the time of occurrence of the event.

Coincidence unit.   It produces a logic pulse if two pulses occur within a specified time interval—the **resolving time**.

Scalar Ratemeter.   This is a pulse counter with/without internal timing circuitry.

Pulse Height Analyzer (PHA).   Incoming linear pulses are sorted into channels on the basis of pulse amplitude.

Multichannel Analyzers (MCA).   These contain 2^n channels (512, 1024, 4096, etc.) with provision to add a count to a given channel each time a voltage pulse of a specific amplitude is detected by a PHA.

Multichannel Scaling (MCS).   Here the memory address of an MCA increases by 1 after a fixed delay time. The channel sweep may be correlated with an external experiment.

Charge Sensitive Amplifier

If $A \gg (C_i + C_f)/C_f$, $V_{out} = -V_{in}$

$$V_{out} = \frac{-A\,Q}{C_i + (A + 1)C_f} = \frac{-Q}{C_f}$$

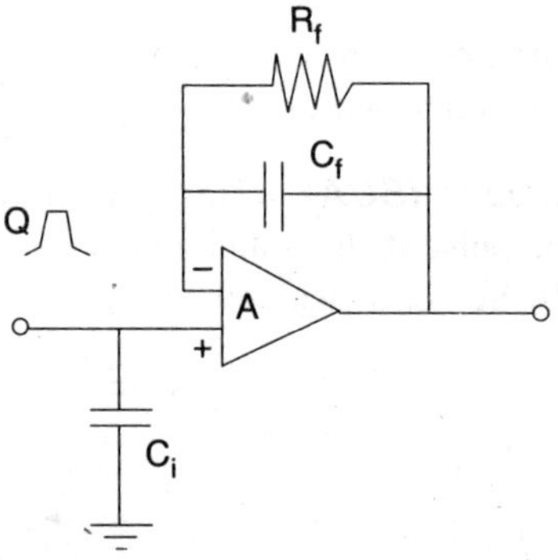

Fig. 5.25   An amplifier in which the output pulse is proportional to the total integrated charge in a pulse at the input terminals (pulse time $\ll R_f C_f$).

REFERENCES AND FURTHER INFORMATION

1. *Burr Brown Integrated Circuits Data Book,* Burr Brown Corporation, 6730 S. Tucson Blvd, Tucson, Arizona 85706, USA.

2. Amplifier Applications Guide (1992), Analog Devices Inc., 1 Technology Way, P.O. Box 9106, Norwood, Massachusetts 02062-9106, USA.

3. Texas Instruments Handbooks—*Linear Circuits (Operational Amplifiers) Data Book Vol. 1, Linear Circuits (Regulators, Comparators, Special Functions) Data Book Vol. 3, The TTL Data Book Vol. 2,* Literature Centre, P.O. Box 809066, Dallas, Texas 75380-9066, USA.

4. *Hewlett-Packard Test and Measurement Catalog (Annual),* Hewlett-Packard (Canada) Ltd., 6877 Goreway Drive, Mississauga, ON L4V 1M8, Canada.

5. Glenn F. Knoll, *Radiation Detection and Measurement,* 2nd ed., John Wiley & Sons Ltd., New York, 1979.

6. *Nuclear Instruments and Systems, EE & G Ortec Catalog,* EE & G Ortec Inc., 100 Maitland Road, Oak Ridge, Tennessee 37830-9912, USA.

7. *LabVIEW™ Graphical Instrumentation and Programming,* National Instruments Corporate Headquarters, 6504 Bridge Point Parkway Austin. TX 794-0100, USA

   Technical support fax: (800) 328-2203, (512) 794-5678; World Wide Web: www:http.natinst.com. .

ANSWERS TO QUESTIONS

(i)

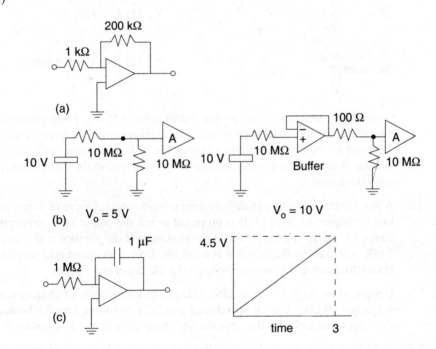

(a)

(b) $V_o = 5$ V $V_o = 10$ V

(c)

(ii) (a) 8 bit → 256 1 V/256 → 0.00396 V

    (b) 12 bit → 4096 1 V/4096 → 0.00024 V

    (c) 8 bit ADC → 1.6 μs total $\Delta f = 1/2T = 313$ kHz

    (d) Square wave → need up to the 9th harmonic → 30 kHz

(iii) 0.04 mV/°C → 4 × 10^{-5} V/°C

    Resolution → 2.5 × 10^{-4}

    This requires 15-bit resolution (1 part in 33768)

(iv) 30 kSa/s implies a frequency limit of 15 kSa/s

    4   fundamental

    12   3rd harmonic

    10   5th harmonic at 20 kHz reflected back to 10 kHz

    2   7th harmonic at 28 kHz reflected back to 2 kHz

DESIGN PROBLEMS

5.1   Design a measurement circuit (in block diagram form) for a direct reading reflectometer for the comparison of diffuse reflection from a test surface to that from a standard.

It is proposed to use the open circuit voltage developed by solar cells to give a measure of the reflected light intensity for uniform incident illumination.

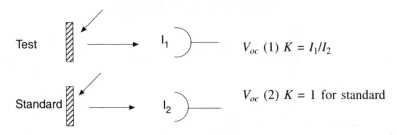

Test I_1 V_{oc} (1) $K = I_1/I_2$

Standard I_2 V_{oc} (2) $K = 1$ for standard

Give a block diagram of the analog circuit you could use which employs a logarithmic amplifier and appropriate multifunction converters and operational amplifier circuits to give a direct output for K. For the solar cells, assume that the photon induced currents i_v are large compared to the thermal reverse leakage current i_o.

5.2   A photomultiplier tube can deliver a maximum output current of 1 mA to a load resistance of 1000 Ω. It is proposed to use the output of a comparator circuit to activate a relay to shield the phototube if the photocurrent exceeds 1 mA. A standard voltage of 5 V is available. Design an operational amplifier circuit to achieve this requirement giving all resistance values.

5.3   Design a two stage low pass filter using individual stages of characteristic frequency 10 kHz. Use an operational amplifier circuit as a buffer between the stages and estimate the characteristic frequency of the combination.

5.4   State four interface requirements that are important for assembling 'smart' instrumentation systems and describe how the IEEE-488 standard system satisfies these requirements. What types of instrumentation systems are likely to benefit most from the development of the standard interface bus?

5.5   Describe one method used to achieve analog to digital conversion, and a typical method for digital to analog conversion. What type of ADC would you use to digitize a 1 kHz (a) sine wave and (b) a square wave with a reasonable accuracy?

5.6   A reverse-biased photodiode having a thermal generation current of 0.1 nA is used as optical sensor. The diode current is measured using a 47 kΩ load resistor connected to a sensitive voltmeter which has an input resistance of 10 MΩ and an input capacitance of 30 pF. The connection is made with a coaxial cable 1.5 m long. The capacitance per metre of cable is 88 pF/m. $1/f$ noise in the analog section of this measurement circuit is negligible.

(a) Assuming that a signal voltage of 2 × thermal noise at 300 K is acceptable for this measurement, what is the smallest signal current that you would expect to measure?

(b) What contribution does shot noise make at this level? What is the total noise in the circuit?

(c) If the first stage of amplification has a gain of 750 and noise figure of 1.3, what is the rms signal to noise voltage ratio after the first stage of amplification.

(d) If the voltmeter is replaced with an analog to digital converter with a conversion time of 10 μs, what is the highest frequency signal you would expect to measure. Use sufficient samples to achieve a reasonable assurance that the waveform is sinusoidal.

(e) If the light signals are rectangular unipolar pulses which have significant rounding of the corners, what would be the highest pulse repetition frequency that you would measure. Explain your reasoning.

5.7 Calculate the Fourier Transform of a triangular waveform of period 40 μs which has the form shown:

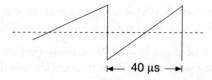

← 40 μs →

Use a spreadsheet to obtain a graphical plot of this series with 5, 11 and 15 harmonics.

5.8 The waveform shown in question 5.7 is to be digitized and filtered using a digital filter of cut-off frequency 250 kHz:

(a) What is the minimum sampling rate you would employ for the digitization.

(b) Sketch the form of the Fast Fourier Transform for the filtered waveform.

(c) What type of A to D converter might you employ for the digitization?

(d) If you examined this waveform in the time domain after filtering using an analog oscilloscope with a frequency response > 1 MHz, what would you expect the waveform to look like and why?

5.9 (a) Outline the differences between the following types of interfaces:

serial

parallel

IEEE-488

What does the concept of 'handshaking' mean in connection with each type of interface.

(b) What would you choose (and why) for the following situations:

(i) Connection between a remote computer acting as a data collection machine and a central network.

(ii) Development of a laboratory system for the measurement of temperature and frequency dependent conductivity. Components of the experiment include a programmable voltmeter and variable frequency oscillator, oven, power controller and digital voltmeter for temperature measurements, and controlling computer.

(iii) Connection of a computer to a local printer.

5.10 The modes of vibration of a wire clamped at each end are measured by using its motion to interrupt a laser beam.

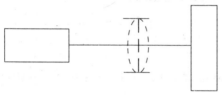

The fundamental frequency observed is around 2.0 kHz, but higher harmonics (up to the 10th) are suspected.

Laser   Wire   Detector

(a) What sampling rate (kSa/s) would you use in order to undertake a spectral analysis of this waveform using a Fast Fourier Transform?.

(b) If a gate time of 2 ms is used, how many periods of the fundamental frequency are examined. At what frequencies will the measurements be made and plotted within an FFT.

(c) Using a set of estimated spectral amplitudes, use a bar chart to show the results of an FFT using the sampling rate calculated in (a).

(d) If the sampling rate is arbitrarily set to 25 kSa/s, and assuming the same set of spectral amplitudes as in (c), use a bar chart to show the results of an FFT.

6

Vacuum Techniques

Vacuum practice enters almost every field of physics and engineering—to study or use matter at low densities, as a tool to isolate other systems or materials from the outside world, or as part of a deposition system to create thin films and coatings.

6.1 UNITS OF PRESSURE MEASUREMENT

The pressure of a gas is defined by the force per unit area it exerts on a surface. The basis of measurement is that of a standard atmosphere of gas at a temperature of 273 K. Two units of pressure (vacuum) are in current practice; the SI unit is the millibar (mbar), the older unit is the torr. The recommended unit is the mbar, although much North American equipment is calibrated in torr.

Units:

> The SI unit of pressure is Pascal (Pa).
> $1 \text{ Pa} = 1 \text{ N/m}^2$; $1 \text{ bar} = 10^5 \text{ Pa}$; $1 \text{ mbar} = 100 \text{ Pa}$
> 1 standard atmosphere = 1013 mbar
>
> 1 standard atmosphere = 760 mm of Hg = 760 torr
> 1 torr = 1 mm Hg = 1.33 mbar
>
> 1 mbar = 0.75 torr

Range of Accessible Pressures

The ranges of pressure which are used in practice and the conventional names applied to them are shown in Fig. 6.1. A rough distinction between the regions lies in the nature of the flow processes resulting from molecular interactions. The mean free path l is the distance a molecule moves freely between collisions. At pressures close to atmospheric, molecules interact strongly with each other, l is comparable to the inter-molecular spacing and the flow is analogous to that of a **viscous** liquid. As the pressure is reduced the mean free path increases and becomes comparable with the size of the container L. **Knudsen** flow has $l \approx L$, while true **molecular** flow occurs when $l > L$. At the lowest pressures, the molecules interact with each other rarely, and the interest lies in the rate at which they impinge on a surface.

The lowest pressure that can be achieved is of the order of 10^{-15} mbar. Pressure levels used for such industrial and laboratory practice lie between 10^{-5} and 10^{-7} mbar, while high vacuum experiments in surface physics are carried out at pressures from 10^{-9} to 10^{-11} mbar. These distinctions lie partly in what is the purpose of creating the vacuum, and partly in what can be easily accomplished by specific equipments.

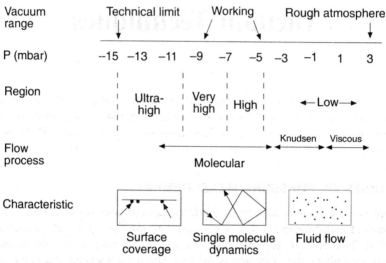

Fig. 6.1   Range and characteristics of vacuum practice.

6.2 CHARACTERISTICS OF VACUUM

The parameters of interest are:

— atomic or molecular density (N) (cm^{-3})
— atom spacing (d) (cm)
— mean free path between molecular collisions (l) (cm)
— time to form a monolayer on a 'clean' surface (t_m) (s)

These are estimated from the perfect gas equation $PV = nRT$ and from simple ideas of kinetic theory. More detailed consideration of kinetic theory generally gives the same form of equation with the addition of a multiplying factor [1]. In keeping with the scale of vacuum apparatus, the calculations are made on a scale of cm.

Perfect gas law. $PV = nRT$, n is the number of moles of gas in volume V, P is the pressure and T is the temperature. The gas constant $R = 8.314 \text{ mol}^{-1}\text{K}^{-1}$.

1 gram molecular weight (mol) of a gas at standard temperature (273.14 K) and pressure (1 atmosphere) (STP) occupies a volume of 22.414 l and contains 6.023×10^{23} molecules. 1 l = 1000 cm^3.

Density of molecules.   (cm^{-3}): (per mbar of pressure)

$$N = \frac{P}{T} \frac{6.023 \times 10^{23} \times 273.14 \text{ cm}^{-3}}{22.414 \times 10^3 \times 1013} = 7.25 \times 10^{18} \frac{P}{T} \text{ cm}^{-3} \text{ mbar}^{-1}\text{K} \quad (6.1)$$

Molecular spacing (cm)

$$d = (N)^{-1/3} \text{ cm} \tag{6.2}$$

Mean free path (*l*) (cm). Molecules of diameter ζ cm collide when their centres approach within a distance ζ cm. The mean free path *l* may be estimated from the volume swept out by a disc of radius ζ moving between collisions spaced distance *l* (Fig. 6.2(a)). On average, no other particle intrudes into this volume during one mean free path. Since there are *N* molecules/cm³,

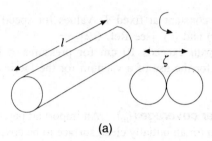

(a)

Molecule	Molecular weight G	Molecular 10^6 diameter ζ cm	$(l \cdot P \times 10^{-3})$ (cm-mbar)
He	4	2.17	18.0
Hg	2	2.66	12.0
N_2	28	3.73	6.1
O_2	32	3.62	6.5
H_2O	18	4.64	3.95
Ar	40	3.64	6.4
Air	29	3.59	3.59

(b)

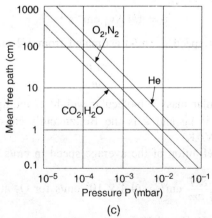

(c)

Fig. 6.2 Mean free path as a function of pressure in mbar for representative gases at $T = 300$ K. (a) Volume swept out by molecules of diameter ζ, *l* is the mean free path. (b) $l \cdot P$ product for various gases. (c) Mean free path versus pressure for various gases.

$$(\pi \zeta^2 l) = 1/N$$

The kinetic theory of gases leads to an additional factor of $1/\sqrt{2}$. Hence

$$l = \frac{1}{(\sqrt{2}\pi \zeta^2 N)} \qquad (6.3)$$

Substituting for N (number density) from eqn. (6.1)

$$l = 3.11 \times 10^{-20} \frac{T}{\zeta^2 P} \text{ cm} \qquad (6.4)$$

The product $l \cdot P$ = constant at fixed T. Values for specific gases at 20°C are shown in Figs. 6.2(b) and (c). [see Ref. 6.2]

The mean free path exceeds 10 cm for pressures $< 10^{-4}$ mbar. This is a significant threshold for the use of a vacuum for the fabrication of thin films by vapour deposition.

Time for monolayer coverage (t_m).   An important parameter in high vacuum work is the time taken for an initially clean surface to be covered with a monolayer of gas. The time for this is t_m (in seconds) $= S/(\beta v)$, where

S (cm^{-2}) is the number of sites in a monolayer
v (cm^{-2}s^{-1}) is the rate at which molecules strike the surface
β is the probability that a molecule sticks to the surface.

The number of sites S is deduced from the atomic structure of the surface and the type of interaction which takes place between it and the impinging molecules. The sticking probability β is generally (0.1–0.2).

On the average within a gas containing N molecules/cm^3 moving in random motion with molecular speed v_a cm/s, approximately 1/4 of the molecules will be moving towards a surface. The molecular flux v at the surface [1] is,

$$v = 1/4 \, N v_a \text{ cm}^{-2}\text{s}^{-1}$$

The average molecular speed v_a (m/s) estimated from $1/2 \, m v_a^2 = 3/2 kT$

$$v_a = \sqrt{\frac{3kT}{m}}$$

where m is the molecular mass = molecular weight G multiplied by the atomic mass unit 1.66×10^{-27} kg and k is the Boltzmann's constant. (1.38×10^{-23} J/°K or 8.62×10^{-5} eV/°K)

A more precise definition of the average speed in cm/s is

$$v_a = 100 \sqrt{\frac{8kT}{\pi m}} \text{ cm/s} \quad (4.4 \times 10^4 \text{ cm/s for O}_2 \text{ at 293 K}) \qquad (6.5)$$

Combining with eqns. (6.1) and (6.5)

$$t_m = \frac{4S}{\beta N v_a} = \frac{4ST}{7.25 \times 10^{18} \beta P} = 3.79 \times 10^{-23} \frac{S}{\beta P} \sqrt{TG} \qquad (6.6)$$

Table 6.1 shows values of these parameters for oxygen at atmospheric pressure, working vacuum, and for ultra-high vacuum respectively. The surface site density $S = 10^{15}$ cm^{-2} and the sticking probability $\beta = 0.2$ for this table.

Experiments to study clean surfaces or non-interacting molecules require pressures less than 10^{-7} mbar so that a reasonable time is available to carry out the measurements. A knowledge of the mean free path l in technical processes such as the vacuum deposition of thin films provides an estimate of the distance a depositing species can travel in a chamber held at a given pressure.

Table 6.1 Oxygen (O$_2$) in Three Pressure Regimes at 293 K

P (mbar)	1013	10^{-6}	10^{-10}
Number of atoms/cm^3 (N)	2.5×10^{19}	2.5×10^{10}	2.5×10^6
Atom spacing ($N^{-1/3}$) cm	3.4×10^{-7}	3.4×10^{-4}	7.4×10^{-3}
Mean free path cm (l)	6.9×10^{-6}	6.9×10^3	6.7×10^7
Monolayer coverage time (s)	1.8×10^{-8}	1.8×10^1	1.8×10^5

Question (i) Calculate the monolayer coverage time for a surface in argon at a pressure of 10^{-9} mbar and 20°C. The surface has a surface site density of 3×10^{15} cm^{-2} and sticking probability of 0.1 for argon.

6.3 APPLICATIONS OF VACUUM

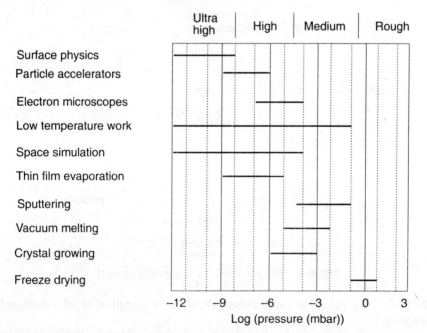

Fig. 6.3 Pressure ranges used for applications of vacuum in industry and in research and development (with permission—Leybold Vacuum Products Inc.).

Figure 6.3 shows the pressure ranges employed in important applications of vacuum in industry and in research and development. These ranges determine the nature of the pumping and measurement systems used.

A comprehensive reference to the material of this chapter is found in *Vacuum Technology, its Foundations, Formulae and Tables* prepared by Leybold Vacuum Products Inc., 5700 Mellon Road, Export, PA 15632 USA.

6.4 VACUUM SYSTEMS

The generic features of a vacuum system are shown in Fig. 6.4. A chamber is pumped by a high vacuum stage **backed** by a lower pressure rotary **mechanical** or adsorption pump. Most high vacuum pumps only operate effectively below 50 mbar. The purpose of the backing pump is to maintain the base pressure of the high vacuum pump, and to undertake the initial evacuation of the system.

The operation of the system is controlled by the **baffle** or **high vacuum** valve, and the **roughing** and **backing** valves. An **inlet** valve allows the system to return to atmospheric pressure.

A liquid nitrogen **cold trap** placed between the chamber and the high vacuum pump condenses water and organic vapours desorbed from the chamber walls and reduces back-streaming of vapours from the pump to the chamber.

Pressures are monitored by the **high vacuum, roughing** and **backing** gauges.

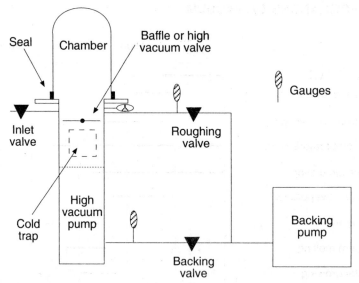

Fig. 6.4 The elements of a vacuum system.

The basic principle to be remembered for the safe operation of all valves and pumps is:

> Always maintain the high vacuum pump at low pressure.
> Never expose the pump to atmosphere.

Pump-down sequence. When the chamber of a vacuum system is open with the high vacuum valve closed, the high vacuum pump is pumped to a low base pressure through the open backing valve. The chamber pump-down sequence is as follows:

1. The **backing valve** is **closed** to seal the high vacuum pump.
2. The **roughing valve** is **opened** to pump the chamber to **10–20 mbar**.
3. The **roughing valve** is **closed** and the **backing valve** is **opened**.
4. The **high vacuum valve** is **opened** to pump the system to low pressure.

To vent the chamber, the high vacuum valve is closed and the inlet valve is opened.

A major source of apparent leaks and long pump-down time is the desorption of water and other organic vapours from the chamber walls. While the effects of this can be reduced by cold traps, it is good vacuum practice to avoid fingerprints by using gloves when working in the chamber, to vent with dry nitrogen gas, and to bake the chamber during pumpdown at a temperature >100°C using heating tapes wrapped around its exterior. Good vacuum chamber construction ensures that no unvented air pockets exist at the bottom of screw threads venting into the chamber.

Construction of vacuum equipment. Vacuum chambers are built of glass, stainless steel or aluminum. The criterion is whether the system has to be baked, must be easily demountable, or is required for ultra-high vacuum. Metal systems are welded or vacuum brazed.

Vacuum seals are an important part of vacuum practice. In medium and high vacuum systems, neoprene or silicon rubber 'O-rings' are used for groove seals. The pressure limitation is 10^{-7} mbar, set by the vapour pressure of the rubber. In ultra-high vacuum systems, copper discs are sandwiched between stainless steel knife edges.

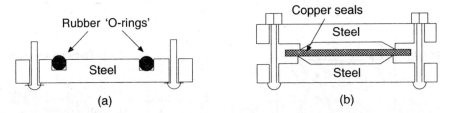

Fig. 6.5 Vacuum seals for demountable flanges: (a) Neoprene or silicon rubber 'O-rings', (b) Copper seals.

The individual components for the design of vacuum systems are summarized in Fig. 6.6. These components are found in commercial vacuum systems. Older gauges such as the McLeod gauge or the mercury manometer based on the relative levels of mercury columns are rarely used in such systems.

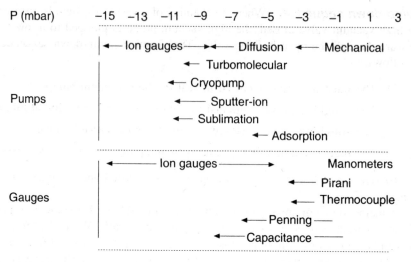

Fig. 6.6 Components available for vacuum system design in various pressure ranges.

6.5 VACUUM PUMPS

Pumps which are found on most conventional vacuum systems are shown in Fig. 6.7.

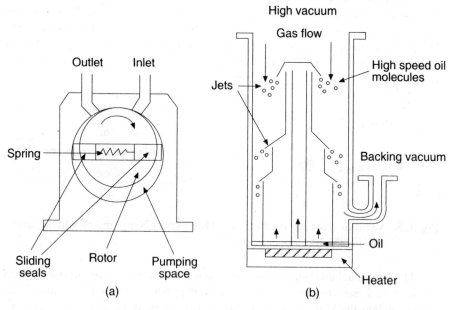

Fig. 6.7 (a) Mechanical rotary pump. (b) Multistage diffusion pump.

Mechanical rotary pump	**Multistage diffusion pump**
10^3 to 10^{-3} mbar	10^{-3} to 10^{-9} mbar
A sliding seal mounted in a rotor sweeps out a volume of gas from inlet to outlet.	A fluid (usually a silicone based oil) is heated so that molecules pass up a central column and emerge from jets at high speed in a downward direction. Gas molecules are entrained in the vapour stream and are carried downwards with the oil to exit at the backing pump connection.

The high molecular speed utilized in a diffusion pump is acquired by careful design of the jet surfaces. These should always be treated with care during cleaning. The highest pumping speed is obtained using light organic pumping fluids. These are susceptible to degradation or 'cracking' from oxidation and the use of stable silicone fluids such as Silicone 704 is almost universal.

A particular problem with diffusion pumps is the **back-streaming** of pump oil into the chamber. This is minimized by careful design of the pump and by the use of water-cooled baffles between the pump and the chamber to ensure no 'line-of-sight' path between the two parts of the system. This problem is almost, but not quite, eliminated by a liquid-nitrogen-cooled cold trap.

Cold traps. The mass spectrum of the gas in a typical vacuum system generally contains evidence of residual air (20% O_2, 80% N_2), a large peak at $M = 18$ due to H_2O, and organic vapours from the pumps. For all high vacuum systems ($P < 10^6$ mbar), the highest cleanliness must be observed. When working within the system gloves should be worn to eliminate fingerprints on internal surfaces. **Ultra-high vacuum systems** are constructed of stainless steel and are baked to 300°C using external ovens in order to eliminate adsorbed gases.

A simple and convenient method of reducing the effect of desorbed vapours or of unwanted components of vacuum pump oil is to have surfaces within the vacuum system which are reduced in temperature. The vapour pressure of various gases is shown in Fig. 6.8. At the temperature of liquid He ($T = 4.2$ K) all gases but helium are frozen out into a solid form. This concept has been implemented in the cryopump (Fig. 6.11) used to create ultra-high vacuum.

In systems where a liquid nitrogen trap is employed ($T = 77$ K), in principle the partial pressure of H_2O is reduced to 10^{-22} mbar. However since degassing of internal surfaces occurs for long times, the partial pressure of water vapour in most conventional vacuum systems remains substantially higher than this.

Specialized Pumps

Modern vacuum equipment utilizes a number of pumping systems which avoid the inherent complexity and backstreaming properties of diffusion pumps.

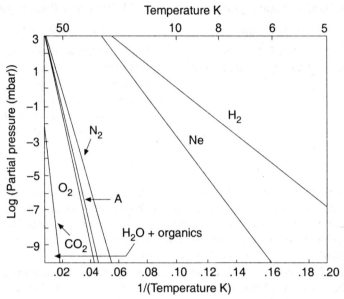

Fig. 6.8 The partial pressure of gases and vapours as a function of temperature.

Turbomolecular pump. This consists of an axial turbine in which a rapidly rotating (45,000 rpm) rotor sweeps gas molecules to an output port maintained at a backing pressure. Turbomolecular pumps are capable of ultimate pressures of 10^{-11} mbar in systems which have been baked to remove all hydrocarbon vapours.

Fig. 6.9 An illustration of a turbomolecular pump. (with permission—Leybold Vacuum Products Inc.).

Adsorption pump. Zeolites are inorganic alumino-silicate 'sponges' having a very large surface area in pores of diameter comparable to the molecular diameter of water, oil and other hydrocarbons. On cooling in liquid nitrogen, adsorption of gas occurs in the pores of the zeolite. The zeolite is regenerated by warming and pumping off the desorbing gas.

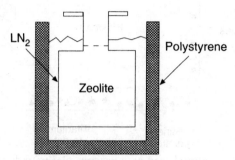

Fig. 6.10   A schematic representation of a zeolite adsorption pump.

Liquid nitrogen cooled semiconductor detectors (Fig. 10.19) employ zeolites to maintain internal vacuum, while adsorption pumps are used as backing pumps for 'oil-free' ultra-high vacuum systems.

Cryopump.   As noted in Fig. 6.8 the partial pressure of all gases and vapours become very low at temperatures close to absolute zero. This ensures excellent vacuum in vacuum systems surrounding liquid helium (4.2 K). A cryopump utilizes a closed cycle helium refrigerator to cool an adsorbing metal surface.

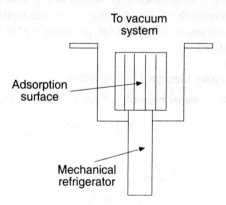

Fig. 6.11   Cryopump using a closed cycle helium refrigerator.

Sputter or getter ion pumps.   Freshly deposited metal films act as efficient 'getters' for gas molecules. Titanium is particularly effective for this purpose. In a **sublimation pump**, a metal is evaporated from a resistance heated source to capture gas molecules as it deposits on surrounding walls. In a **sputter ion pump** a gas discharge is initiated near a titanium electrode and ions hitting the surface cause Ti films to build up elsewhere within the pump.

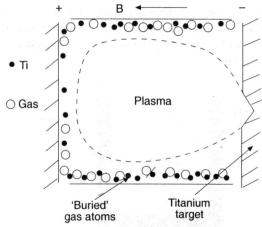

Fig. 6.12 Sputter ion pump based on a gaseous plasma induced within a titanium enclosure.

The pumping action results from the incorporation of gas atoms within the titanium layer. The action is continuous since gas ions are buried within successive layers of Ti. A magnetic field is employed to increase the number of collisions between gas molecules and electrons and hence to maintain the discharge to lower pressures.

6.6 VACUUM GAUGES

In the two stage vacuum system of the type shown in Fig. 6.4, there is a need to indicate the **backing pressure** of 10–50 mbar at the mechanical pump, and to make a precise measurement of the **ultimate pressure** within the high vacuum chamber (10^{-5}–10^{-11} mbar).

Pirani or Thermocouple Gauge (1–500 mbar)

The gauge measures the temperature of a heated filament. The electrical input

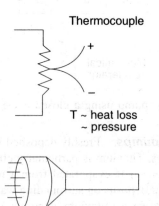

Fig. 6.13 Backing pressure gauges.

power is constant, and the filament temperature is determined by heat loss due to thermal conduction in the gas. This is a measure of pressure. In a **Pirani** gauge, the filament temperature is measured as resistance $R(T)$. In a **thermocouple** gauge the temperature is measured directly.These gauges are non-linear.

Penning Cold Cathode Gauge (10^{-3} to 10^{-9} mbar)

A high voltage (≈ 2 kV) is applied between a central cathode and a grounded cylindrical anode. The current flowing in the resulting gas discharge is a measure of the pressure.

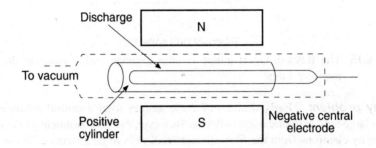

Fig. 6.14 Penning gauge utilizing current flow through a low pressure plasma.

In order to extend the range, a magnetic field is applied using cylindrical permanent magnets to cause electrons to undergo helical motion in the discharge. This increases the probability of ionization. The cold cathode gauge is a robust, general purpose gauge for use at medium and high vacuum. It has no heated filament to burn out if the gauge is exposed to high pressures. The reading depends on the nature of the gas being pumped and is calibrated for a given gas.

Capacitance Gauge (10^{-4}–10^{-7} mbar)

A capacitance gauge measures absolute pressures by sensing very small deflections of a metal diaphragm (Fig. 3.40). The reference side of the sensor is pumped to high vacuum and is permanently sealed. A capacitor is formed between two electrodes, one of which is fixed and one of which is attached to the diaphragm. When the pressure changes, the diaphragm moves and this is recorded as a change in capacitance. These sensors provide highly accurate measurements of pressure that are independent of the type of gas.

Ionization Gauge (10^{-5} to 10^{-13} mbar)

A heated filament produces a stream of electrons which ionize gas molecules. The positive ion current is measured between a central fine wire biased negatively and a circular ring or grid systems of anodes.

The copious stream of electrons allows efficient ionization of the gas and the ion current is a sensitive measure of the pressure.

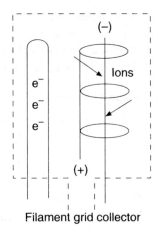

Filament grid collector

Fig. 6.15 The Bayard-Alpert gauge is both a gauge and a pump at the lowest pressure range.

X-ray problem. Early designs of these gauges used a central positive anode and a large area cylindrical ion collector. However, the bombardment of the positive anode by electrons from the filament caused X-rays to be produced. These X-rays cause photoelectrons to be emitted from the ion collector. The photoelectrons flow to the positive anode **away** from the ion collector.

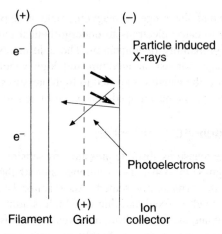

Fig. 6.16 Photoelectron generation leading to spurious currents in an ionization gauge.

The total current to the collector is

$$I \text{ (total)} = I \text{ (ions)} + I \text{ (photoelectrons)}$$

and an apparent increase in pressure is recorded. The effect is proportional to the area of the ion collector exposed to the X-ray flux.

The ionization current is reduced by inverting the system and by making the ion collector a fine wire of very small surface area. This problem and its solution is an excellent example of how unexpected physical phenomena may have to be considered in a given instrumentation problem.

The flow of charge within an ionization gauge causes adsorption of gas molecules on the surfaces of the electrodes and of the surrounding glass envelope. This is employed as a pump in the ultra-high vacuum range of pressures.

6.7 PUMPING SPEED FOR A VACUUM SYSTEM

Factors which affect the pumping speed of a vacuum system are shown in Fig. 6.17. A vacuum pump is connected to a chamber of volume V by means of a pipe of known dimensions. The original pressure is P_o, gas enters through leaks or desorption from the walls, and the pump can reach an ultimate pressure of P_u.

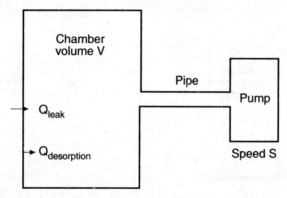

Fig. 6.17 Factors which affect the pumping speed of a vacuum system.

The following quantities are defined:

Quantity of gas G (PV). Since $PV = nRT$, where n is the number of gram-moles in the volume V,

$$PV = \frac{G}{M} RT$$

where G is the mass of gas in the volume and M is the gram-molecular weight. Thus

$$G = PV \cdot \frac{M}{RT}$$

If the gas composition (M) and temperature (T) do not change, the (PV) product and the mass of gas are analogous.

Pumping Speed S

For a pump having a volume flow of gas at intake pressure P, the pumping speed is

$$S = \left(\frac{dV}{dt}\right)_P \, 1\,\text{s}^{-1}$$

Throughput Q. The mass flow rate (g/s) of gas through an aperture or into a pump is defined as the throughput. This can be written either in terms of G or the (PV) product as shown by

$$Q_m = \frac{dG}{dt} \quad Q_{PV} = \frac{d(PV)}{dt} = P\left(\frac{dV}{dt}\right)_P = PS$$

if P and V are constant at the intake side of the pump (Q_{PV} mbar-l s^{-1}), and S is constant over the range of pressures concerned.

Pumping speed for mechanical and high vacuum pumps. The pumping speed as a function of intake pressure is shown in Fig. 6.18 for a rotary mechanical pump and for a diffusion pump respectively.

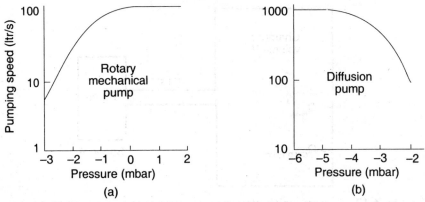

Fig. 6.18 Pumping speed as a function of pressure for (a) a rotary mechanical pump, (b) a diffusion pump.

The physical reason why the pumping speed varies with pressure in this manner for the two pumps can be rationalized if the mechanisms shown on page 176 are reviewed. The rotary pump works best if the gas acts like a viscous liquid, while the diffusion pump operates well only if molecular flow enables molecules to be entrained in the high speed flow of vapour.

Question (ii) A calibrated leak of 1.05×10^{-2} mbar-l s^{-1} is admitted to a vacuum system. If the pressure after an extended period of pumping is 5×10^{-5} mbar, what is the pumping speed of the pump?

Pumping Time

From the definition of throughput and assuming that a given pump is operating at a pumping speed which is independent of pressure, the pressure at time t for a pump of speed S and ultimate pressure P_u evacuating a volume V is given by

$$\frac{dG}{dt} = \frac{d(PV)}{dt} = -\left(\frac{S}{V}\right)(P - P_u)$$

Solving for a pressure change from P_o to P

$$(P - P_u) = (P_o - P_u) \exp -\left(\frac{S}{V}\right) t$$

If $P_u \ll P$ then

$$P = P_o \exp -\left(\frac{S}{V}\right) t$$

The form of this curve as a function of time is shown in Fig. 6.19. At higher pressures, a plot of log P versus t is a straight line of slope $-(S/V)$. At low

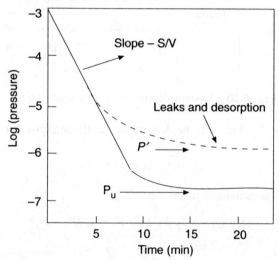

Fig. 6.19 Pressure as a function of time: (a) Assuming a pump of ultimate pressure P_u and no leaks. (b) In the presence of leaks and desorption.

pressures, the pressure tends towards the ultimate pressure of the pump P_u. In practice leaks and desorption from the walls causes an apparently stable higher pressure P'. Question (ii) illustrates a method of calculating the size of the leak. The time dependence of the pressure during pump down is an important characteristic of a vacuum system. It should be monitored carefully and the reason for any changes noted.

Question (iii)   A chamber of volume 2000 l is to be pumped by a diffusion pump based system of effective pumping speed 21 l/s from 10^{-3} to 10^{-6} mbar. The ultimate pressure of the pump is 8×10^{-7} mbar. (a) Calculate the pump down time, (b) a real system is likely to take longer than this, why?

Effective pumping speed—conductance.   Pumping speed graphs such as those shown in Fig. 6.18 are characteristic of the pump at its intake port. In a real system, the pump is connected to the chamber through valves, cold traps or pipes. Such connections will reduce the pump speed to a lower **effective** value S_e. Gas will also flow into the system from leaks, desorption from the walls, and from cavities such as screw threads arising from mechanical construction. The systems then becomes as described in the following:

Pump of speed S_p; intake pressure P_1

Tube conductance C; input pressure P_2

Equivalent speed S_e; intake pressure P_2

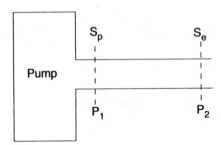

Fig. 6.20 System design for a pumping system.

Conductance *C.* This may be defined as the **throughput per unit pressure difference**

$$Q = Q_{PV} = C(P_2 - P_1)$$

Since no gas is accumulated in the system

$$Q = Q_{PV} = S_p\, P_1 = S_e\, P_2 = C\,(P_2 - P_1)$$

$$\frac{Q}{C} = \frac{Q}{S_e} - \frac{Q}{S_p}$$

$$\frac{1}{S_e} = \frac{1}{C} + \frac{1}{S_p}$$

The form of these equations suggests an analogy with electrical quantities:

Pressure   = voltage
Throughput   = current
Conductance   = conductance

The value of C for a given component depends on its shape and the pressure range. Formulae for important shapes in vacuum technology for molecular flow are given in Table 6.2 [see Ref. 6.1].

The conductance of a pipe varies as the **cube** of its diameter, while the conductance of an aperture varies as its area. The effect of bends in a pump is to change its effective length by a substantial factor. The practical implications of this table are that for maximum conductance, vacuum connections should be as **wide** and as **short** as possible. This is the reason that high speed high vacuum pumps have large apertures, and that vacuum system design minimizes the complexity of the piping arrangements.

Table 6.2 Conductance for Vacuum Components Operating under Conditions of Molecular Flow

Component	Conductance (l/s)
Tube length L cm diameter d cm $L > 10d$	$12.1 \dfrac{d^3}{L}$
Aperture of area A cm^2	$11.6\ A$
n bends in a pipe of diameter d cm total length L cm	$L_{\text{eff}} = L + 1.33\ nd$

Question (iv)   A vacuum chamber of volume 120 l is pumped through a line 20 cm long and 2.5 cm in diameter with one bend. The pump has a nominal pumping speed at its intake of 100 l/s. How long does it take to pump the chamber from 1013 mbar to 10^{-3} mbar if the ultimate pressure $P_u = 8 \times 10^{-4}$ mbar.

Leak Testing

In the event of a leak in a vacuum system, after all joints have been checked, the following procedures can be carried out.

1. If the pressure remains >200 mbar, check all O-rings and soldered joints.

2. If the pressure is in the range 1–40 mbar, carefully spray the outside of the system with an organic fluid of low flammability—methanol, for example. The diffusion rate of molecules through the leak will increase and a pressure rise will be noted when the fluid covers the hole.

3. For smaller leaks, a **leak detector** is invaluable. This is a mass spectrometer detector tuned to detect helium ions. This is attached to the vacuum system and helium gas from a cylinder is directed to the outside of the system. The small helium ions pass into the leak and are detected within the chamber.

6.8 THIN FILM TECHNIQUES

Thin film technology is the basis of integrated circuit fabrication, and is used to manufacture thin film devices ranging from electronic capacitors to low temperature Josephson junctions. The technology has widespread industrial applications, particularly in the coating of plastic components for automobiles. The major techniques using physical vapour deposition are shown in Table 6.3. All of these methods employ vacuum methods, either to eliminate the effect of a background gas, to set the pressure of a gas to allow a reaction or deposition to proceed, or to stabilize a glow discharge.

Table 6.3 Physical and Chemical Processes for Thin Films

Technique	Power source	Materials
Vacuum Evaporation	Thermal Electron beam	metals single elements oxides
Sputtering	DC plasma discharge RF plasma discharge RF magnetron plasma	metals and alloys insulators oxides
Plasma	RF glow discharge	Amorphous silicon
Chemical Vapour Deposition (CVD)	Vapour phase chemical reactions at high temperatures	Transparent oxides metals semiconductors
Molecular Beam Epitaxy (MBE)	Molecular source	Semiconductors GaAs

The fabrication of materials as thin films has the following advantages:

1. Microcircuit fabrication can be achieved
2. Economy in material use
3. Non-equilibrium alloys and crystal structures may be created.

Related processes which are important fall under the heading of **thick film** fabrication. This includes the use of pastes prepared by mixing powdered metals with appropriate organic binders and wetting agents. The pastes are applied to a surface either by painting or screening and the paint is dried and fired to produce a conducting layer. Such techniques can be extended to oxide and insulating layers.

Deposition Procedures

Thermal evaporation. A tungsten spiral Fig. 6.21(a) or flat molybdenum 'boat' Fig. 6.21(b) is heated electrically. Pieces of metal placed on the spiral melt and wet the surface at high temperature. Evaporation occurs at higher power inputs to the spiral.

A difficulty of the metal alloying with the tungsten spiral is avoided by an **electron beam source.** A beam of electrons is accelerated (1 A at 6 kV) and is focussed onto a pressed powder target held in a water cooled copper hearth as shown in Fig. 6.21(c). The material acts as its own crucible and contamination from the source is avoided.

Thermal evaporation has the advantages that it can be easily carried out, is rapid, and is good for metals and single elements.

A **disadvantage** of thermal evaporation is that the films tend to be non-stoichiometric, although this can be avoided by **flash evaporation** in which powder is dropped onto a heated tungsten ribbon.

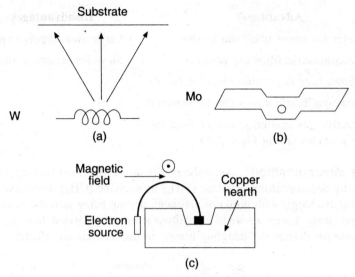

Fig. 6.21 Thermal evaporation with various forms of power input.

Radio-frequency (RF) and DC sputtering. A DC or RF plasma is established near a target. The plasma gas is normally argon for metal targets and argon/oxygen mixtures for oxide targets.

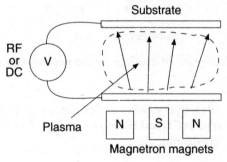

Fig. 6.22 Ions from a gas discharge impact onto target surface resulting in material transfer to the substrate and to other parts of the chamber.

A **DC discharge** can be used with conducting targets, but insulating targets charge up due to ion bombardment and ions are unable to reach the target. In this case, a **radio-frequency discharge** is employed so that electronic and ionic bombardment on successive half cycles of the exciting waveform maintain the target neutral (or at least at a controlled DC voltage above ground potential).

The light electrons in the plasma are ineffective in the sputtering process and serve to 'short-circuit' the ion current. However electron current is necessary to maintain the discharge at low sputtering pressures. In **magnetron sputtering** a magnet is placed behind the target to confine the electrons to a circular path away from the substrate while increasing their collision frequency with ions at low pressure. Sputtering has the following advantages and disadvantages.

Advantages	**Disadvantages**
• Better for compounds and oxides	• Large area targets required
• Stoichiometric films are possible	• Slow for ceramics and oxides
• Allows high deposition rates for metals	
• Non-equilibrium alloys can be obtained	
• Reactive gas sputtering can be used to form oxides (Zn in O_2 = ZnO)	

Plasma decomposition.   Amorphous silicon or diamond-like coatings are prepared by decomposition of silane (SiH_4) or methane (CH_4) respectively in an RF plasma discharge, with boron or antimony doping being introduced as borane (BH_3) and SbH_5. Large amounts of hydrogen are incorporated into the film to compensate for defects or 'dangling bonds' in the amorphous structure.

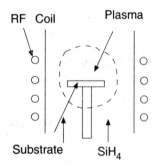

Fig. 6.23   Plasma deposition of coatings by the decomposition of silane in a gas discharge.

Spray pyrolysis.   Reactive vapours are combined on a heated substrate. An example is the reaction of $SnCl_4$ and H_2O to form SnO_2. This is a highly transparent oxide used to form 'conducting glass'.

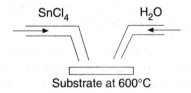

Fig. 6.24   Vapour phase reaction on a heated substrate.

Chemical vapour deposition (CVD).   Complex metallorganic compounds are reacted at low pressure on the surface of a substrate. The reaction rate may be enhanced by the use of plasmas in 'plasma enhanced CVD' (PECVD). CVD is being actively pursued for the fabrication of GaAs films, metal coatings and oxide films.

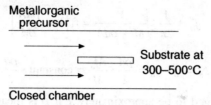

Fig. 6.25 Vapour phase chemical reactions in chemical vapour deposition (CVD).

Molecular beam epitaxy (MBE). Developments in molecular sources now allow films of controlled composition to be built up **epitaxially** on single crystal substrates.

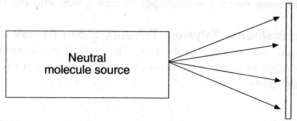

Fig. 6.26 Molecular construction of layers using molecular beam epitaxy.

The crystal structure in the film follows the underlying substrate. This technique is currently very expensive, but has significant potential for future VLSI circuit fabrication.

Post-deposition annealing. Most deposition processes introduce significant disorder into the crystal structure. For example, freshly deposited metal films of gold or aluminum are generally soft and easily abraded. Post-deposition annealing (30 min at 110°C) serves to remove gas incorporated during the deposition and to remove lattice defects. In metal films this causes the film to become hard, durable and adhesive. Oxide systems show the growth of specific crystal structures upon annealing.

6.9 FILM THICKNESS MONITORS

It is desirable to monitor the thickness of a film as it is being deposited. The standard instrument is a quartz crystal monitor. The piezoelectric quartz transducer forms part of an oscillator circuit whose frequency is monitored. As the film deposits on the crystal, the mass of the crystal increases and the frequency changes from its initial value given by,

$$f_o = \frac{1}{2\pi} \sqrt{\frac{Y}{M}}$$

where Y is the elastic constant and M the mass. If a film of thickness d and density ρ is deposited over area A, the mass increases by ρdA

$$f_o' = \frac{1}{2\pi} \sqrt{\frac{Y}{M + \rho dA}} = f_o \left[1 - \frac{\rho dA}{2M} \right]$$

$$\Delta f = f_o' - f_o = \frac{\rho d}{2} \cdot \frac{A f_o}{M} = \text{constant} \cdot \frac{\rho d}{2}$$

The constant is arranged to be approximately 1 if ρ is measured in g/cm^3 and d is measured in angstroms.

6.10   FILM THICKNESS MEASUREMENTS

Optical and mechanical methods are used to determine the film thickness after deposition. In some cases it is necessary to etch a step into the film.

Mechanical methods: Talystep, Talysurf, Sloan Dektak.   A mechanical probe is moved over an edge in the film. The vertical mechanical movement is determined—often by a piezoelectric or electrodynamic output from a transducer attached to the probe.

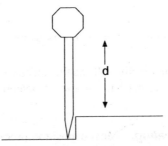

Fig. 6.27   Stylus or profilometry methods for thickness measurement.

Optical interference methods: multiple beam interferometry.   Observation of **interference colours** on white light illumination can be used to estimate the thickness of a dielectric layer. If light reflects normally from the top and bottom surfaces of a film separated by distance d, and taking account of a phase change of $\lambda/2$ on reflection at the first surface, an interference maximum will occur when, $2dn = (m + 1/2)\lambda$ where n is the refractive index of the film, and m is the order

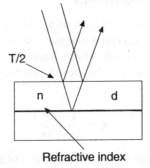

Fig. 6.28   Generation of interference colours in layers.

of the interference. For example, as the thickness of a ZnS layer ($n = 2.37$) increases from zero, the colour of the film changes sequentially. The wavelength of the maximum wavelength for $m = 0$ is given by $d = \lambda/4n$. The colour progression corresponds to the film thicknesses shown below.

Colour	m = 0				m = 1			
	Blue	Green	Yellow	Red	Blue	Green	Yellow	Red
Thickness nm	42	50	61	68	126	150	180	200

REFERENCES AND FURTHER INFORMATION

1. S. Dushman, *Scientific Foundations of Vacuum Technique*, 2nd ed., John Wiley, New York, 1962. (While this is an old book, it provides a wealth of quantitative and calculated data.)

2. *Product and Vacuum Technology Reference Book*, Leybold Inc., Leybold Vacuum Products Inc., 5700 Mellon Road, Export, Pennsylvenia 15632-8900 USA.

3. *Handbook of Thin Film Technology*, Edited by L.I. Maissel and R. Glang, McGraw-Hill Book Company, New York, 1970.

4. *Handbook of Thin Film Process Technology*, Edited by D.A. Glocker and S.I. Shah, IOP Publishing Inc., Boston, 1995.

ANSWERS TO QUESTIONS

(i) 1.23×10^5 s

(ii) 210 l/s

(iii) (a) 811 s, (b) because of desorbed gases and virtual leaks

(iv) 2470 s

DESIGN PROBLEMS

6.1 You have a vacuum system of total volume 100 litres. It is connected to a pump of speed 5 l/s and ultimate pressure 5×10^{-4} mbar. It becomes necessary to redesign the system to include a pipe of conductance 0.3 l/s. How much longer will it take for the pressure to fall from atmospheric (1013 mbar) to 10^{-3} mbar? Assume that the pumping speed remains constant.

6.2 A vacuum chamber of volume 500 l is to be pumped from a pressure of 30 mbar to 10^{-6} mbar. A diffusion pump of pumping speed 600 l/s and ultimate pressure 5×10^{-7} mbar is available and this must be attached to the chamber by a pumping line 20 cm long and 10 cm diameter. If the conductance (l/s) of such a line is

$$C = 12.1 \times d^3/L$$

where d and L are the diameter and length of the line measured in cm,

(a) How long will it take to pump the chamber assuming no gas leaks or desorption?

(b) If gas leaks to the chamber at a rate of 2×10^{-4} mbar l/s, what is the lowest pressure that will be attained?

(c) In the case of (b), what will be the pumping speed required in a rotary pump used to maintain a backing vacuum of 30 mbar for the diffusion pump?

6.3 It is required to set up a vacuum system for an experiment in which organic contamination of the sample should be eliminated. The pressure range is from 10^{-5} to 10^{-7} torr, and the temperature range of the experiment is 0–600°C. Design such a system, clearly indicating what kind of equipment you would require and why?

6.4 It is required to pump out a vacuum chamber of volume 1.5 m^3 from a roughing pressure of 1 mbar to a pressure of 2×10^{-6} mbar in a time of 3 minutes. Pumping arms of 10 cm diameter and 20 cm long allow connections of pumps. What pumping speed do you require in pumps having an ultimate pressure of 1×10^{-7} mbar? How might you ensure that the pump-down time is always the minimum?

6.5 It is required to develop a vacuum interlock to be able to pass samples into an experimental chamber held at a pressure of 5×10^{-7} mbar. The sequencing of the valves to undertake the pump down of the interlock is to be done under automatic control. It is proposed to use a roughing pump of pumping speed 100 l/s and ultimate vacuum 0.5 mbar to take the system from 1013 mbar to 1 mbar, followed by a turbomolecular pump of pumping speed 500 l/s and ultimate vacuum 5×10^{-7} mbar to pump from 1 mbar to 8×10^{-7} mbar. If the size of the interlock chamber is $10 \times 10 \times 20$ cm, what is the rate that samples can pass through the interlock?

Describe the pressure gauges that you would use to monitor and control the sequence.

6.6 It is desired to study the diffusion into the bulk of the materials of a surface layer of gold deposited on a polymer sheet $(CH_2)^n$. The proposal is to use an accelerated-based technique in which a 10 μm thick polymer sheet is fastened to a temperature controlled surface within a vacuum system. A thermal source is used to evaporate a layer of gold 1 μm thick on the surface of the polymer. Without breaking vacuum, a 4 MeV proton beam is used to measure the depth distribution of gold using Rutherford backscattering.

During the experiment, the vacuum should be maintained at a pressure of 5×10^{-6} mbar, the polymer temperature should be variable from 77K to 500°C and the peak gas load to the system is $Q(= pv) = 4 \times 10^{-4}$ mbar-l/s. Samples will be maintained at specific temperatures. The concentration of gold as a function of distance from the surface is to be measured after a given time interval.

Devise an experimental chamber and pumping system to be located on an accelerator beam line which could undertake these measurements. Briefly describe the function and method of operation of all equipment required and be as clear as possible regarding the type and speed of the vacuum system, the method of temperature measurement, and the nuclear detector and measuring technique. Credit will be given for a reasonably detailed drawing of the layout of an experimental chamber which would work in practice.

6.7 A quartz crystal monitor indicates a change in frequency of 1600 Hz when an aluminium film of density 2.7 gm/cm^3 is deposited on its face. Describe how such a monitor works and determine the film thickness in Å, assuming the normal conditions for such crystals. If the quartz crystal is 0.2 mm thick and the density of quartz is 2.3 gm/cm^3, estimate the starting frequency of the oscillator.

7

Optical Instruments

Optical instrumentation is used not only in the laboratory but in a wide range of industrial applications. Equipment based on lasers is widespread, and the optical fibre has become the mainstay of high speed ground communications. However, since much of the optical instrumentation used in the laboratory is concerned with spectroscopy, this section will proceed from spectroscopy to other applications. Much of the terminology is common.

7.1 SPECTROSCOPIC INSTRUMENTATION

Optically induced transitions between energy levels of atoms and molecules are detected in three major ways:

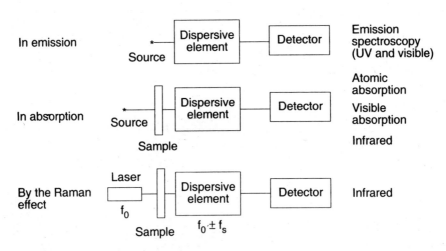

Fig. 7.1 Methods of light spectroscopy.

Emission spectroscopy observes light emitted from excited atoms or molecules, absorption spectroscopy measures light absorbed from an illuminating source, while Raman spectroscopy examines changes in the wavelength of incident light caused by its interaction with vibrational frequencies within a molecule. In all cases, it is necessary to record the wavelength of the light which is emitted or absorbed. Some form of dispersive element is required for this purpose.

196

7.2 VISIBLE AND INFRARED SPECTROSCOPY

Much laboratory spectroscopy is carried out in the ultraviolet, visible and infrared regions of the spectrum utilizing light having a wavelength from 200 nm to 100 μm. Sources of radiation for this region are found in excited gas discharges. Modern ultraviolet spectroscopy utilizes synchrotron radiation sources in which radiation is emitted through the acceleration of charged particles (Fig. 9.6).

The elements of spectrometers are:

Sources	White light thermal sources line sources (discharge lamps) lasers
Dispersive elements	Monochromators Spectrometers Spectrographs Interference filters
Detectors	Photographic plates Photomultipliers Photoconductive cells Photodiodes

Details of many of these optical detectors have been discussed in Chapter 3.

The wavelength dependence of the output of any optical system is a function of the wavelength dependence of its constituent elements. In the above spectrometers, the output signal $X(\lambda)$ which is recorded on a chart recorder involves not only the input signal $S(\lambda)$, but also the wavelength dependence of the detector $D(\lambda)$, the dispersive element $A(\lambda)$ and the light source $L(\lambda)$. Thus

$$X(\lambda) = S(\lambda) \cdot D(\lambda) \cdot A(\lambda) \cdot L(\lambda)$$

all of which have to be evaluated before an absolute value of $S(\lambda)$ can be deduced. This makes absolute measurements of spectroscopic information a non-trivial task.

As with other types of instrumentation, the availability of specific technology determines the 'easy' range of experimentation. Figure 7.2 shows these limiting factors.

Practical spectroscopy—Limiting factors. The optical spectrum ranges from 100 nm to 100 μm. A major factor in the accessible wavelength range is the optical transmission of materials used in the optics of instruments. Thus optical glass transmits from about 310 nm to 3 μm, while more expensive quartz is required from 200 nm to about 50 μm. Below 200 nm, air absorbs and vacuum spectrographs are required; in the far infrared, lenses are made of sodium chloride. These are not only fragile but are damaged by atmospheric moisture. The necessary protective chambers make such instruments expensive.

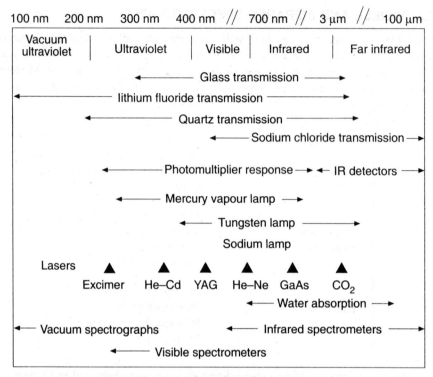

Fig. 7.2 Limiting factors and available technology for the design of optical instruments.

Sodium:	589.0 and 589.6 doublet
Mercury:	257.6, 365.0, 404.7, 435.8, 547.1, 579.1, 690.1
He–Ne:	632.9

Fig. 7.3 Wavelengths of some convenient monochromatic light sources in nm.

7.3 SPECTROMETER DESIGN

Parameters of interest are the intensity of light passed by a spectrometer and the degree to which it can resolve the light into separate wavelengths.

Dispersive elements for the separation of a continuous (white light) spectrum usually depend on the fact that light rays of different wavelengths take different paths when they interact with the element. A basic spectrometer therefore includes:

- entrance and exit slits to define the respective beam directions.
- a dispersive element such as a grating or prism
- focussing elements such as lenses or mirrors

The spectrometer performance is defined in terms of quantities which are generally measured using a monochromatic light source such as a sodium discharge lamp. This emits at $\lambda = 589.0$ nm. The output is recorded on the output scale of the instrument.

Instrumental Profile

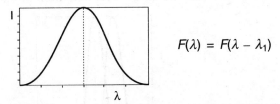

$$F(\lambda) = F(\lambda - \lambda_1)$$

Fig. 7.4 The instrumental profile is a plot of the spectrometer output as a function of wavelength when the source is monochromatic light of wavelength λ_1.

Resolving Power $R = \lambda/\Delta\lambda$

The smallest range of wavelengths $\Delta\lambda$ which can be resolved by the spectrometer is defined by either of the following two methods.

(a) By the half width at half maximum ($\Delta\lambda$) of the spectrometer output when a monochromatic source of wavelength λ is viewed.

(b) As the wavelength difference ($\lambda_2 - \lambda_1$) recorded when viewing monochromatic sources of wavelength λ_1 and λ_2 such that the combined instrument profile has a dip of 20% between the maxima.

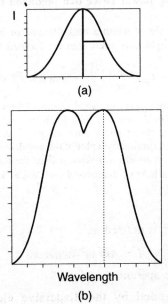

Fig. 7.5 Two methods of defining the resolving power of a spectrometer.

These definitions are arbitrary, but they agree substantially with the **Rayleigh Criterion** for the angular resolution of closely spaced objects.

For objects of size close to that of the wavelength of light, diffraction causes the image to appear as a series of maxima and minima. The images of two closely spaced objects therefore consist of overlapping diffraction patterns.

The Rayleigh Criterion states that two objects can be resolved if the **first maximum** of one image lies on the **first minimum** of the second.

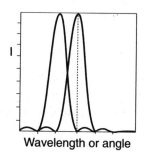

Fig. 7.6   Overlapping diffraction patterns for two sources at the Rayleigh limit.

The theoretical angular limit of resolution θ_r for images formed by line sources of radiation of wavelength λ is $\theta_r = \lambda/d$ radians, while for a circular aperture

$$\theta_r = 1.22\lambda/d \text{ radians}$$

where θ_r is the angular separation of two objects of diameter d. For the Rayleigh Criterion, the instrument profile falls to 50% between the peaks. The 20% criterion corresponds to a resolving power **twice** that predicted by the Rayleigh Criterion.

Question (i)   For a single slit of width a back-illuminated by light of wavelength λ, the intensity of light I_θ at angle θ in the forward direction is

$$I_\theta = I_m \left(\frac{\sin \alpha}{\alpha} \right)^2, \quad \text{where } \alpha = \frac{\pi a}{\lambda} \sin \theta$$

(a) Use a convenient computer or spreadsheet program to plot the angular distribution (in degrees) of light intensity along a line perpendicular to the axis of a long slit of width a for values of a/λ from 3 to 10. (b) Establish the Rayleigh Criterion by a plot of the angular direction of the first minimum (in radians) versus λ/a. (c) What will be the form of the image if the slit is replaced by a hole of diameter d with $d/\lambda = 5$?

Luminosity L

The light-gathering power is defined as

$$L = A\Phi \text{ m}^2\text{-steradians}$$

A – area of the entrance aperture

Φ – solid angle subtended by the dispersive element at the entrance aperture.

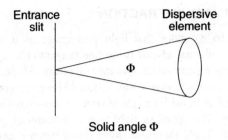

Fig. 7.7 Factors determining the luminosity L.

The luminosity is the optical 'gain' of the instrument and R is its 'bandwidth'. $R \times L$ is similar to the gain-bandwidth product of an electronic amplifier. It is a property of the system, and an improvement in one aspect is normally accompanied by a reduction in the other. For example, the resolving power of a spectrometer will be partly determined by the accuracy with which the angular directions of the input and exit beams are defined. This resolving power can be improved by narrowing the entrance slit, but this reduces the luminosity and $R \times L$ remains constant. The resolving power is also set by intrinsic properties of the dispersive element (page 207). Only a different type of dispersive element improves the performance. This is shown in Fig. 7.8.

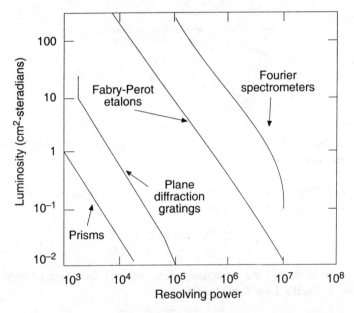

Fig. 7.8 Performance limits of different types of optical spectrometers.

Question (ii) Sodium light is focused on a set of linear slits at the entrance to a spectrometer 1 cm in length and 0.02 cm wide. The dispersive element is of area 5 cm × 5 cm placed 10 cm distant from the slits with its centre line on a line joining the slits and the element. Calculate the luminosity.

7.4 REFRACTION AND DIFFRACTION

Huygens' Principle (1678) states that light propagates as a series of spherical wavelets. The position of a **wavefront** is the line tangential to a set of wavelets. A **ray** is drawn in a direction perpendicular to the wavefront. All points on a wavefront can be treated as a source of secondary wavelets and Huygens' construction assumes that wavelets are drawn at equal intervals of time to show the distance travelled by a wavelet in that time. The speed of light v in a medium of permeability μ and permittivity ε is $1/\sqrt{\mu\varepsilon}$. Light therefore slows down when it enters a medium for which the relative permittivity $\varepsilon_r > 1$.

Refraction. Light crosses an interface between media (1) and (2). The initial ray direction is at θ_1 to the perpendicular.

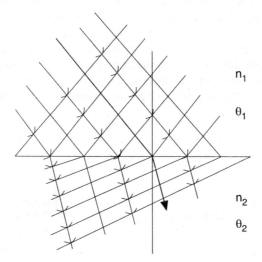

Fig. 7.9 Refraction of a ray towards the normal as light enters a medium in which its speed is reduced.

The **speed of light** in each medium is v_1 and v_2 respectively ($v_2 < v_1$).
By a simple geometrical construction

$$\frac{\sin \theta_1}{\sin \theta_2} = \frac{v_1}{v_2} = \frac{n_2}{n_1} = \sqrt{\frac{\varepsilon_1}{\varepsilon_2}}$$

where $n = c/v$ is termed the refractive index, and c is the speed of light in free space. This is **Snell's Law** for refraction.

$$\boxed{n_1 \sin \theta_1 = n_2 \sin \theta_2}$$

In passing from a **rare** to a **dense** medium a ray of light is refracted **towards** the normal.

The **optical path length** of light in a medium is nd, where d is the physical distance travelled by light in a medium of refractive index n.

Diffraction. Light is reflected from a set of point sources or mirrors spaced distance d apart. Wavelets are drawn at equal intervals of time.

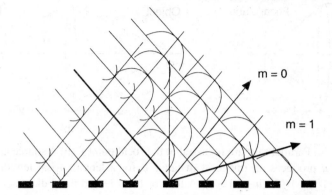

Fig. 7.10 Diffraction from a periodic set of reflecting sources spaced at distance d. The wavelets build up in certain directions θ which are denoted by the integer m.

The wavelets in the **diffracted** wavefronts combine only along certain directions. By geometry, these directions are at angle θ such that

$$d \sin \theta = m\lambda$$

where m is an integer which defines the **order** of the diffraction.

The intensity of diffracted rays for higher orders falls rapidly for larger values of m.

Diffraction patterns can also be generated from light sources or illuminated apertures transmitting light.

7.5   LENSES AND REFRACTIVE OPTICS

Convex and concave lenses focus light by the refraction of light at a curved surface. The deviation of the refracted light is assessed by reference to the normal to the surface. A ray passing through the centre of the lens is undeflected. A ray parallel to the axis passes actually or virtually through the focal point f.

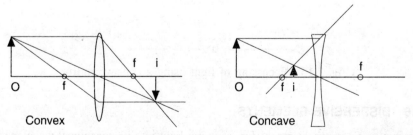

Fig. 7.11 Ray paths to calculate the position of the image in convex and concave lenses.

In most of the methods of instrumentation, the convex lens is of great importance either to focus parallel light to a point, or to image a source on to a detector.

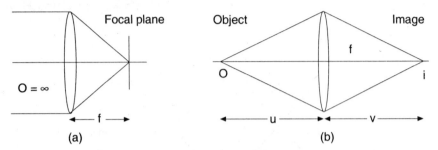

(a)   (b)

Fig. 7.12 (a) Parallel rays are focussed at the focal point f. (b) A source at distance u produces an image at distance v for a lens of focal length f. $1/u + 1/v = 1/f$ if $(u + v) > 4f$. A lens of focal length f and diameter a has an f-number of f/a.

Lens aberrations occur because the focal length of a simple lens varies with the perpendicular distance of an incident ray from the axis along the centre of the lens (**spherical aberration**) or with the wavelength of light (**chromatic aberration**). High quality lenses are made with a combination of lens and glass types in order to reduce the effect of these aberrations.

Because a lens has to transmit light, the material of which it is made restricts the wavelength response of the system in which it is employed. Conventional glass optics have a wavelength limit of about 310 nm. Quartz or fused silica extends the wavelength range to the ultraviolet (< 254 nm). Front surface aluminized concave mirrors of radius R have the least restriction on wavelength response.

The focal length f of a concave mirror is $f = R/2$

The image position is given by $1/u + 1/v = 2/R$

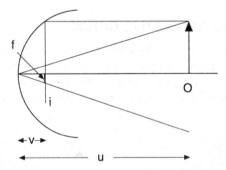

Fig. 7.13 Focussing of light using a concave mirror.

7.6 DISPERSIVE ELEMENTS

Dispersive elements are means for resolving light into its constituent wavelengths. The resolution $(\lambda/\Delta\lambda)$ by which two wavelengths differing by $\Delta\lambda$ can be separated is set partly by the **geometry** of the optical system (the accuracy with which incident and exit angles can be defined), and partly by the **theoretical maximum resolving power** of the element.

Prisms

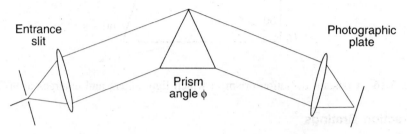

Fig. 7.14 Optical paths through a prism spectrometer.

The angular deviation of a ray for a prism of angle ϕ is given in Appendix 6. Dispersion arises because the refractive index n varies with wavelength λ. The variation can be approximated by Cauchy's equation $n = [A + B/\lambda^2]$, where A and B are constants for a particular glass.

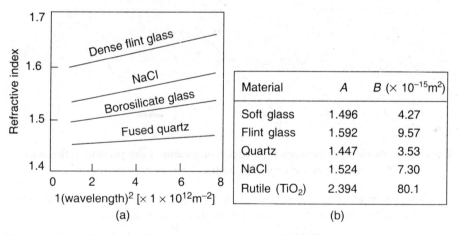

Material	A	B ($\times$ 10^{-15}m^2)
Soft glass	1.496	4.27
Flint glass	1.592	9.57
Quartz	1.447	3.53
NaCl	1.524	7.30
Rutile (TiO$_2$)	2.394	80.1

(a)   (b)

Fig. 7.15 (a) Variation of refractive index with λ. (b) The table gives the constants of the Cauchy equation for λ from 355 nm to 1 μm.

Constant deviation prism. In a conventional prism spectrometer the angular deviation in the beam direction for a given wavelength is complex and requires adjustment of the viewing direction for each λ. In a **constant deviation prism** the incident and refracted beams always differ in direction by 90°.

The dispersion in λ is measured by simply rotating the prism about its axis. This makes it particularly suitable for industrial applications having fixed exit and entrance slits.

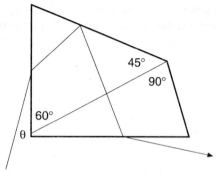

Fig. 7.16 Constant deviation prism in which light enters and emerges at 90°.

Diffraction Gratings

For general purposes, the most widely used dispersive element is the diffraction grating. It is normally used in the reflection mode and consists of a series of rulings spaced by distance d on the surface of a plane medium. Common gratings have 3000–8000 lines/cm

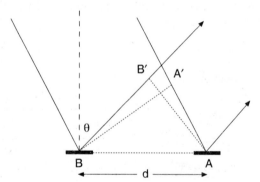

Fig. 7.17 Interference between light beams originating from periodic reflectors or transmitters in a diffraction grating. The source spacing is d.

For constructive interference at angle θ_r for a grating spacing d any two optical paths must differ in phase by an integral number of wavelengths.

Path difference $BB' - AA' = m\lambda$

$$\boxed{d \sin \theta_r - d \sin \theta_i = m\lambda}$$

where m is an integer and θ_i and θ_r are the angles of incidence and reflection respectively. If θ_i is kept constant, the angular dispersion with λ can be calculated by differentiating with respect to λ.

$$d \cos \theta_r \, d\theta_r = m \, d\lambda$$

so that

$$\frac{d\theta_r}{d\lambda} = \frac{m}{d \cos \theta_r}$$

The dispersion of a prism therefore increases with order ($m > 1$). If θ_r is small, $\cos \theta_r = 1$, and the dispersion becomes highly linear ($d\theta = m/d \cdot d\lambda$). The linearity of the wavelength scale with angle is an advantage of the grating over the prism. The primary disadvantage of the diffraction grating is that the beam energy is spread over several orders. This problem is alleviated by **blazing** a grating.

Blazing of a grating. The grating is ruled or 'blazed' such that each ruling is at angle ϕ to the plane. This causes the beam energy to be focused into a particular order m without changing the resolution. The effect is wavelength dependent and gratings are blazed for a particular spectral region.

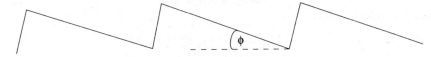

Fig. 7.18 Blazing of a grating by ruling it at an angle ϕ to the plane to focus light of a wavelength range into a specific order m.

Plastic replica gratings. Inexpensive, large area gratings are made for a variety of purposes by solidifying polymer solutions onto a grating ruled into a stainless steel plate.

Question (iii) Calculate the angular deviation for the first order diffraction pattern for a mercury line at 365.0 nm using a grating with 5000 lines/cm and light at normal incidence.

Theoretical Maximum Resolving Power for the Prism and Grating

For both the diffraction grating and the prism, the **practical** resolving power is affected by the definition of the incident and exit beam angles, i.e. by the exit and entrance slit widths.

However, the **theoretical** or maximum resolving power can be calculated by noting that the Rayleigh Criterion (Fig. 7.6) applies. If images are formed at a particular angle for wavelength λ, the minimum angular separation for a second image at angle $\theta + d\theta$ corresponding to $\lambda + d\lambda$ is such that the first maximum of one image falls on the first minimum of the second image.

In the following section, a comparison of the maximum resolving power expected for prism and diffraction grating spectrometers is developed. In both cases, the resolving power increases with the size of the dispersive element.

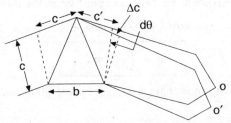

Fig. 7.19 Refracted images for wavelengths λ and $\lambda + \Delta\lambda$.

Prism.   The Rayleigh criterion for a line source states that the minimum angular separation $d\theta_{min}$ for a beam of width a is

$$d\theta_{min} = \lambda/a$$

The optical path lengths for rays of wavelength λ through the **tip** and the **base** of the prism forming an image at O are identical at

$$c + c' = nb$$

For a wavelength $\lambda' = \lambda + \Delta\lambda$, the new image is at O' and the path length is

$$c + c' + \Delta c = (n + \Delta n)b$$

$$d\theta_{min} = \lambda/a = \Delta c/a$$

Thus $\lambda = (\Delta n)\, b$ and

$$\boxed{R = \frac{\lambda}{\Delta\lambda} = b\,\frac{dn}{d\lambda}}$$

Using the Cauchy equation (Sec. 7.6), the resolving power R of a soft glass prism of width $b = 5$ cm is $R = 3400$ at $\lambda = 500$ nm.

Grating.   A **maximum** will occur when the optical paths from two sources differ by an integral number of wavelengths.

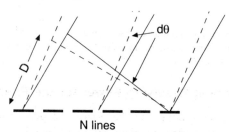

Fig. 7.20   Diffracted rays for λ and $\lambda + \Delta\lambda$ with angular separation $d\theta$.

If the grating has a total of N lines, for a maximum to be obtained for the extreme rays for diffraction order m,

$$D_{max} = mN\lambda$$

The next **minimum** will occur if the path difference between the central ray and the two extreme rays differ by $\lambda/2$,

$$D_{min} = mN\lambda + \lambda$$

To be resolved, the maximum for $(\lambda + \Delta\lambda)$ must lie at the minimum for λ

$$mN(\lambda + \Delta\lambda) = mn\lambda + \lambda$$

$$\boxed{R = \frac{\lambda}{\Delta\lambda} = mN}$$

The first order   resolving power for a grating having 3000 lines/cm and a width $b = 5$ cm is

$$R = mN = 1 \times 3000 \times 5 = 15{,}000$$

Interference Filters

Interference filters are planar elements with excellent practical application as narrow band filters. Light reflected from the upper and lower faces of a single dielectric layer of refractive index n and thickness d undergoes interference. As the light passes from a rare to a dense medium, a phase change of $\lambda/2$ also occurs. For light at normal incidence, a maximum reflected intensity therefore exists for a wavelength λ.

$$2nd = (2m + 1)\frac{\lambda}{2} \quad \text{where } m = 0, 1, 2, 3 \ldots$$

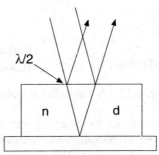

Fig. 7.21 Interference by reflection from two surfaces.

The resolution is improved by fabricating semi-transparent metal layers on each side of the dielectric film. The $\lambda/2$ phase change at the front surface then does not apply.

The high resolution occurs because of multiple reflections within the dielectric layer. Only wavelengths which accurately give rise to an interference maximum remain in phase after successive reflections. The maximum intensity occurs at

$$2nd = m\lambda, \text{ where } m = 0, 1, 2, 3, \ldots.$$

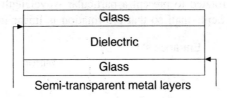

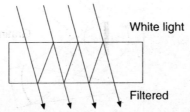

Fig. 7.22 Multiple reflection interference filters using semi-transparent metal layers each side of a dielectric layer. The bandpass of the filter can be < 10 nm.

Fabry-Perot Etalon

An **etalon** consists of two parallel semi-transparent metal films acting to form an optical cavity. The reflected or transmitted light is influenced by the interference between multiply reflected beams such that the maximum response is given by $\lambda = 2d$.

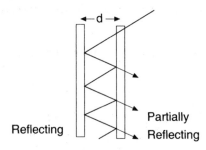

Fig. 7.23 A Fabry-Perot etalon.

A spectrum can be traced out by changing the spacing of the etalon in a controlled manner. This type of measurement is extensively utilized in Fourier transform infrared spectroscopy (FTIR). The intensity or time variations of an irradiance pattern made using polychromatic radiation can be converted to wavelength response using Fourier transform methods [see Ref 7].

7.7 SPECTROMETER DESIGN

Grating Spectrometers

The Czerny-Turner mounting is used for many grating spectrometers. Light is focused on the grating using concave front-surface mirrors.

The grating is rotated to present a particular wavelength to the exit slit. The diffraction angle is kept small so that the rotation is linear in λ.

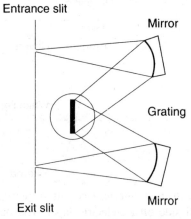

Fig. 7.24 Czerny-Turner grating spectrometer.

Absorption Spectrometers

An important measurement in spectroscopy is to determine the absorption of light by a sample as a function of wavelength.

The absorption of a sample is recorded by measuring a spectrum with and without the sample in the beam. This requires measuring the difference between two rapidly changing functions which leads to substantial errors and inconvenient measurements.

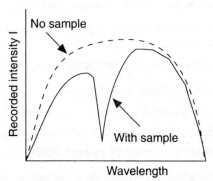

Fig. 7.25 Absorption recorded by measurement of the spectrometer output with (I) and without (I_0) a sample of thickness x in position. The linear absorption coefficient μ is given by $I = I_0\, e^{-\mu x}$.

Double Beam Absorption Spectrometers

Commercial spectrometers use a **double beam** system in which light passing through a **sample** compartment is compared with that passing through a **reference** cell. This is shown in Fig. 7.26. The beam emitted by the monochromator or spectrometer is split into two paths, one of which is passed directly through a reference compartment to a detector, while the other passes through the sample to a second detector. The outputs from the two detectors are compared electronically, so that the recorder shows the difference between them. The output is recorded in terms of the optical density $D = \log_{10}(I_0/I)$ (page 81)

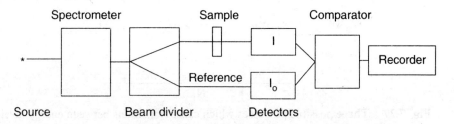

Fig. 7.26 Double beam absorption spectrometer which eliminates the wavelength dependence of the spectrometer components by a comparison technique. The output is normally recorded as optical density $D = \log_{10}(I_0/I)$.

7.8   LASERS

Lasers [**L**ight **A**mplification by **S**timulated **E**mission of **R**adiation], have been an important additions to physics instrumentation since 1960. Originally predicted by Einstein in the 1920's, the effect remained a minor piece of theory until 1957 when first the microwave M(icrowave) ASER and then the LASER was demonstrated. The laser is now an important tool in communications, medicine, optical imaging and manufacturing.

The **requirements** for LASER action include:

1. a system in which an inverted population distribution can be created in a set of energy levels;

2. energy feedback—generally by the use of a resonant cavity.

The **characteristics** of a LASER source include:

1. a coherence in which all wavefronts are in phase,

2. large output power (eigher pulsed or continuous),

3. a strongly collimated beam of emitted light.

The third characteristic is probably the most striking feature of a laser source. However, collimation is not a fundamental property of laser action, but arises as a result of the form of the resonant cavity used in most sources. The basic requirements for laser action are first reviewed, and then physical systems in which laser action is possible are summarized.

The Einstein Coefficients

Consider a system of two energy levels in which the levels contain N_1 and N_2 atoms respectively. The population of each state can change through absorption and emission of radiation of frequency v. Einstein described the mechanism for how these transitions can take place in terms of a set of coefficients A and B.

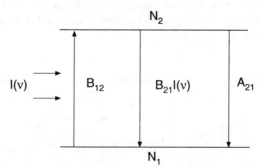

Fig. 7.27   **Three** possible processes which cause transitions between energy levels are absorption (B_{12}), spontaneous emission (A_{21}) and induced emission (B_{21}).

The transition probabilities for these processes are related to the number of atoms in each state and to the strength of the incident light intensity $I(v)$.

Process	Transition Probability
1. spontaneous emission	$A_{21}N_2$
2. absorption from a radiation field of intensity $I(v)$	$I(v)B_{12}N_1$
3. induced emission stimulated by the radiation field $I(v)$	$I(v)B_{21}N_2$

In equilibrium, the rate upward must equal the rate downward

$$I(v)B_{12}N_1 = A_{21}N_2 + I(v)B_{21}N_2$$

and the equation may be rearranged in the form:

$$I(v) = \frac{A_{21}N_2}{B_{12}N_1 - B_{21}N_2}$$

In **thermodynamic equilibrium** $N_2 \ll N_1$ since the distribution of atoms among the levels will be given by the classical Maxwell-Boltzmann distribution

$$\frac{N_2}{N_1} = \exp\left(-\frac{hv}{kT}\right)$$

and assuming the $B_{12} \sim B_{21}$

$$I(v) = \frac{A_{21}N_2}{B_{12}N_1} = \frac{A_{21}}{B_{12}} \exp\left(-\frac{hv}{kT}\right)$$

This is unremarkable and is consistent with classical physics. However if a non-equilibrium system can be arranged in which $N_2 \gg N_1$ the emission intensity becomes very large since the above exponent becomes positive. This may be regarded as a state of **negative** thermodynamic temperature. Different types of lasers utilize different methods to implement this 'top-heavy' or inverted population of energy levels. The principle is generally to absorb energy in one set of easily accessible states, to transfer this energy rapidly to the upper of the set of states in which the laser action can occur, and to de-excite the lower states as rapidly as possible. This process is illustrated by three important types of lasers.

Laser Systems

Helium-neon lasers (He-Ne). Excited spectroscopic states of helium and neon gases occur at similar energies. In a gas mixture excited by an electrical discharge, energy absorbed by He ions is preferentially transferred to N_2 neon ions by a process of resonance transfer. The neon atoms de-excite between two excited energy states with the emission of a photon over a characteristic recombination time. The neon atoms in the lower level return to their ground state very rapidly so that $N_2 > N_1$.

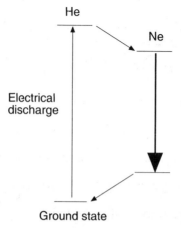

Fig. 7.28   The characteristic 'blue' colour of a gas discharge in helium transfers to the 'pink' glow of neon. De-excitation within neon leads to an inverted population.

***Solid state lasers*: Al$_2$O$_3$: Cr or YAG : Nd.**   A host oxide such as aluminium oxide (Al$_2$O$_3$) or yttrium aluminium garnet (YAG) contains a small concentration of a transition metal (Cr) or rare earth impurity (Nd). The laser transition occurs within the impurity states.

Strong resonance transfer processes couple the impurity to the oxide lattice to give $N_2 \gg N_1$.

Initial excitation or 'pumping' takes place by optical absorption in the host lattice.

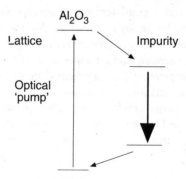

Fig. 7.29   Laser transitions between impurity states in the band gap of Al$_2$O$_3$.

***Semiconductor lasers*: GaAs.**   In a forward biased *pn*-junction, minority carriers are injected at the edge of the depletion layer (see Appendix 3).

In the depletion region both electrons and holes are present. Under strong forward bias injection of holes and electrons occurs into the depletion region and the laser condition $N_2 \gg N_1$ is achieved. In **direct band gap** semiconductors such as GaAs, radiative recombination takes place. The light is emitted at the edges of the depletion region.

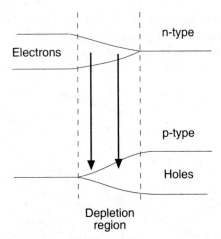

Fig. 7.30 Laser action by minority carrier injection in a forward biased semiconductor junction.

Optical Feedback—The Resonant Cavity

Concentration of the energy with the medium into a set of coherent wavefronts occurs within an optical cavity. The cavity is formed by two mirrors, one of which is semi-silvered to pass a fraction of the light within the medium. Photons are emitted in all directions, but only those passing along the length of the cavity induce further stimulation of radiation. These build up coherent wavefronts precisely aligned along the length of the cavity. This leads to the characteristic highly collimated 'beam' of the laser.

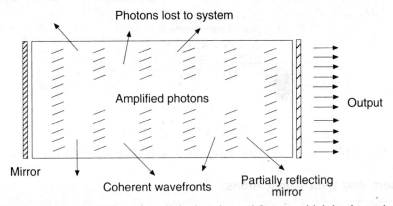

Fig. 7.31 The construction of an optical cavity and factors which lead to coherent wavefronts.

Practical cavities. Gas lasers such as He-Ne and CO_2 use the physical arrangement shown in Fig. 7.31. Since various modes of oscillation and polarization can occur within the cavity, careful adjustment of the mirrors is required and in critical applications, polarizing mirrors are used to maintain the plane of polarization of the emitted light in one direction. Inexpensive He-Ne lasers often do not have this provision and the plane of polarization may fluctuate. This can give rise to unexpected signal variations in experiments involving polarization or reflection.

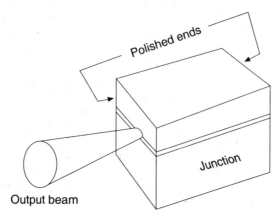

Fig. 7.32 The structure of a semiconductor laser. The output is shaped to link to an optical fibre.

Semiconductor lasers are important in optical communications. The 'cavity' shown in Fig. 7.32 is produced by polishing parallel faces on the semiconductor crystal containing the *pn*-junction. The emission is then primarily along the plane of the junction.

Solid state and dye or vapour lasers use flash lamps for optical pumping, the latter controlling the wavelength of the emission by changes of the composition of the dyes within the optical cavity. The form of such pulsed lasers is illustrated in Fig. 7.33.

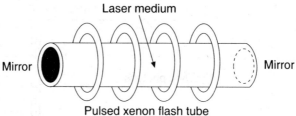

Fig. 7.33 Cavity and flash lamp for a solid state, dye or vapour laser.

Lasers and their Applications

Table 7.1 summarizes major classes and applications of lasers.

7.9 FIBRE OPTICS

Optical wave guides, or optical fibres, allow light to be guided over large distances. They have now become the basis for most land-based communications systems. A waveguide consists of a fine glass fibre which confines light by total internal

Table 7.1 Major Classes, Special Properties and Applications of Lasers

Type	Primary Wavelength (µm)	Tunable	Pulse or cw	Power cw/peak (W/J)	Comments
He-Ne	0.63 1.15, 3.39	no	cw	0.5–50 (mW)	Laboratory/industry metrology/ instrumentation
Argon	0.49 0.52	no	cw pulse	2.0 10	Pump laser Cutting
CO_2	10.6	no	cw pulse	10^4 10^9	Industrial high power laser Cutting/welding/heat treating
Gas dynamic	10.6	no	pulse	10^5	Military lasers using gas expansion in shock tubes.
Chemical	1–10	yes	pulse	10^{11}	Produced by chemical reactions. Tunable by choice of reaction.
Excimer	0.38	no	pulse	10^5	Ultraviolet laser which may be electron beam pumped. Laser fusion research, light source for high definition photolithography in the semiconductor industry.
Ruby	0.69	no	pulse	10^3	Solid state lasers excited with a xenon flash lamp.
Nd:YAG	0.9	no	pulse	10^3	Laboratory pulsed laser.
Semi-conductor GaAs	0.9	no	cw	0.1	Communications laser.
Dye	0.4–0.9	yes	pulse	10^3	Tunable/pumped with flash lamp or nitrogen laser.

reflection, and which has been designed to minimize the scattering of light from the surface of fibre. The simplest model of a fibre is shown in Fig. 7.34.

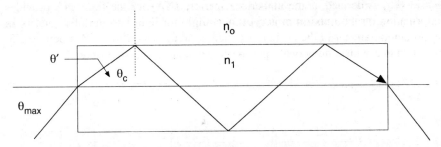

Fig. 7.34 Ray path through an ideal optical fibre.

Light which enters the perpendicular end of the fibre is refracted at the entering face, and at the side of the fibre. If the angle of incidence at the side is greater than the critical angle i_c, the light is totally internally reflected. Simple geometry then implies that all further reflections along the fibre will also be internally reflected.

The largest angle of incidence θ_{max} at the entrance to the fibre is given by Snell's Law ((Fig. 7.9), page 202).

Question (iv)   (a) Calculate the angle to the normal for a refracted ray of light entering a block of glass at an angle of 30° to the normal if the refractive index $n = 1.5$. (b) If light is incident on an interface from within a material of refractive index 1.5, complete internal reflection will occur when the output beam is effectively refracted by > 90°. What is the condition for the 'critical angle θ_c' and what is its value for this medium?

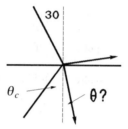

If the refractive index of the glass of an optical fibre is n_1 and that of the surrounding medium is n_o, the condition for complete internal reflection at the glass/air interface within the fibre is

$$n_1 \sin \theta_c = n_o \sin 90$$

$$\sin \theta_c = n_o/n_1$$

At the entrance:

$$n_o \sin \theta_{max} = n_1 \sin \theta' \quad \text{with} \quad \theta' = 90 - \theta_c \text{ and } \sin \theta' = \cos \theta_c$$

Hence

$$\boxed{n_o \sin \theta_{max} = n_1 \cos \theta_c}$$

Numerical Aperture

$n_o \sin \theta_{max}$ is defined as the numerical aperture (NA) for the fibre. NA provides information on the **maximum acceptance angle** for light into the fibre, and is an important parameter in the design of optical systems using fibres. It is pointless to use a source with a numerical aperture larger than that of the fibre.

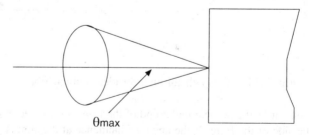

θmax

Fig. 7.35   The numerical aperture determines the maximum acceptance of light into a fibre.

Question (v) What is the largest angle at which light can enter a plastic rod of refractive index 1.31 and propagate along the rod? What is the value of the numerical aperture?

Fibre Cladding

Because light interacts with dirt, scratches and other imperfections at the surface of the fibre, good optical fibres cannot be prepared using the simple glass/air interface, but the light must be reflected at an interior interface created by combining glasses having different refractive indices n_2 (for the inner core) and n_1 (for the outer cladding). Depending on the method of manufacture, the interface may be either a **graded** or a **step** change in refractive index across the radial direction. Graded fibres are fabricated by drawing molten glass from two coaxial containers at high speed through an aperture, step index fibres are prepared by chemical vapour deposition of the cladding onto the core.

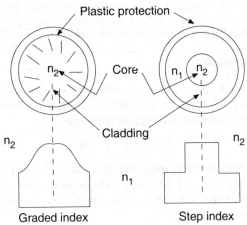

Fig. 7.36 Schematic representation of graded and step index fibres.

The effect of the cladding is to steer the light **within** the fibre.

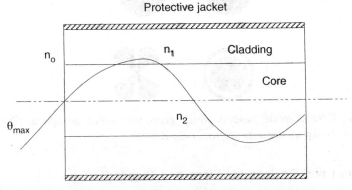

Fig. 7.37 Light is steered by the interface between the core of refractive index n_2 and the cladding of refractive index n_1 in an optical fibre.

For this structure:

$$n_2 \sin i_c = n_1 \sin 90; \quad n_o \sin \theta_{max} = n_2 \cos i_c$$

Taking the square of the numerical aperture NA, and noting that $\cos^2 \theta = 1 - \sin^2 \theta$

$$(NA)^2 = n_o^2 (\sin \theta_{max})^2 = n_o^2 \cdot \frac{n_2^2}{n_o^2} \frac{(1 - n_1^2)}{n_2^2}$$

$$\boxed{\text{Numerical Aperture NA} = (n_2^2 - n_1^2)^{1/2}}$$

This depends only on properties of the fibre. A convenient experiment to determine many important fibre properties is to measure the transmission of the fibre as a function of the angle of incidence to a perpendicular face.

Question (vi)   The angular width of the beam incident on an optical fibre is measured by the incident angle at which the fibre output has decreased to 10% of its maximum value is $\pm 17.2°$. If the central core has a refractive index of 1.61, what is the refractive index of the cladding?

Modes of Propagation

Maxwell's equations govern the propagation of light in an optical fibre. Different modes propagate depending on the dimensions of the fibre. A critical parameter is the *V*-number, or characteristic waveguide parameter.

$$V = 2\pi/\lambda \cdot a \text{ NA}$$

where λ is the wavelength of propagation, a is the radius of the core, and NA is the numerical aperture. If $V < 2.405$, only a **single mode** will propagate, while for $V > 2.405$ **multiple modes** will exist. Some types of modes which are propagated in a fibre are illustrated in Fig. 7.38.

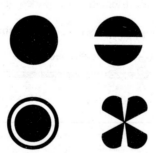

Fig. 7.38   Optical mode patterns which propagate within an optical fibre. After Newport Instruments Inc. (see Ref. 1).

Equipment and Devices for Fibre and Planar Optics

An excellent set of fibre optic experiments and examples of mode propagation are provided in materials from Newport Instruments (page 221).

Fibre optic connectors. A major drawback of optical fibres compared to conventional metal wire is the difficulty of making connections. In practice, the ends of a fibre are made very smooth and perpendicular and butted against its neighbour. Various types of connector are now available to hold the fibres together.

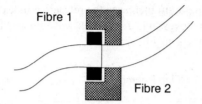

Fig. 7.39 Connector for fibre optic cables.

Fibre optic sensors. If the cladding of two fibres is removed and they are glued together, an 'evanescent' wave associated with the optical energy in the core couples energy from one fibre to another.

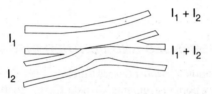

Fig. 7.40 Coupling of optical energy between two light guides which are placed in close proximity. The coupling is through the evanescent wave associated with each fibre.

Fibre optic couplers provide a means of producing a fibre optic interferometer.

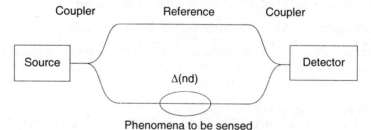

Fig. 7.41 A fibre optic interferometer utilizing two directional couplers.

Such interferometers are very sensitive to changes in the optical path length between the two paths. If one fibre is used as a reference, the other can be mechanically strained, or used as a temperature or pressure sensor. It is also possible to coat the surface of the fibre with a piezoelectric or magnetic material in order to form a sensor for electric and magnetic fields.

Planar optics. In the quest for **integrated optics** in which functions now carried out by electronics are undertaken purely in the optical domain, planar optical waveguides are of interest. In principle these can be achieved by the fabrication of high refractive index films on glass or silicon substrates. However, scattering of light in such films currently remains a problem in achieving low loss guides. The diffusion of patterns of titanium Ti into single crystal lithium niobate has achieved the most success. The titanium increases the refractive index of the lithium niobate. In order to insert and recover light from such waveguides, TiO_2 (rutile) prisms are used in the manner shown in Fig. 7.42.

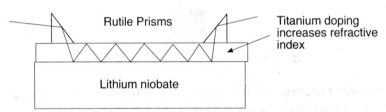

Fig. 7.42 The use of high refractive index rutile (TiO_2) prisms to insert and extract light into a dielectric film prepared on a substrate.

REFERENCES AND FURTHER INFORMATION

1. Newport Corporation, *Newport Optics Catalog*, 1791 Deer Ave, Irvine, California. 92714, USA Newport Instruments Canada Corp., 2650 Meadowvale Blvd. Unit 3, Mississauga, Ontario. ON L5N 6M5 Canada
   Newport have laboratory workbooks and kits of experiments for fibre optics/ general optics.

2. Oriel Corporation, Vol. I—*Tables and Positioners,*

   Micropositioners, Optical Mounts, Vol. II—*Light Sources,*

   Monochromators, Detection Systems, Vol. III—*Optics and Filters.*

3. Edmund Scientific Co., *Edmund Scientific Catalog,* 101 E. Gloucester Pike, Barrington, New Jersey, 08007-1380. USA; Efston Science Inc., 3350 Dufferin St., Toronto, ON M6A 3A4 Canada.

4. Auriga Fibre Optics Ltd., *Auriga fibre optic component and accessory catalog*, 60 Doncaster Ave., Unit 6, Thornhill, ON L3T 1L5 Canada Ontario.

5. An important source of information on the optical properties of many materials and systems is found in various editions of the *American Physics Handbook,* American Institute of Physics, McGraw Hill.

   G.W.C. Kaye and T.H. Laby, *Tables of Physical and Chemical Constants,* Longman, London and New York, 1986.

6. F.L. Pedrotti, S.J and L.S. Pedrotti, *Introduction to Optics*, Prentice Hall, Inc., Englewood Cliffs, New Jersey 07632 (1987).

ANSWERS TO QUESTIONS

(i)

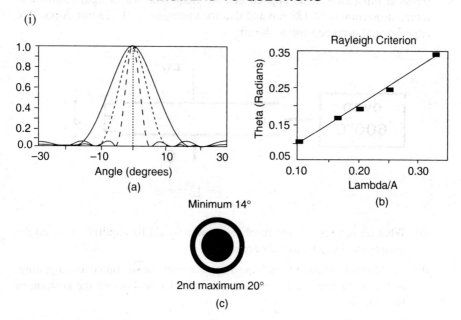

Minimum 14°

2nd maximum 20°

(c)

(ii) 5×10^{-7} m^2-steradians

(iii) 10.5°

(iv) (a) 18.7°, $n_1 \sin \theta_c = 1 \sin 90$, (b) $\theta_c = 41.8°$

(v) 57.8°, 0.85

(vi) 1.57

DESIGN PROBLEMS

7.1 You require to display the d lines of sodium ($\lambda_1 = 589.0$ nm; $\lambda_2 = 589.6$ nm) on a screen placed 30 cm from the dispersive element. The line separation should be at least 1 mm on the screen.

(a) Show a possible optical arrangement which would do this using a diffraction grating ruled with 8500 lines/cm and light at normal incidence.

(b) If the light source is 20 cm from the grating, what must be the exit slit width if the angular widths of the diffracted lines are to be no broader than 1/10th of their separation?

7.2 It is proposed to design a system for deuterium (^{2}H) separation from a mixed deuterium/hydrogen atomic beam using optical enhancement to increase the yield of the process. A high intensity synchrotron radiation source of ultraviolet radiation will be used. The beam of atoms originates in an oven held at 600°C and the separation will be done in vacuum. It is proposed to use optical absorption leading to selective ionization of one of the species, followed by physical separation in an applied electric field. The wavelength required to ionize deuterium is 91.188 nm and that for hydrogen is 91.213 nm. A possible experimental arrangement is shown:

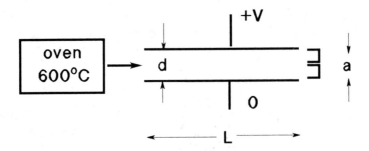

(a) What photon energy and resolving power would be required to set up the conditions for selective absorption?

(b) Could this resolution be achieved using a continuous source and a grating, and if so, design the grating, choose the order, and define the position of the exit slits.

(c) Estimate the beam speed and suggest dimensions, voltages, etc. for a trial design. Justify your choice by simple calculations.

7.3 In a development program for the production of power from nuclear fusion, it is required to measure the temperative of a plasma using the temperative broadening of the 615.43 nm spectral line from hydrogen. This entails a measurement of the breadth of the spectral emission line. The atoms are excited by an electric discharge through the gas which is leaked into the vacuum chamber through a leak valve. At low pressures, the line width (FWHM) of a spectral line from a gas of atomic mass M at temperature T (K) is

$$\Delta\lambda = 10^{-6} \sqrt{\frac{T}{M}} \cdot \lambda \text{ (nm)}$$

(a) Design an apparatus to make this measurement if the operating temperature of the arc is about 10,000 K. Draw a sketch of a workable optical system and determine the characteristics of a grating spectrometer capable of making the measurement.

(b) Could a phase sensitive detector (lock-in amplifier) be usefully employed in this system. How would you employ it, and what advantages would you gain by using such a device?

The following question may be assisted with a computer simulation.

7.4 Design a prism to have an angle of minimum deviation of close to 40°. Prisms having angles of 30°, 45° or 60° are available in media of refractive indices of 1.310, 1.517 or 1.617. State the angle and material.

7.5 A proposal has been made to develop an optical technique to aid in the avoidance of icebergs by liquefied natural gas tankers using the Davis Strait and the Beaufort Sea in Northern Canada. The proposal is based on the experimental observation by a diver that large volumes of undersea light appear to originate from the direction of icebergs. It is proposed to mount an optical detector consisting of 0.2 m diameter photomultiplier capable of detecting a light level of 10^{-13} W on the keel of the ship.

Determine whether this idea is feasible for the detection of icebergs of approximate horizontal area 30×30 m at a distance of 1 km from the ship.

(a) Under daylight conditions where the solar irradiance with the sun approximately vertical is 1 W/m^2.

(b) At night where the irradiance is 2×10^{-4} W/m^2.

(c) If a laser beam is projected from the ship in the forward direction and of which a fraction of 0.16 is isotropically scattered from the iceberg.

(d) If the project is not feasible at 1 km, what is the maximum distance at which an iceberg could be detected at night?

This problem requires some careful modelling. Information that you may wish to utilize is the following:

Light is attenuated in seawater according to the relation $I = I_o \exp(-\mu x)$, where $\mu = 0.06$ m^{-1} for arctic seawater.

The iceberg ice contains many small fissures, cracks and bubbles which cause the light to be scattered isotropically. Assume that natural light captured by the iceberg propagates uniformly in all directions. Vertical surface light entering the water may be scattered $90°$ into the horizontal direction. The amount is given by the volume scattering coefficient $\beta(90)$ defined by $dI(90) = \beta(90) I \, dV$ where $dI(90)$ is the light scattered at $90°$ from a volume element dV due to vertical illumination intensity I. For seawater $\beta(90) = 0.06$.

7.6 (a) A step index optical fibre 1 m in length is used to transmit parallel light collected by a lens of diameter D from a gas discharge and focused on the end of the fibre. The index of refraction of the core and sheath of the fibre is 1.65 and 1.35 respectively. If the intensity of light in the gas discharge corresponds to 2 mW/m^2, reflection losses at the faces of the lens and fibre are about 6%, and the linear attenuation coefficient for light in the fibre is .05 m^{-1}, (i) If a lens of focal length $F = 5$ cm is used, what is its diameter D, (ii) what is the maximum power transmitted to the end of the fibre?

(b) After a further 3% loss at the exit interface, the light intensity is measured by a photodiode having a conversion efficiency of 0.5 A/W. If the resulting current is measured using a 100 kΩ load resistance connected to an oscilloscope having an input impedance of 1 MΩ in parallel with 30 pF by a coaxial cable 50 cm in length. What is the minimum signal to noise ratio you might expect from the measurement. The cable characteristics are 88 pF/m.

(c) If the gas discharge is excited by the 60 Hz mains, comment briefly on all sources of noise within the system and on methods to enhance the signal to noise ratio.

7.7 (a) An optical fibre is constructed of two glasses of refractive index 1.46 and 1.31 respectively. It is desired to transmit parallel light coming from an area of an illuminated plasma which is 3 cm in diameter. Design a simple optical system to focus the light into the fibre. What will be the luminosity of this arrangement?

(b) If the plasma is AC driven at a frequency of 400 Hz show how a lock-in detector may be used to increase the signal to noise ratio of the signal recorded by a photodiode viewing light emerging from the end of the fibre.

8

X-ray Measurements

Electromagnetic radiation has the characteristics both of a wave of wavelength λ and frequency v and a photon of energy hv, where h is Planck's constant $(6.63 \times 10^{-34}$ J.s) (see Sec. 3.6). The X-ray region of the electromagnetic spectrum corresponds to wavelengths of 0.08 to 100 Å, where 1 Å = 0.1 nm, and photon energies of 0.1 to 150 keV. Two important phenomena are linked with this region:

1. the interaction of X-rays with the inner electron shells of atoms. This leads to the analytical tool of X-ray fluorescence.

2. the diffraction of X-rays by crystal lattices. This is important for the study of crystallographic structures.

8.1 THE ELECTRON STRUCTURE OF ATOMS

In order to understand the energies measured in X-ray fluorescence, the energy levels for electrons in atoms are briefly reviewed.

For a singly charged atom such as the hydrogen atom, the Bohr theory of the atom assumes that atoms consist of a massive, positively charged central nucleus surrounded by an electron which can take up various spherical orbits. The angular momentum of the electron is quantized in units of $(h/2\pi)$. An electron in an orbit has an **allowed** energy characterized by a quantum number n, where n is the number of quantized units of angular momentum $(h/2\pi)$. All other energies are **forbidden**.

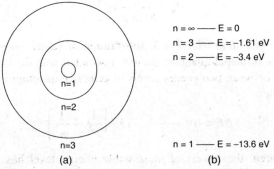

$n = \infty$ —— $E = 0$
$n = 3$ —— $E = -1.61$ eV
$n = 2$ —— $E = -3.4$ eV

$n = 1$ —— $E = -13.6$ eV

(a)   (b)

Fig. 8.1 The Bohr model for the hydrogen atom: (a) Represents circular orbits. (b) Shows the allowed energies for the electron (n is the principal quantum number).

227

The mathematical form of the orbit is called the **wavefunction**, which describes the probability of finding the electron in a particular region of space. Figure 8.1 shows spherical orbits of radius r, and an energy level diagram showing the energies for a single electron of charge $(-q)$ orbiting a positively charged nucleus.

The energy scale has a zero (the vacuum level) which represents the energy which an electron would have at an infinite distance from the nucleus. The energy of an electron within the atom is then **negative** with respect to the vacuum level. If the classical Bohr theory is extended to multi-electron atoms, and the effects of Z electrons of mass m and charge $-q$ orbiting an atom of mass M and charge $+Zq$ taken into account as a reduced mass μ, the energies of the allowed electron levels are given by

$$E\,(eV) = h\nu = -\frac{\mu Z^2 q^4}{8\varepsilon_o^2 n^2 h^2}$$

Z atomic number
q electronic charge 1.60×10^{-19} C
ε_o permittivity of free space 8.85×10^{-12} F/m
h Planck's Constant 6.63×10^{-34} J-s
μ reduced mass of the atom and electron $\mu = mM/(m + M)$
m electron mass 9.11×10^{-31} kg
M atom mass $\approx$ atomic mass $A \times 1.67 \times 10^{-27}$ kg
c velocity of light 3.00×10^8 m/s

The frequency of radiation resulting from energy transitions between such levels is often expressed

$$\nu = cRZ^2 \left[\frac{1}{n^2} - \frac{1}{m^2} \right]$$

where R is the Rydberg constant $(1.097 \times 10^7$ m$^{-1})$, the reduced mass is approximated by the electron mass, and n and m are the principal quantum numbers of the levels concerned.

$$R = \frac{mq^4}{8\varepsilon_o^2 h^3 c} \ \text{m}^{-1}$$

For a hydrogen atom with $Z = 1$, substitution of values gives $E_n = -13.6/n^2$ eV, where n is the principal quantum number associated with the level. The energy difference in eV between two energy levels of principal quantum number n and m respectively is then

$$\Delta E = h\nu = \frac{hc}{\lambda} = -13.6 \left[\frac{1}{n^2} - \frac{1}{m^2} \right]$$

For **hydrogen**, the lowest or most stable energy level has $n = 1$ and the smallest energy required for an electron to make a transition from this level to a state with $m = 2$ is $\Delta E = 10.2$ eV. From $n = 2$ to $m = 3$, $\Delta E = 1.89$ eV, and from $n = 3$ to $m = 4$, $\Delta E = 0.66$ eV. These energies correspond to the long wavelength limit of optical absorption and emission spectra for hydrogen termed as the Lyman ($n = 1$), Balmer ($n = 2$) and Paschen ($n = 3$) series respectively. These wavelengths are in the ultraviolet and visible regions of the electromagnetic spectrum.

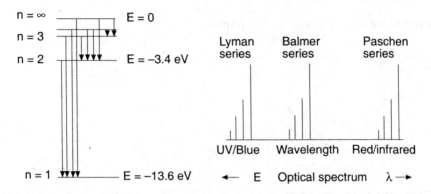

Fig. 8.2 Emission spectra for hydrogen: (a) The atom is excited so that the electron enters an orbit with a principal quantum number > 1. (b) Electromagnetic radiation of wavelength $hc/\Delta E$ is emitted when the electron returns to an orbit of quantum number $n = 1$ (Lyman), $n = 2$ (Balmer) and $n = 3$ (Paschen series) respectively.

Question (i)   What are the optical wavelengths corresponding to the first four terms of the Balmer series for hydrogen?

8.2 MULTIELECTRON ATOMS

This simple model of circular orbits is not sufficient to describe the properties of real atoms, particularly for higher values of Z. The postulates of quantum mechanics must be used to calculate more details of the energy levels. The results are expressed in terms of a set of quantum numbers which define the type and units of angular momentum which an electron may have. The unit is $h/2\pi = \hbar = 1.054 \times 10^{-34}$ J-s. The nature of each quantum number is summarized as follows:

Principal quantum number, n (1, 2, 3...).   The **total energy** of the electron in an atom of atomic number Z is the number derived in the Bohr theory. The inequality is shown because other quantum numbers change the energy slightly.

$$E(n) \approx E(Z)/n^2$$

Orbital quantum number, ℓ (0, 1, 2...(n − 1)).   This defines the orbital angular momentum of the electron, and in simple terms is a measure of the elliptical nature of the orbit. $\ell = 0$ is the lowest energy state.

Magnetic quantum number, m, (−(ℓ) through 0 to +(ℓ), (2ℓ + 1) states).   Angular momentum is a vector and therefore has a direction in space.

The component of ℓ along any axis is quantized in units of $\hbar$. An external magnetic field defines a specific direction in space and the component of angular moment is quantized with respect to that direction.

Example: $\ell = 2$; $-2\hbar$ to 0 to $+2\hbar$ (5 states)

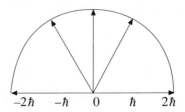

Spin quantum number, *s* ($s = \pm 1/2$).   The electron has a spin angular momentum of $m_s = \pm 1/2\hbar$. This moment may be aligned with respect to other moments (e.g. orbital angular momentum ℓ) within an atom to give a set of total angular momentum states $j = (\ell + s)$.

Total quantum number, *j* ($j = \ell + s$).   If **(j)** is the vector total of all forms of angular momentum characteristic of a particular state. j is aligned with respect to any axis in components from $(\ell + s)$ to $(\ell - s)$:

Spectroscopic notation.   The energy states for the electron are labelled in terms of their total energy quantum number and the letter *s, p, d, f* referring to the angular momentum vector ℓ.

ℓ	0	1	2	3	4
n					
1	1s				
2	2s	2p			
3	3s	3p	3d		
4	4s	4p	4d	4f	
5	5s	5p	5d	5f	5g

Fig. 8.3   Spectroscopic states corresponding to values of n and ℓ.

Each of the states have $(2\ell + 1)$ magnetic sub-levels. Each sub-level can have two electrons with spin $+1/2$ and $-1/2$ respectively. The total number of states within each set of levels is $2(2\ell + 1)$. Thus *s* states contain 2 electrons, *p* states accommodate 6 electrons, and *d* states have 10 electrons. This is written $1s^2 2s^2 2p^6 3s^2 3p^6$ for a representative atom (argon). For small Z the states are filled sequentially with electrons to form the light elements. However, after the 3p states are filled the 4s states fill before the 3d. This gives rise to the transition metal series of elements. Similar anomalies occur for the 5f and 6f levels.

8.3 X-RAY FLUORESCENCE—LINE SPECTRA

Optical spectra involve transitions between energy levels associated with the outer valence electrons of a multi-electron atom. Transitions to levels closer to the nucleus involve higher energy photons in the X-ray region of the spectrum. It is possible to estimate what photon energies (eV) are involved in X-ray transitions in a manner similar to that used for the optical spectra of the hydrogen atom.

For a *K* transition ($n = 1$) one electron is removed from the *K* shell. The nuclear charge *Z* is shielded by **one** charge

$$h\nu_K = hcR(Z - 1)^2 \left[\frac{1}{1^2} - \frac{1}{m^2} \right]$$

For an *L* transition ($n = 2$) one electron is removed from the *L* shell. The nuclear charge is shielded by **nine** charges,

$$h\nu_L = hcR(Z - 9)^2 \left[\frac{1}{2^2} - \frac{1}{m^2} \right]$$

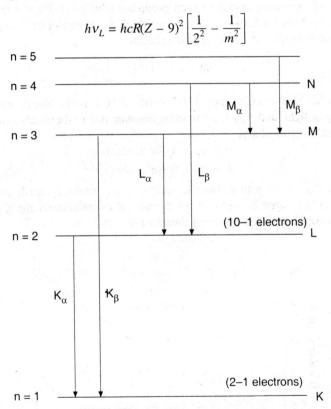

Fig. 8.4 Transitions leading to the α and β series of X-rays.

To observe X-ray transitions it is necessary to excite the atom to produce vacancies in internal or low-lying states so that electrons from outer states can make transitions into them. The atom may be excited by bombarding it with high energy electrons obtained by direct electrical acceleration, or through excitation by low energy X-ray radiation from a radioactive source.

When a vacancy is created in a $1s$ state of a multi-electron atom of atomic number Z, the energy of an electron in that state can be estimated by noting that the core of the atom consists of Z positive charges and the one remaining negative electron from the $1s$ state. The effective atomic charge is then $+(Z - 1)q$. 10 electrons occupy states having principal quantum numbers $1s$ and $2s$ and $2p$. If a vacancy is created in the $2s$ or $2p$ states, the effective atomic charge is $+(Z - 9)q$.

Like the optical spectrum, the X-ray spectrum consists of groups of lines. In emission, the radiation results from upper level electrons falling into inner shell vacancies, the energy released being emitted as an X-ray photon of energy given by $h\nu = \Delta E$. The probability for a photon to be emitted is known as the **fluorescence yield**. The origin of the lines may be deduced from a Bohr-like electronic structure. X-rays resulting from transitions to states with principal quantum number $n = 1$, 2, and 3 are termed K, L and M X-rays respectively. Transitions of increasing energy in each group are labelled α, β and γ respectively. In general, the **Moseley Law** is predicted in which the energy of an X-ray transition is proportional to $(Z - \sigma)^2$ through the relation

$$\nu = cR(Z - \sigma)^2 \left[\frac{1}{n^2} - \frac{1}{m^2} \right]$$

where R is the Rydberg Constant (1.097×10^7 m^{-1}), Z is the atomic number, c is the velocity of light, and $\sigma_{K,L}$ is a shielding constant due to the number of electrons shielding the nuclear charge.

$$\sigma = \sigma_K \sim 1 \text{ for } K \text{ shells}$$
$$\sigma = \sigma_L \sim 9 \text{ for } L \text{ shells}$$

In practice, $\sigma_{K,L}$ vary with Z, but the variations are relatively small and become significant only at large Z. A plot of the theoretical expression for the K_α emission lines is compared with experimental data in Fig. 8.5.

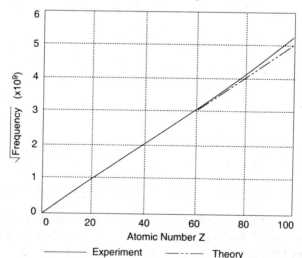

Fig. 8.5 A Moseley plot for K_α emission lines for the elements. The dashed line is the theoretical expression, the full line represents experimental data.

A plot of $(v)^{1/2}$ versus Z is known as a Moseley Plot. It provides a method of analysis or confirmation of the presence of a particular element.

Question (ii) Calculate the K_α X-ray photon energy and wavelength for (a) iodine ($Z = 53$) and (b) copper ($Z = 29$).

8.4 FINE STRUCTURE

A detailed examination of X-ray emission lines shows a fine structure of doublet and triplet lines etc. These are designated $K_{\alpha 1}$, $K_{\alpha 2}$, $K_{\beta 1}$, $K_{\beta 2}$ etc. The intensity of $K_{\alpha 2}$ is generally 0.5–0.53 $K_{\alpha 1}$.

The structure in the spectrum arises from the various values of the total magnetic moment j for the shells concerned, and the selection rules that govern transitions between the levels.

Selection rules: $\Delta\ell = \pm 1$; $\Delta j = 0, \pm 1$

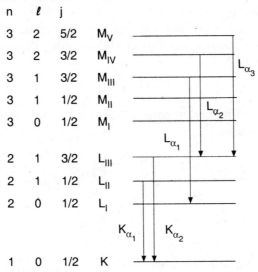

Fig. 8.6 Electron transitions leading to fine structure.

K_α shell transitions are doublets $K_{\alpha 1}$ and $K_{\alpha 2}$, L_α shell transitions are triplets L_I, L_{II} and L_{III}. The nominal position of the K_α emission lines (Appendix 4.2) is generally a weighted average of the $K_{\alpha 1}$ and $K_{\alpha 2}$ line positions.

8.5 ABSORPTION AND EMISSION PROCESSES

Emission requires the removal of a K or L shell electron and the transition of an outer shell electron into the inner state.

Absorption requires the removal of an inner electron from the atom to a higher vacancy or from the atom itself if all upper states are filled. The **absorption edge** occurs at a slightly higher energy (wavelength slightly lower) than that for the emission line.

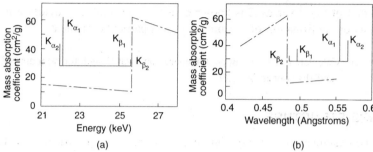

Fig. 8.7 Relationship between the *K* absorption edge (graph of μ/ρ cm^2/g vs *E* or λ) and *K* emission spectra (inset) for silver (a) vs energy (b) vs wavelength.

In general, the K_α linear attenuation coefficient μ decreases by about a factor of 6 at an absorption edge. Since the attenuation law $I = I_o \exp(-\mu x)$ is similar for all electromagnetic radiation (see Fig. 3.32), the attenuation of a layer of material of thickness *x* for wavelengths above and below the absorption edge changes by a factor of $\exp(-6) = 0.0025$ (see Sec. 9.5).

The position of X-ray emission lines and absorption edges are tabulated (see *Handbook of Chemistry and Physics,* 66th ed., CRC Press, pp. E139–183 and Appendix 4.2 respectively).

X-ray Filters

In order to produce a **monochromatic** source of X-rays it is necessary to filter the emission from a source.

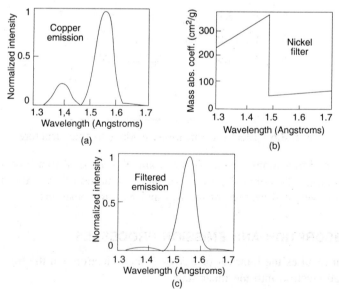

Fig. 8.8 Creation of a filter for Cu K_β X-rays: (a) Experimental spectrum of Cu emission taken using a solid state X-ray detector with a resolution of about eV. (b) Absorption spectrum for Ni. (c) Spectrum after filtering by 0.05 mm thick Ni sheet.

To filter the K_α from the K_β radiation from copper, say, an excellent filter material would be an element whose absorption edge lay between the K_α and the K_β emission lines.

This may be selected by choosing a material with an atomic number less than that of the target material.

Thus for copper ($Z = 29$) K_α radiation, nickel ($Z = 28$) is used as a filter.

Copper emission	$K_{\alpha 1}$ — 1.5405 Å; $K_{\alpha 2}$ — 1.5443 Å; K_β — 1.3926 Å
	$Z = 29$; $K_{\dot{C}u}$ absorption edge — 1.380 Å
Nickel absorption	$Z = 28$; K_{Ni} absorption edge — 1.488 Å

In terms of **energy**:

Cu: $K_{\alpha 1} = 8.0478$ keV; $K_{\alpha 2} = 8.0278$ keV; $K_\beta = 8.9029$ keV,

Ni: K_{Ni} absorption edge = 8.333 keV

Attenuation coefficient for mixtures. The linear attenuation coefficient for a compound or a mixture can be calculated through summing the contributions to the **mass** attenuation coefficient

$$\frac{\mu}{\rho} = \sum_i w_i \left(\frac{\mu}{\rho}\right)_i \frac{cm^2}{g}$$

w_i = Weight fraction of element i present
(μ/ρ) = Mass attenuation coefficient (cm^2/g) of element i

Continuous absorption spectrum. Between absorption edges radiation is absorbed approximately such that

$$\mu/\rho \sim Z^3 \lambda^3 \text{ cm}^2/\text{g}$$

8.6 PRODUCTION OF X-RAYS

X-ray emission spectra can be produced by excitation of target atoms by charged particle bombardment, or by the absorption of electromagnetic radiation.

Electron bombardment. Conventional X-ray machines employ the bombardment of a cooled target (usually Mo or Cu) by an electron beam accelerated through a potential difference V.

This process produces:

— a **continuous** spectrum

— **characteristic** X-ray lines

The slowing (deceleration) of the charged electrons in the target causes the emission of **Bremsstrahlung** radiation having a continuous wavelength spectrum over the X-ray region (Fig. 9.6).

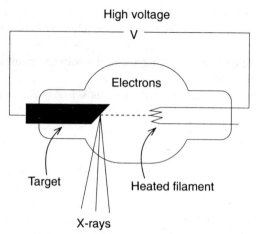

Fig. 8.9 High voltage X-ray tube in which an electron beam strikes a target.

The maximum energy photon that is created is determined by the incident electron energy (eV) such that

$$hv_{max} = \frac{hc}{\lambda_{min}} = qV$$

where q is the charge of the electron, and V is the accelerating voltage.

A spectrum for a Mo target is shown in Fig. 8.10. For clarity, characteristic K_α and K_β emission lines for molybdenum are included only on the 40 kV spectrum.

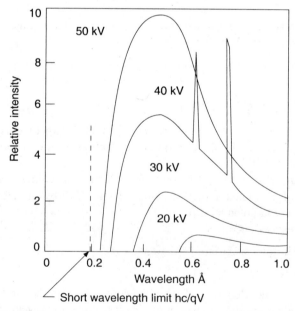

Fig. 8.10 The X-ray emission spectrum for electrons of various energies exciting a molybdenum target.

X-ray and ultra-violet emission occurs whenever electrons are accelerated. When this occurs under circular motion in a synchrotron, the resulting synchrotron radiation source is an intense UV and X-radiation source.

Question (iii)   What is the short wavelength limit for X-rays generated in a copper anode operated at 20 kV?

The excitation of characteristic X-rays for analytical purposes is termed **X-ray fluorescence**. Solid state X-ray detectors (Fig. 10.20) can provide simultaneous analysis of the entire spectrum. Important methods for producing such X-rays in practice are as follows.

Electron beam microprobe (electron excitation).   A finely focused electron beam of diameter 2–10 μm is used to excite characteristic X-rays over a microscopic area. The background continuum is unwanted. All emission lines are detected simultaneously using a solid state X-ray detector.

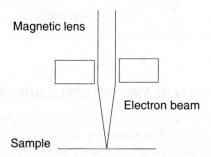

Fig. 8.11   Electron beam X-ray excitation for materials analysis.

Proton induced X-ray emission (PIXE) (proton excitation).   A beam of protons from an accelerator excites characteristic X-rays. Because protons slow less rapidly than electrons, the background Bremsstrahlung continuum is reduced for protons and the detection sensitivity for elemental X-rays is thereby increased.

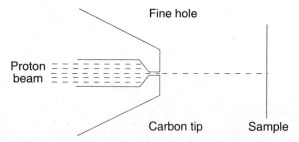

Fig. 8.12   X-ray excitation by a fine beam of protons from a particle accelerator.

X-ray fluorescence (X-ray excitation).   A conventional X-ray generator or a radioactive source such as ^{125}I which emits a 27.4 kV X-ray can be used to excite characteristic X-rays. This is a standard analytical tool in the materials industry.

The simultaneous nature of the analysis for a complex compound using a X-ray fluorescence is shown for a metallic alloy (brass) coated with silver.

X-ray emission line energies for the analysis of such spectra are listed in Appendix 4.2.

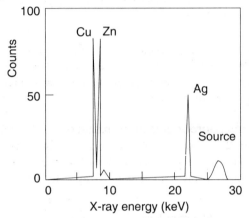

Fig. 8.13 Simultaneous measurement of characteristic X-rays from silver-coated brass.

8.7 X-RAY DIFFRACTION AND CRYSTALLOGRAPHY

A crystal lattice consists of planes of atoms spaced at regular intervals. Radiation falling on a crystal surface is scattered by individual atoms in different crystal planes separated by plane spacing d. The scattered waves will undergo constructive interference when the path difference between interfering rays differs by an integral number of wavelengths, i.e., when $2d \sin \theta = m\lambda$, where m is the order of diffraction. The angle θ is defined in Fig. 8.14. For $m = 1$, since $d \approx 4$ Å, $\sin^{-1} \lambda/2d$ is significant when $\lambda = 0.5$–10 Å. This range of wavelengths corresponds to the X-ray region.

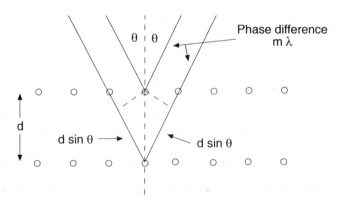

Fig. 8.14 Diffraction of X-rays by crystallographic planes spaced by distance d.

Interference maxima. In three-dimensional crystals, many plane spacings are found, each of which will give rise to a maximum when the interference equation is satisfied. These maxima (or high intensities of reflected radiation at specific angles) are characteristic of the structure of the crystal and can be used to identify it.

Absent reflections. The absence of a reflection can also identify a type of crystal structure. Consider a face-centred cubic structure:

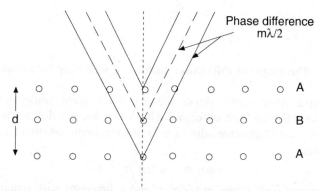

Fig. 8.15 The origin of 'absent' reflections in diffraction from a face-centred cubic lattice. Atom planes 'A' are at the top and bottom of the unit cell. Atom planes 'B' are located on each face. All planes have the same number of atoms and therefore reflect equally.

Atom planes A constitute the top and bottom of the unit cell separated by distance d. A maximum is expected at an angle when $2d \sin \theta = \lambda$. However, the crystal structure also contains a set of atom planes B at $d/2$. In physical space atoms B will be located between atoms A, but since the reflecting power of the plane depends primarily on the number of atoms present, they are shown aligned for simplicity. At angle θ, these planes will produce radiation scattered with a phase difference of $\lambda/2$ with respect to the top and bottom faces, and hence a minimum rather than a maximum occurs at this angle. For example, the **absence** of the anticipated reflection can distinguish between the simple cubic and the face- or body-centred cubic structures (Fig 8.25).

In the following discussion, for experimental simplicity, we will primarily examine **cubic** structures and concentrate on the use of powder diffraction techniques for analysis and crystal structure determination.

More general techniques and structures are reviewed by H. Lipson and H. Steeple: (Ref. 1) or B.D. Cullity (Ref. 2).

Diffraction by Three-dimensional Crystals

Suppose we consider a beam of X-rays falling on a plane of atoms in which the interatomic distance is a. Each atom can be regarded as a source of wavelets radiating into three dimensions.

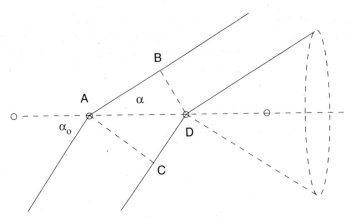

Fig. 8.16   The origin of diffraction maxima in scattering by a regular lattice.

A simple construction shows that for an incident angle α_o, diffraction maxima occur when the path difference $AB - CD = h\lambda$, at a diffraction angle of α. Since the emission is into three dimensions this maximum occurs in a cone of half angle α, where

$$a \cos \alpha - a \cos \alpha_o = h\lambda$$

In a crystal with planes in the x, y, and z directions with spacings a, b and c respectively, two equations must be satisfied in addition

$$b \cos \beta - b \cos \beta_o = k\lambda$$
$$c \cos v - c \cos v_o = \ell\lambda$$

each of which corresponds to a cone of X-rays in space. A maximum will be observed only when the three equations are satisfied simultaneously. This corresponds to a point or to a line where the three cones of diffracted radiation intersect. For a complex crystal and radiation of a single wavelength, the number of such directions is very small. Special experimental arrangements have therefore to be made to obtain a diffraction pattern containing useful information. The two most commonly used methods are described below:

Laue method.   This method is often used in the preparation and mounting of single crystal specimens for other experiments in which a knowledge of the crystallographic orientation is essential.

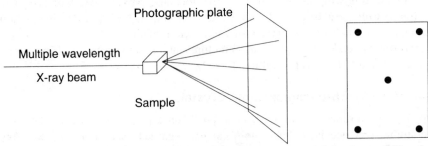

Fig. 8.17   Experimental arrangement and diffraction pattern for Laue diffraction.

The incident beam contains a continuum of wavelengths $I(\lambda)$. 'Spot' maxima are observed at all wavelengths for which the diffraction equations are satisfied at angles characteristic of the crystal structure. A crystal having a cubic structure oriented with a plane perpendicular to the incident beam direction will give rise to a square pattern of maxima on a photographic plate placed in the location shown. In order to orient a crystal, the crystal orientation is adjusted until an initially asymmetric pattern of spots becomes symmetric. The main features are:

> Multiple wavelength diffraction
> Diffraction from single crystals
> Orientation of crystal axes in space

Debye-Scherrer method. A monochromatic beam of X-rays (say K_α emission from a copper target at $\lambda = 1.5412$ Å filtered using a Ni filter) is used to illuminate a finely powdered sample. All orientations of the crystallites are present and the X-ray beam will select out those which are correct to give a maximum at a particular angle. Since the crystallites are oriented randomly, the maxima will occur as cones about the beam axis. The actual record is the intersection of these cones with a photographic film or to the plane of movement of an X-ray detector.

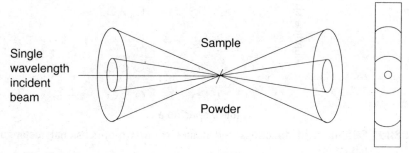

Fig. 8.18 Experimental arrangement and diffraction pattern for Debye-Scherrer powder diffraction.

Powder diffraction is of importance for analysis and material identification. Its usefulness lies in the fact that both the plane spacings and the crystal structure (cubic, tetragonal, etc.) can be determined. This combination may be compared with published data in order to identify an unknown substance. Careful measurements of lattice spacing as a function of temperature or applied stress can provide information on various material constants.

The main features are:

> Single wavelength diffraction
> Powder diffraction and analysis

Some interesting variants of powder diffraction can be found. The method can be used for a single crystal if the crystal is rotated during the measurement. If the sample is a thin film on a substrate, the effect of the substrate can be reduced by using an X-ray beam which has been carefully collimated into a flat beam. This beam is caused to enter the rotating sample at a fixed glancing incident angle (0.5–4°) so that only the film material scatters X-rays. The technique is known as **glancing angle X-ray diffraction**.

Crystallographic Directions and Planes—Miller Indices

In order to discuss the results of X-ray diffraction, a nomenclature is required to classify crystal directions and planes. For simplicity these may be illustrated using a two dimensional rectangular lattice of lattice spacing a and b.

The full lines show [directions] within the lattice. The dotted lines show various (planes) of atoms.

Both direction and planes may be defined by taking intercepts along axes drawn along the principal directions.

The direction along the x-axis is [100], while the cube diagonal is [111].

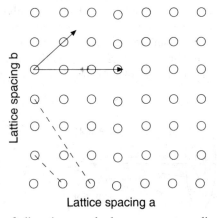

Fig. 8.19 Definition of directions and planes on a two-dimensional rectangular lattice.

Examples of sets of planes in two dimensions are (a, b) or $(2a, 3b)$. However, since planes along the principal axes on this scheme are $(0a, \infty b)$ or $(\infty a, 0b)$, a better nomenclature is to use the **inverse** of the intercept. A procedure to define the **Miller Indices** of a plane is as follows:

1. Take the intercepts of the plane on the major axes in units of lattice spacing.
2. Invert
3. Rationalize fractions to the lowest set of integers.
4. Express as $(hk\ell)$

Example:	Planes	(a)	(b)
	Intercepts	1a, 1b	2a, 3b
	Inverse	1/1, 1/1	1/2, 1/3
	Integers	1, 1	3, 2
	Indices	(11)	(32)

The principal planes of the cubic lattice are

(001) – parallel to *x-y* plane
(010) – parallel to *x-z* plane
(111) – major diagonal plane

Plane spacing.   An important result of defining the Miller Indices is that the plane spacing for a set of planes defined by a set of indices can be calculated. This is most easily done for a cubic lattice, but more complex expressions can be derived for other lattices.

Consider the set of (21) planes, the intersection on the *x*-axis = *a*/2
In general, for (*hkl*) planes
Intercepts: *x*-axis = *a/h*
       y-axis = *b/k*
       z-axis = *c/l*
In two dimensions, the distance from the plane to the origin

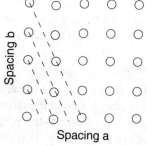

Spacing b

Spacing a

Fig. 8.20   Planes on a two-dimensional lattice.

$$d = a/h \cdot \cos \alpha$$

$$d = \left(\frac{a}{h}\right) \cdot \frac{b/k}{\sqrt{\left(\frac{a}{h}\right)^2 + \left(\frac{b}{k}\right)^2}}$$

$$= \frac{ab}{\sqrt{k^2 a^2 + b^2 h^2}}$$

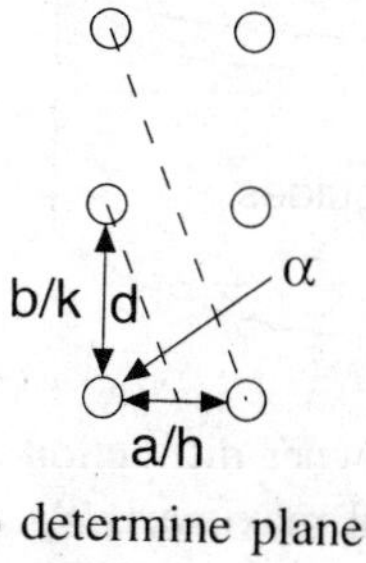

b/k $\downarrow$ d

α

a/h

Fig. 8.21   Construction to determine plane spacing for (*hk*) planes.

If the lattice is square, $a = b$ and the equation becomes

$$d = \frac{a}{\sqrt{h^2 + k^2}}$$

This can easily be extended to three dimensions for a cubic lattice to give

$$d_{hkl} = \frac{a}{\sqrt{h^2 + k^2 + l^2}}$$

and the Bragg equation $2d_{hkl} \sin \theta = \lambda$ may be written for a cubic crystal

$$\sin^2\theta = \frac{n^2\lambda^2}{4a^2} (h^2 + k^2 + l^2)$$

and this gives all possible angles at which a maximum is observed for the (hkl) planes for a cubic crystal.

Similar calculations can be made for other types of lattice to give a series of analogous expressions (see H. Lipson and H. Steeple, (Ref. 1).

Question (iv)   The cubic lattice of MnO has a lattice constant of 4.426 Å. What are the plane spacings for the (001), (111) and (211) planes.

Powder Diffraction Spectra

Spectra may be obtained with a Debye-Scherrer camera or a powder diffractometer. In the camera a strip of film is placed around the circular body of the camera, and conical beam guides direct the X-ray beam to the powders sample stuck to a fine glass rod at the centre of the camera. This rod is rotated during the measurement. After exposure, the film is removed and developed in the dark. The pattern obtained results from cones intersecting a plane. The angle of each maxima is determined from the relationship $\theta = S/4R$, where S is the distance between symmetrical maxima about the central spot and R is the camera radius.

In a powder diffractometer, the powder sample is mounted on a flat plate and is placed in the sample holder. A goniometer mechanism rotates the sample through

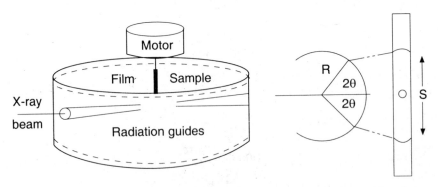

Fig. 8.22   A Debye-Sherrer powder diffraction camera of radius R. The distance between symmetrical maxima is S.

θ, while the X-ray detector rotates through 2θ. The detector output is recorded on a chart or computer as a function of reflection intensity versus 2θ.

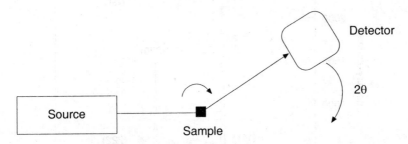

Fig. 8.23 A powder diffractometer. The crystal rotates about the centre line, while the X-ray detector moves at twice the angular speed.

Analysis of X-ray powder diffraction patterns. The following section outlines the methods used to analyze and characterize materials by X-ray diffraction, primarily using powder diffraction. The most common target used in X-ray sources is copper with a K_α wavelength defined as 1.5412 Å. After the data has been recorded, the values of $\sin\theta$, $\sin^2\theta$, plane spacing d calculated from $2d\sin\theta = \lambda$, and intensity are calculated for each line.

Question (v)   X-ray data is taken using a chromium anode ($\lambda_{Cr} = 2.289$ Å). If the spectrum has a line at ($2\theta = 45.4°$), what would be the equivalent (2θ) line position for a copper anode ($\lambda_{Cu} = 1.5412$ Å)?

Powder diffraction file.   The powder diffraction pattern of every compound is distinctive in terms of both line positions and intensities. This information has been systematized by the JCPDS International Centre for Diffraction Data. All published diffraction spectra are included in a data file. This file is available as a card file, in book form as a search manual, and most recently on CD-ROM. Examples of the data available are shown below with courtesy of the publishers of the Powder Diffraction File (Ref. 3).

The search manual gives a listing for the d-values and intensities of the 8 most intense lines in order of line intensity. The technique for using the file is to start with the most intense line, check the listing for the anticipated compound to see that the first three lines are present, and then compare the intensities and d-values for all other lines. Full data can then be obtained from the index card for the compound in the diffraction file. An example of the search procedure for a compound (sodium chloride) is shown below:

The diffraction pattern was recorded using a diffractometer with a copper target with a K_α wavelength of $\lambda = 1.5412$ Å.

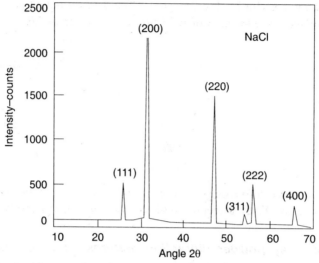

Fig. 8.24 Diffraction spectrum for NaCl using Cu K_α radiation.

The *d*-values computed for lines of decreasing intensity are:

Intensity order	1	2	3	4	5	6
d-value:	2.829	1.996	1.629	3.272	1.411	1.706
Search manual:	2.82	1.99	1.63	3.26	1.41	1.705
Compound:	NaCl					

General analysis for cubic structures. For a simple cubic lattice of spacing *a*, diffraction maxima occur for the (*hkl*) planes of the crystal at angles given by

$$\sin^2\theta = \frac{n^2\lambda^2}{4a^2}\,(h^2 + k^2 + l^2)$$

If the lattice is simple cubic, the maxima which occur are determined by the allowed values of ($h^2 + k^2 + \ell^2$) as *h*, *k* and ℓ take all values starting with 0. The allowed values are illustrated by the first line of Figure 8.25. Mathematically, certain numbers are absent from the series (7, 15,...). These 'missing' numbers are given by (8*m* − 1), where *m* is an integer.

$\sin^2\theta$	$h^2+k^2+l^2$	0	1	2	3	4	5	6	8	9	10	11	12	13	14	16	17	18	19
Simple cubic	All h,k,ℓ																		
Body-centred cubic	h+k+ℓ even																		
Face-centred	h,k,ℓ all even all odd																		
Missing numbers																			

Fig. 8.25 Allowed planes for three cubic lattices.

The effect of absent reflections due to additional planes of atoms in the face-centred cubic and body centred cubic structures is to remove reflections from certain planes.

Face-centred cubic:	h, k, and ℓ must be all even or all odd
Body centred cubic:	$h + k + \ell$ must be even.

The experimental technique to determine the lattice spacing is to calculate $\sin^2\theta$ for all lines in the diffraction spectrum and to fit them into a series set by the allowed values of h, k, and ℓ. This is shown for NaCl in Fig. 8.26.

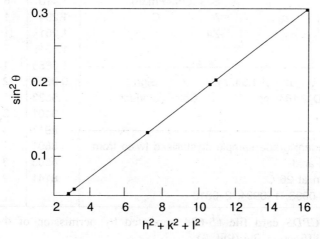

Fig. 8.26 $\sin^2 \theta$ versus $h^2+k^2+\ell^2$: slope gives $a = 5.642$ Å.

When a fit has been achieved and the type of crystal structure has been identified, the slope of a plot of $\sin^2 \theta$ versus $(h^2+k^2+\ell^2)$ gives a value for the constant of proportionality $(\lambda^2/4a^2)$. This enables a value for a to be determined. The JCPDS diffraction card file for NaCl is shown in Fig. 8.27.

Expressions for $\sin^2 \theta$ for lattices of different crystal systems, and detailed procedures for fitting data for those lattices are given in Chapter 5 of Lipson and Steeple.

X-ray density

$$\text{Density} = \frac{\text{Mass of the unit cell}}{\text{Volume of the unit cell}}$$

$$= \frac{\Sigma Z}{6.023 \times 10^{23} \times V}$$

$$= 1.66020 \, \Sigma \, Z/V \text{ g/cm}^3$$

where ΣZ is the sum of the atomic weights of the atoms in the unit cell, and V is its volume in Å.

5-0628

d	2.82	1.99	1.63	3.258	NaCl		
I/I_1	100	55	15	13	Sodium chloride		(Halite)

Rad. CuKα λ 1.5404 Filter Ni	d Å	I/I_1	hkl
Dia. Cut off Coll.	3.258	13	111
I/I_1 G.C. Diffractometer D corr abs.	2.821	100	200
Ref. Swanson and Fuyat, NBS Circ. 539, II, 411953	1.994	55	220
	1.701	2	311
	1.629	15	222
Sys. CUBIC S.G. O^5_H–Fm3m	1.410	6	400
a_0 5.640$_2$ b_0 c_0 A C	1.294	1	331
α β γ Z4	1.261	11	420
Ref: ibid	1.1515	7	422
	1.0855	1	511
εα n = β 1.542εγ Sign	0.9969	2	440
2V D 2.164 mp Color colorless	.9533	1	531
Ref: ibid	.9401	3	600
	.8917	4	620
An ACS reagent grade sample crystallised twice from	.8601	1	533
hydrochloric acid	.8503	3	622
X-ray pattern at 26°C	.8141	2	444
Replaces 1–0993, 1–0994, 2–0818			

Fig. 8.27 *JCPDS card file #5-628, reprinted by permission of the Powder Diffraction file (Ref. 3).*

Question (vi)   NaCl (Z_{Na} = 23, Z_{Cl} = 35.5) has a face-centred cubic structure with a lattice constant of 5.642 Å. What is its density in g/cm^3?

Information Available from Diffraction Patterns

Accurate measurement of cell dimensions.   X-ray diffraction can provide information on changes in lattice constant in the study of alloys, or in the effects of thermal expansion, mechanical stress, or defect generation under radiation bombardment. The highest accuracy in lattice constant measurements is achieved at large θ.

The Bragg equation ($2d \sin \theta = \lambda$) may be written $d = \dfrac{\lambda \cosec \theta}{2}$

Differentiating: $\delta d = - \lambda \cosec \theta \cot \theta \; \delta\theta/2$
$$\delta d = - d \cot \theta \; \delta\theta$$

Since $\cot \theta \to 0$ as $\theta \to 90°$, for a given $\delta\theta$ the relative error in the lattice parameter d becomes very small when θ exceeds 80°.

In defect studies, comparison of the X-ray density with the density determined by, say, Archimedes method provides information on the relative volume of voids or defects.

Crystallinity. A single crystal gives well resolved spots on a photographic plate taken by the Laue method. Polycrystalline material shows a pattern of full circles. Intermediate patterns (spotty circles) are good indicators of the crystalline state.

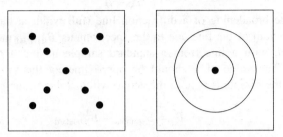

Fig. 8.28   X-ray Laue patterns of diffraction rings and spots characteristic of polycrystalline and single crystal material respectively.

Amorphous materials. These are characterized by diffuse rings corresponding to the average set of spacings present in the disordered structure.

This information is expressed in terms of the radial distribution function $\phi(r)$ which is a measure of the number of atoms at radius r from any other atoms.

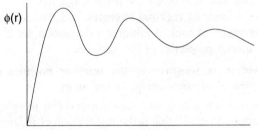

Planar spacing r

Fig. 8.29   Radial distribution function representing the distribution of inter-atom spacings in an amorphous material.

Temperature effects. The peak positions for a given pattern change as the lattice constants change under thermal expansion. The vibrations of the atoms about their mean positions in the lattice smear out the diffracting planes and lead to a decrease in the **intensity** of a given line. This occurs rather than the line broadening which may be expected. This decrease in intensity at angle θ is described by the Debye-Waller factor B_T.

$$I = I_o \exp\left[-2B_T \left(\frac{\sin \theta}{\lambda} \right)^2 \right]$$

Expressions for B_T are given in Cullity (Ref. 2) but may be approximated (within 20% for many metals) up to 900 K by $1.3T/A$, where T is the temperature (K) and A is the atomic weight.

Grain size. As the particle size t of a powder decreases, the line width of X-ray diffraction lines increases. A quantitative estimate of the particle size t in the nanometer size range may be obtained from,

$$B = \frac{0.9\,\lambda}{t\,\cos\theta}$$

where B is the broadening of a diffraction line (full width at half maximum) at angle θ. The width of the line due to the spectrometer $B_{standard}$ must be evaluated by comparison with a line from a standard sample having a reasonable large particle size. The value of θ should be approximately the same. If the width measured for the sample is B_{sample}, the actual width B due to the grain size is then given by

$$B^2 = B^2_{sample} - B^2_{standard}$$

8.8   NEUTRON DIFFRACTION

Quantum mechanics shows that particles such as neutrons also have a wave-like behaviour such that a particle having momentum p has a corresponding wavelength λ such that $\lambda = h/p$. If the energy of the particle is $E = p^2/2m$, $\lambda = h/\sqrt{2Em}$.

Significant diffraction effects are found with *neutrons*. As noted earlier, the wavelength required to give appreciable diffraction angles can be estimated from Bragg's Law as $\sin\theta \approx \lambda/2d$. For $d = 4$ Å, $\lambda \approx 0.2\text{–}4$ Å. For neutrons $m = 9.1 \times 10^{-31}$ kg, and $E = 0.021$ eV for $\lambda = 0.2$ nm.

This is in the range of **thermal energies**, and thermal neutron diffraction is a significant experimental tool. Additional information from neutron diffraction arises from the special properties of the neutron:

1. The neutron is sensitive to the **nuclear** position rather than to the distribution of electron charge in the atom.

2. The neutron has a magnetic moment and can therefore detect **magnetic structure** in a crystal, that is the distribution of magnetic moments within a larger crystallographic structure.

3. The neutron is more sensitive to nuclear movement and therefore is a good probe of **vibrational spectra** in crystals.

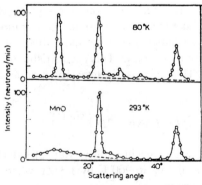

Fig. 8.30   Neutron diffraction spectra for MnO above and below a temperature range where magnetic ordering takes place (from C.G. Shull and J.S. Smart, Phys. Rev., **76**, 1256 (1949)).

An example of the value of neutron diffraction is seen in studies of the magnetic structure of manganese oxide MnO.

MnO is paramagnetic at room temperature, but develops an ordered antiferromagnetic structure at 80 K. This is shown by added peaks at low temperature in the neutron diffraction spectrum which can be indexed to a unit cell twice as large as that at 293 K.

REFERENCES AND FURTHER INFORMATION

1. H. Lipson and H. Steeple, *Interpretation of X-ray Powder Diffraction Patterns*, Macmillan, London, 1968.

2. B.D. Cullity, *Elements of X-ray Diffraction*, 2nd ed., Addison-Wesley Publishing Co. Inc., Reading Mass., 1978.

3. *JCPDS Diffraction File:* compiled by the International Centre for Diffraction Data, 1601 Park Lane, Swarthmore, Pennsylvania, USA.
   This file is available as a series of cards, or on magnetic disc. A convenient laboratory compendium is in the form of two books.
   Inorganic Phases—Alphabetical Index (Chemical and Mineral Name).
   Inorganic Phases—Search Manual (Hanawalt).

4. *Table of Isotopes*, 7th ed., Michael Lederer and Virginia S. Shirley (Eds.), John Wiley, & Sons Inc., New York, 1978, Appendix 3, Table 10.

5. *Chemical Rubber Handbook*, Various Editions (Table of X-ray Mass Absorption Coefficients).

6. Catalogues from the Rigaku Corporation:
   Rigaku USA/Inc., 199 Rosewood Drive, Northwoods Business Park, Danvers, Massachusetts 01923, USA.
   Rigaku Canada, 34 Berczy St, Aurora, Ontario L4G 1W9 Canada.

ANSWERS TO QUESTIONS

(i)   656, 486, 434, 410 nm

(ii) (a) $K_\alpha(I)$ — 27.48 keV, $\lambda = 0.045$ nm

      (b) $K_\alpha(Cu)$ — 7.96 keV, $\lambda = 0.156$ nm

(iii)   0.061 nm

(iv)   (001) 4.426 Å, (111) 2.555 Å, (211) 1.807 Å

(v)   30.1°

(vi)   2.61 g/cm^3

DESIGN PROBLEMS

8.1 An X-ray powder diffraction photograph taken using a camera of 28.06 mm radius and Cu K_α radiation ($\lambda = 1.542$ Å) gives the following *D* values between successive pairs of lines on flat X-ray film.

Ring #	1	2	3	4	5	6	7	8	9	10
D mm (± .2)	38.1	44.2	63.8	76.5	80.4	97.0	111.4	114.2	134.2	156.6

What type of lattice does the sample have and what are the lattice constants? What is the element?

8.2 For X-ray cancer therapy it is necessary to achieve an X-ray beam for which the flux is uniform across the beam. A directed beam of X-rays from a Bremsstrahlung target has a flux which is not uniform, but which reduces with distance from the central beam axis according to the law

$$f_r = f_o e^{-r}$$

where r in cm is measured from the central axis. It is therefore necessary to use some form of filter to make the flux more uniform.

Design a filter using an appropriately shaped lead block which will produce a beam of uniform flux over a circular diameter of 2.0 cm. What is the value of the flux after the filter? (The mass attenuation coefficient is 0.050 cm^2/g for Pb which has a density of 11.34 g/cm^3.)

8.3 Describe an X-ray machine having a Cu target and operating at 40 kV. Discuss the energy spectrum produced by such a machine and a method of making the beam more nearly monochromatic so that it can be used with a powder camera.

For Cu K_α = 0.154 nm K_β = 0.139 nm

8.4 Discuss the application of X-ray fluorescence to the identification of chemical elements. Make a diagram of a typical set up and describe the limitations of the method when fluorescing radiation of energy 27 keV is used.

8.5 Use an energy level diagram to explain the origin of the K_α, K_β and L X-rays for an atom of atomic number Z. If the levels are described by a modified Rydberg equation, calculate the energy of the K_α X-ray for calcium (Z = 20).

8.6 Explain why a Ni filter can be used to isolate the Cu K_α radiation from the Cu K_β X-rays emitted by a copper X-ray target.

8.7 Explain how to generate an almost monochromatic X-ray beam of approximately 7 keV using an X-ray tube.

8.8 Describe briefly a camera for obtaining X-ray powder photographs for the analysis of crystal structure. Explain why some reflections are missing from certain types of crystal structure.

In an experiment to analyze the crystal structure of metallic copper (atomic weight. = 63.5, density 8.9 × 10^3 kg/m^3) X-rays of wavelength 1.5412 Å were incident on a copper foil and the first four strong Bragg reflections were found at the following Bragg angles:

$$\theta = 21.87, \ 25.45, \ 37.47 \ \text{and} \ 45.58$$

Show that the data are consistent with a face-centred cubic lattice and calculate the lattice constant.

9

Radioactivity and Matter

Radioactive nuclei and nuclear particles play a major part in modern technology, in power plants and radio-isotopes: from smoke detectors to food sterilization. The properties of such nuclei and the manner in which they decay are reviewed. The interaction of nuclear particles with matter and the relevance of such interactions to instrumentation and analysis is discussed.

9.1 NUCLEAR PROPERTIES

Radioactive nuclei are unstable with time and decay to other species with the emission of particles such as electrons, positrons or α-particles. This is often followed immediately by the emission of high energy electromagnetic photons termed X- or γ-rays. Heavy nuclei such as ^{235}U are naturally radioactive, while radioactive nuclei of lighter elements are prepared by bombardment of stable nuclei by neutrons in a nuclear reactor, as a result of nuclear fission, or by other particles in an accelerator. A nucleus of chemical element X is described by:

$$\begin{array}{ll} & X \text{ is the chemical symbol} \\ {}_Z^A X & A \text{ is the atomic mass} \\ & Z \text{ is the atomic number} \end{array}$$

The **atomic number** Z is the number of positive charges within the nucleus. Since the number of positive charges is matched by the number of electrons in the resulting atom, Z determines the chemical properties of the atom. The **atomic mass number** A describes the number of nucleons. The mass is predominantly built up of charged protons and uncharged neutrons. The number of neutrons varies in different **isotopes** of a nucleus which have different % **abundance.** Naturally occurring radioisotopes exist and are important for nuclear power, but generally it is unstable radioactive isotopes produced by irradiation which are used for technical purposes. Appendix 4.1 lists important laboratory isotopes.

The atomic mass is calculated in terms of the atomic mass unit (amu) or (u). A scale based on the ^{12}C atom is used. By definition:

$$\text{Mass of } {}^{12}C \text{ atom} = 12.0000 \text{ amu}; \quad 1 \text{ amu} = 1.661 \times 10^{-27} \text{ kg}$$

An atom of atomic number Z contains a nucleus and Z electrons of mass m_e bound together with an atomic binding energy E_{be}; therefore,

$$\text{Atomic mass} = \text{Nuclear mass} + Z\,(m_e) - E_{be}$$

where E_{be} is expressed in mass units through $E = mc^2$

9.2 NUCLEAR AND ATOMIC PARTICLES

Table 9.1 lists laboratory nuclear and atomic particles of interest in instrumentation and analysis.

Table 9.1 Important Particles in Nuclear Instrumentation

Name	Symbol	Charge (q)	Rest mass (M_p)	Energy	Source	Applications
Proton	p	+1	1	0–4 MeV	Accelerators	PIXE
Neutron	n	none	1.006	Slow < 10 keV Fast → MeV	Reactor Reactor Reactions	Activation Analysis Crystallography Power generation
Electron	e^-	−1	1/1840	0–300 keV → 50 MeV	Heated wires E Fields Accelerators	Electronics Microscopy X-ray generation
β^--particle	$\beta^-(e^-)$	−1	1/1840	0.01–5 MeV	Nuclear decay	Thickness monitor
β^+-particle	$\beta^+(e^+)$	+1	1/1840	0.01–5 MeV	Nuclear decay	Positron imaging Interacts with e^- to create two γ-rays of energy 512 keV
α-particle	$\alpha\,(^4_2\text{He})$	+2	4	4.5–8 MeV	Nuclear decay	Nuclear reactions
Deuteron	$d(^2\text{H})$	+1	2	0–10 MeV	Accelerators	Nuclear reactions
γ-ray	γ	None	None	20 keV–MeV	Nuclear decay	Thickness monitor
X-ray	X	None	None	1–100 keV	Electron impact Excited atoms	Crystallography Radiation damage
Fission fragments	Nucleus name	Several	> 100	50–100 MeV	Nuclear fission	
Heavy ions	Ion	1	Ion	0–2 MV	Accelerators	Ion implantation
Neutrino Antineutrino	ν $\bar{\nu}$	None	None	0.01–5 MeV	β decay	Nuclear decay

While the particles have a fundamental role in the structure of the nucleus, the table emphasizes the role of the particle in external measurements. It includes information on the particle charge and rest mass relative to the proton charge ($Q = 1.61 \times 10^{-19}$ C and $M_p = 1.67 \times 10^{-27}$ kg), the energy normally expected in a laboratory experiment, sources and applications.

9.3 RADIOACTIVE DECAY

When an unstable nucleus decays, it may do so by several routes, or by a sequence of decays through a line of daughter products.

Single decay. A radioactive decay used for laboratory purposes is an Americium ^{241}Am alpha particle source produced by irradiation in a nuclear reactor. This decays to Neptunium ^{237}Np with the emission of an alpha particle—a helium nucleus.

$$^{241}_{95}\text{Am} \rightarrow \,^{237}_{93}\text{Np} + \,^4_2\text{He}$$

The **branching ratio** for the decay notes that out of 100 decays,

86 decays result in an α-particle with $E_\alpha = 5.486$ MeV (86%)
13 decays result in an α-particle with $E_\alpha = 5.441$ MeV (13%)

The probability for decay per unit time is a constant and is denoted by the **decay constant** λ. λ is a property of the nucleus and is independent of the environment.

For $N(t)$ nuclei at time t, the number which decay in time t

$$\frac{d\,N(t)}{dt} = -\lambda\,N(t) = -\frac{1}{\tau}\,N(t)$$

where $\tau = 1/\lambda$ is the lifetime of the nucleus. Hence,

$$N(t) = N_0 e^{-\lambda t} = N_0 e^{(-t/\tau)}$$

where N_0 is the number of nuclei at time $t = 0$

The **half life** $T_{1/2}$ is the time required for half the nuclei in a collection to decay

$$N = 1/2\,N_0 = N_0\,e^{-\lambda T_{1/2}}; \qquad \ln 1/2 = -\lambda T_{1/2}$$

$$\boxed{T_{1/2} = 0.693/\lambda = 0.693\,\tau\,\text{(s)}}$$

The **strength** or **activity** $c(t)$ of a source is the number of decays that take place per second.

$$c(t) = -\frac{dN}{dt} = \lambda N = \lambda N_0\,e^{-\lambda t}$$

Units:	Curie =	Ci = 3.7×10^{10}	decay/s
S.I.	Becquerel =	Bq = 1	decay/s

Question (i) (a) How many radioactive nuclei are present in a 10 µCi laboratory source of ^{125}I? ^{125}I decays with the emission of a 35.5 keV X-ray with a half-life $T_{1/2}$ of 56 days. (b) What will be the activity of the source after 1 year?

Parallel decay paths. If alternate modes of decay exist each with decay constant λ_i, the total number of active nuclei $N(t)$ decreases as $dN/dt = -(\lambda_1 + \lambda_2 +...)$. The total activity thus has a half-life

$$\boxed{T_{1/2} = 0.693/(\Sigma\,\lambda_i) = 0.693/\Sigma\,(1/\tau_i)}$$

Sequential decay paths. If the decay (daughter) products are radioactive, they also can decay to give intermediate concentrations of products N_A, N_B,...

$$A \xrightarrow[\lambda_A]{} B \xrightarrow[\lambda_B]{} C \xrightarrow[\lambda_C]{} \dots \text{etc.}$$

$$\frac{dN_A(t)}{dt} = -\lambda_A N_A(t) \tag{1}$$

$$\frac{dN_B(t)}{dt} = \lambda_A N_A(t) - \lambda_B N_B(t) \tag{2}$$

$$\dots \qquad \dots \qquad \dots \qquad \text{etc.}$$

From equations (1) and (2),

$$N_A(t) = N_A(t{=}0)\, e^{-\lambda_A t}$$

$$\frac{dN_B(t)}{dt} = \lambda_A N_A(t{=}0)\, e^{-\lambda_A t} - \lambda_B N_B(t)$$

Guessing a solution of the form shown below (where a and b are constants and $N_B(t) = 0$ at $t = 0$)

$$N_B(t) = N_A(t)\,(a e^{-\lambda_A t} + b e^{-\lambda_B t})$$

$$N_B(t) = \frac{\lambda_A}{\lambda_B - \lambda_A}\, N_A(t{=}0)\, [e^{-\lambda_A t} - e^{-\lambda_B t}] + N_B(t{=}0)\, e^{-\lambda_B t}$$

The activity $c_B(t)$ of $B = \lambda_B N_B$ for any value of N_B ($t = 0$) is then

$$c_B(t) = \frac{\lambda_A \lambda_B N_A(t{=}0)}{\lambda_B - \lambda_A}\, [e^{-\lambda_A t} - e^{-\lambda_B t}] + \lambda_B N_B(t{=}0)\, e^{-\lambda_B t}$$

Thus the first decay 'feeds' the second, and the number of nuclei $N_B(t)$ has the form shown in Fig. 9.1. The conditions when (a) $N_B = 0$ at $t = 0$, and when (b) $N_B > 0$ at $t = 0$ are shown separately.

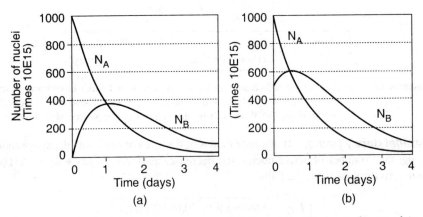

Fig. 9.1   Radioactive decay by a two stage decay path, shown as the number of daughter nuclei N_B as a function of time $\lambda_A = 1$ day^{-1}, $\lambda_B = 0.98$ day^{-1}. (a) $N_A(0) = 10^{18}$, $N_B(0) = 0$, (b) $N_A(0) = 10^{18}$, $N_B(0) = 5 \times 10^{17}$.

Question (ii) ^{228}Ra decays with a half-life of 6.7 years to ^{228}Th which has a half-life of 1.9 years. (An intermediate with a half-life of 6 hrs can be ignored). What is the maximum activity of the thorium product as a function of the time after chemical separation of 10^6 atoms of the radium isotope. After what time does it occur? What is the ratio of the thorium/radium activities at long time?

Build-up of radioactivity due to nuclear reactions. The related problem of the growth of activity of a daughter product resulting from a nuclear reaction is discussed for neutron activation analysis (pages 271 and 311).

9.4 DECAY MODES FOR RADIOACTIVE NUCLEI

Radioactive nuclei decay into more stable nuclei of lower total mass. The decay results in the emission of particles or radiation with characteristic energy. In some cases following the nuclear reaction, X-rays are emitted as a result of changes within electron levels of the atom. The mass difference between the initial and final nuclei reflects both the mass of the products and the energy emitted in the transition. Decay modes for nuclei are summarized in handbooks such as the Handbook of Physics and Chemistry (Chemical Rubber Company). The major modes are as follows:

α-decay: Emission of a helium nucleus $^4_2$He

$$^A_Z X \rightarrow {^{A-4}_{Z-2}} Y + {^4_2} \text{He}$$

Heavy nuclei ($Z > 80$) are unstable against the spontaneous emission of alpha particles. The escape process can be modelled in terms of the penetration of a potential barrier by helium nuclei. Typical alpha particle energies E_α are in the range 4.5–8 MeV.

Spontaneous Fission (SF): Creation of fission fragments

This results from the fission of a heavy nucleus into two medium sized nuclei with or without accompanying light particles. The mass distribution of the fragments is asymmetric (Fig. 9.2), when a Californium ^{252}Cf nucleus fissions into two smaller nuclei plus several neutrons.

The following observations are made regarding spontaneous fission:

- spontaneous fission only happens in very heavy nuclei
- neutrons are generally also emitted
- fissionable nuclei are usually unstable against competing alpha decay. The relative probability of fission and of α-decay can be estimated from the respective half-lives (e.g. 85 year and 2.65 year respectively for a ^{252}Cf nucleus).

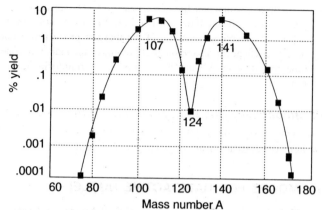

Fig. 9.2   Distribution of mass fragments arising from the spontaneous fission of ^{252}Cf.

β^--Decay: Emission of a negative electron β^-

β^- and β^+-decay results from a reversible change of a neutron into a proton within the nucleus.

$$^{A}_{Z}X \quad \rightarrow \quad ^{A}_{Z+1}Y^* + \beta^- + \bar{\nu}$$

$$^{60}_{27}Co \rightarrow ^{60}_{28}Ni \ + \ \beta^- + \bar{\nu}$$

- In order to satisfy conservation laws, β^--decay is always accompanied by the emission of an **antineutrino** $\bar{\nu}$ which is hard to detect.

- The energy is shared by the daughter nucleus, the β^--particle, and the antineutrino. The energy distribution for the β-particles is continuous with an upper limit or **endpoint energy** E_β. The distribution for β^- from ^{64}Cu is shown in Fig. 9.3(a).

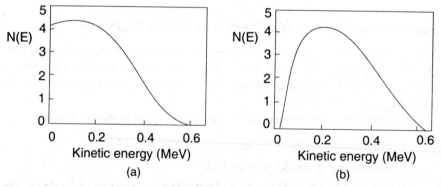

Fig. 9.3   Energy spectrum of β-particles emitted from ^{64}Cu. The end-point energy is around $E_\beta = 0.6$ MeV. (a) β^- (b) β^+. Differing Coulomb interactions between the β^- and the β^+ and the positive nucleus cause the differences between the distributions.

The daughter nucleus may be left in an excited state which decays by gamma emission. The gamma ray is virtually instantaneous with the β^- decay.

β^+-decay: Emission of a positive electron or positron β^+

$$_Z^AX \rightarrow _{Z-1}^AY^* + \beta^+ + v$$

$$_{11}^{22}Na \rightarrow _{10}^{22}Na + \beta^+ + v$$

- β^+-decay is always accompanied by a neutrino v
- The energy is distributed between the daughter nucleus, the positron and the neutrino so that the positron energy is continuous upto E_β (Fig. 9.3(b)).
- Since the positron is the anti-particle of the electron, it is often captured by an electron as soon as it is emitted and two **annihilation gamma rays** of 511 keV are produced ($2m_ec^2 = 1022$ keV) where m_e is the mass of the electron.

Electron Capture: e^-_K or e^-_L capture

The nucleus captures an orbiting negative electron—most often from the electronic K shell since the overlap of the electron with the nucleus is greatest for this shell. This process competes with positron emission and leaves the nucleus in an excited state.

$$_Z^AX + e^-_K \rightarrow _{Z-1}^AY + v$$

$$_{11}^{22}Na + e^- \rightarrow _{10}^{22}Ne + v$$

- Simultaneous gamma emission may be observed
- A neutrino v is emitted with a defined energy
- **No** 511 keV annihilation gamma rays are created
- The process is important if it is energetically unfavourable for the nucleus to decay by positron emission
- X-rays, characteristic of the final atom are emitted when electrons fill the vacancy in the K-shell.

Positron emission and electron capture can occur in the same nucleus.

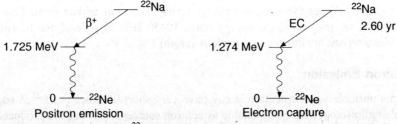

Fig. 9.4 Decay modes for ^{22}Na which result from positron emission and electron capture respectively. The branching ratio for ^{22}Na decay is β^+—90.5%, EC—9.5%.

Electron Emission: Internal conversion or Auger electrons e⁻

The energy of an excited nucleus is transferred directly to an electron in the atomic levels of the atom which is then ejected with significant energy. Filling of the vacancy in the electron levels may then result in the subsequent emission of X-rays.

$$_Z^A X^* \rightarrow {}_Z^A Y + e^-$$

No change in atomic number Z occurs. The electron arises from the following two processes.

Auger processes.   An atom in an excited state de-excites by emitting an outer (valence) electron instead of an X-ray (see page 306). This process is always competitive with X-ray emission.

Internal conversion.   An excited nucleus decays by ejection of an inner K or L shell electron rather than by gamma emission.

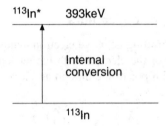

Fig. 9.5   The decay of ^{113}In by internal conversion.

The relative probability for an X-ray photon rather than an Auger electron is called the **fluorescent yield.**

Gamma (γ) Ray Decay

An excited nucleus decays through the emission of electromagnetic radiation (photons) of specific energies (normally $E > 40$ keV). The energy of γ-rays reflect specific nuclear states or reactions. γ-emission often occurs after β decay.

Characteristic X-ray Emission

Interactions between the electronic energy levels of an atom and an excited nucleus produce X-ray photons in the energy range 10–50 keV. In general, the transitions are characteristic of the product nucleus (atom).

Neutron Emission

Nuclei unstable against neutron decay have very short lifetimes ($\sim 10^{-22}$ s) so that stable radioisotopes are not available as neutron sources. Neutrons are produced by a nuclear reaction in which a nucleus such as Be interacts with alpha particles.

$$_2^4 \text{He} + {}_4^9 \text{Be} \rightarrow {}_6^{12} \text{C} + {}_0^1 n$$

Typical neutron sources are shown in Table 9.2. The half life of the neutron source is the half life of the alpha source. The neutrons are emitted with energy of 5–7 MeV and are slowed by embedding the source either in water or paraffin.

Table 9.2   Neutron Source Reactions

Source	$T_{1/2}$	# of neutrons per decay	E_α (MeV)
^{226}Ra + Be	1600 yr	502	several
^{241}Am + Be	432 yr	80	5.48
^{242}Cm + Be	162 days	118	6.1

Other Radiation Sources

Particle accelerators.   Accelerate charged particles (p, d ions) and are important for ion implantation in semiconductors and for the preparation of radioisotopes.

Nuclear reactors.   These are a source of neutrons for isotope production (n) particularly for medical applications. Highly radioactive waste is created in nuclear power generation.

Synchrotron radiation.   Electrons accelerated in a circular path give rise to electromagnetic radiation called Bremsstrahlung. This has a continuous energy distribution in the X-ray and ultraviolet and is used as a high intensity source for materials analysis.

Fig. 9.6   The generation of Bremsstrahlung radiation from a curved electron beam.

Cosmic rays.   These are very high energy particles from space.

Question (iii)   Write down the equations which express the following decay schemes. Use Appendix A4.2 (p. 344) to identify the atoms concerned.

(a) $^{90}_{38}$Sr [β^-]

(b) $^{55}_{26}$Fe [EC]

(c) $^{228}_{88}$Ra [α]

(d) $^{137}_{55}Cs [\beta^-; \gamma(0.663 \text{ MeV})]$

(e) $^{64}_{29}Cu$ [EC, 43%]

$[\beta^+, 19\%]$

$[\beta^-, 38\%]$

9.5 INTERACTION OF RADIATION WITH MATTER

The interaction of radiation or charged particles with matter is important for the use of the radiation for particular purposes—in radiation damage and the design of radiation shielding, and in materials analysis (Chapter 11).

Photons and Matter—γ-ray attenuation

X-rays and γ-rays are electromagnetic radiations with photon energy from a few keV to MeV. Optical radiation is attenuated in a medium according to an expression of the form

$$I(x) = I(0) \ e^{-\mu x}$$

where μ is a linear attenuation coefficient (cm^{-1}) and x is the thickness of the material in cm.

For γ-radiation the same expression applies, but it is more often written in terms of a mass attenuation or absorption coefficient μ/ρ (cm^2/g), where ρ is the density and ρx (g/cm^2) is the mass thickness.

$$I(x) = I(0) \ e^{-\mu/\rho.\rho x}$$

ρx can be interpreted as the mass/unit area of a slab of thickness x. The mass thickness ρx is useful because high energy photons interact with the total electron density Ze. As shown in Fig. 9.7(a) for photon energies > 100 keV, μ/ρ has a smoother variation with Z independent of crystal structure than has μ alone. In the X-ray range ($E < 25$ keV) the effects of transitions between the K and L electron shells (Fig. 8.4) which give rise to sharp absorption edges must be taken into account (Fig. 9.7(b)).

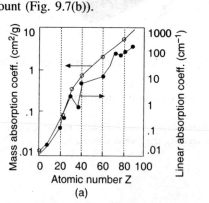
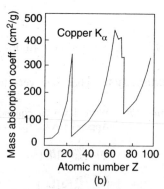

Fig. 9.7 (a) The linear (μ) and mass (μ/ρ) absorption coefficients of the elements as a function of Z for 100 keV photons. (b) Absorption coefficient of 8.041 keV Cu K_α photons as a $f(Z)$.

Values of μ/ρ as a function of photon energy and atomic number Z are given in Appendices 4.2 and 4.3. 8.041 keV Cu K_α radiation with a wavelength of $\lambda = 1.5412$ Å is commonly used for X-ray diffraction. It is useful to relate Fig. 9.7(b) to the emission spectrum and energy level diagram of Cu (Figs. 8.4 and 8.8).

Interaction cross-section σ. When a photon traverses a unit area of a medium, the absorption is related to the number of interactions the photon makes within the medium.

As radiation crosses a slab of thickness Δx its intensity decreases by ΔI from $I(E, x)$ to $I(E, x + \Delta x)$.

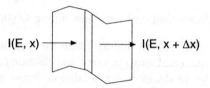

Fig. 9.8 Radiation absorbed in a slab thickness Δx in medium of total thickness x.

Absorption	=	Interaction probability	×	Photon flux	×	Number of atoms in slab thickness Δx
$\Delta I(E)$	=	$\sigma(E)$	×	$I(E)$	×	$\dfrac{\rho N_0 \Delta x}{A}$

where $\sigma(E)$ is the **atomic cross-section** (cm^2) for an interaction, N_0 is Avogadro's number, A is the atomic weight, ρ is the density (g/cm^3).

Rearranging and integrating for a finite thickness $\displaystyle\int_0^x \frac{d\,I(E)}{I(E)} = -\sigma(E)\,\frac{\rho N_0}{A}\,dx$

Intensity after distance x, $\displaystyle I(x) = I(0)\,e^{-\frac{\sigma(E)\,\rho N_0}{A}x} = I(0)\,e^{-\mu x}$

$$\mu(E) = \text{Linear attenuation coefficient} = \frac{\sigma(E)\rho N_0}{A}\ \text{cm}^{-1}$$

$$\frac{\mu(E)}{\rho} = \text{Mass attenuation coefficient} = \frac{\sigma(E) N_0}{A}\ \text{cm}^2/\text{g}$$

Mass attenuation coefficients as a $f(E, Z)$ are graphed in Appendix 4.3. For an absorber or compound with multiple elements i of mass fraction w_i, the total mass attenuation coefficient is given by the expression as shown on page 235.

Interactions of Photons with Matter

Photons interact primarily with electrons in the atom. The mechanisms are as follows.

Photoelectric effect. All of the photon energy is absorbed by a single atom such that an electron is ejected from an internal energy level.

Compton scattering.   A photon is elastically scattered by a single electron with only a fraction of the photon energy being transferred to the electron. The direction of the photon is changed by the collision. The paths of the photon and electron are as shown.

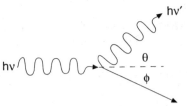

Fig. 9.9   Photon and electron paths during Compton scattering.

Pair production.   The photon interacts with the Coulomb field of the nucleus and all its energy is converted into a positron and electron pair $(\beta^+ \beta^-)$. Recombination of these particles produces two annihilation γ-rays each with an energy of 511 keV.

The dominant interaction depends on the photon energy E and on the atomic number Z of the absorber. For $Z > 40$ and $E < 0.1$ MeV the photoelectric effect (σ_{pe}) is strongest. For $Z > 40$ and $E > 10$ MeV, pair production (σ_{pp}) is of most importance. At intermediate values of Z and E, Compton scattering (σ_{cs}) is large. The total interaction cross-section is the sum of the individual cross-sections.

$$\sigma_T(E) = \sigma_{pe}(E) + \sigma_{cs}(E) + \sigma_{pp}(E)$$

Energy Transfer to a Medium by Photons

The effects of radiation on a material result from the energy which is absorbed. The energy transferred in each process is illustrated in Fig. 9.10(a). Since only part of the input energy is deposited through the Compton effect, the **attenuation** of the photon flux is not a good measure of the region over which energy is **absorbed** when Compton scattering is present. The **Compton absorption cross-section** $\sigma_{ca}(E)$ is defined as $\sigma_{ca}(E)\, E_\gamma = \sigma_{cs}\, \Delta E_\gamma$ where ΔE_γ is the **average energy** transferred to the absorber. Two coefficients are therefore used when Compton scattering is present, one of which is used to define the probability of interaction, the other for the probability for energy transfer.

$$\text{Attenuation coefficient } \mu(E) = \frac{\rho N_0}{A} [\sigma_{pe}(E) + \sigma_{cs}(E) + \sigma_{pp}(E)]$$

$$\text{Absorption coefficient } \mu_a(E) = \frac{\rho N_0}{A} [\sigma_{pe}(E) + \sigma_{ca}(E) + \sigma_{pp}(E)]$$

The two coefficients are graphed for different elements and γ-ray energies in Appendix 4.3. Comparison shows that the attenuation coefficients differ for $20 < Z < 60$ and $0.5 < E < 1$ MeV. The effect on the distribution of photon flux (attenuation) and absorbed energy (absorption) is shown in Fig. 9.10(b) for an absorber of $Z = 60$ and photon energy 0.5 MeV.

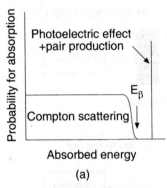
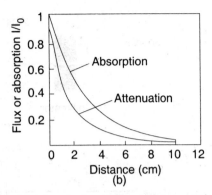

Fig. 9.10 (a) The spectrum of energy absorbed by electrons from photons under Compton scattering, photoelectric absorption and pair production. (b) Photon attenuation and absorption as a function of distance for $Z = 60$ and $E = 0.5$ MeV showing that energy is absorbed over larger distances than may be expected from simple consideration of attenuation.

The peak corresponding to the photoelectric effect and to pair production assumes that the ejected electron and the annihilation radiation respectively are absorbed in the material. In practice, the coefficient which is used depends on whether the interest is in the number of photons penetrating a material and the distance to which they penetrate (radiation shielding, X-ray fluorescence), or the effect that the radiation has on the material (radiation dose, carrier generation).

Question (iv) The mass attenuation coefficient for lead for 1 MeV γ-rays is 0.07 cm^2/g and the density is 11.3 g/cm^3. Assuming linear geometry, what thickness of lead must be used around a source to reduce the source strength by a factor of 100, Answer in terms of (a) linear thickness (b) mass thickness.

Question (v) In a radiation damage experiment a 1 cm^2 area and 0.5 cm thick crystal of CaF$_2$ of density 3.18 g/cm^3 is irradiated for 10 min by 663 keV γ-rays from a ^{137}Cs source having an intensity 2 mW/cm^2. If the mass absorption coefficient is 0.028 cm^2/g, what is the total energy absorbed by the crystal?

Gamma Ray Attenuation: Measurement and Design

The attenuation of gamma rays in various thicknesses of matter can be used to measure thickness. An understanding of the attenuation of radiation in materials is necessary for the design of protective shielding to minimize the effects of radiation. The variable energy loss and changing photon direction in Compton scattering must be considered in such designs. In measurement, the initial and final beam direction must be well defined by lead collimators and the material thickness should not be too large. In shielding, the effect of multi-angle scattering may lead to the 'build-up' of the intensity beyond that expected from simple assumptions. Figure 9.11(a) shows an example of good geometry for attenuation measurements. Figure 9.11(b) shows how scattering causes I to build up above the value predicted by the exponential attenuation formula.

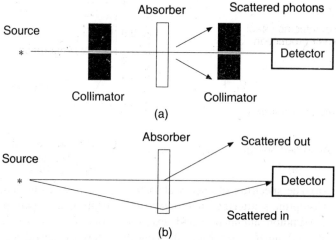

Fig. 9.11 (a) Good measurement geometry for γ-ray attenuation. (b) The build-up factor in radiation shielding.

When build-up occurs, the intensity reaching the detector is

$$I = B(x) \, I_0 \, e^{-(\mu/\rho)\,(\rho x)}$$

where $B(x)$ is the build-up factor. $B(x)$ depends on the geometry and is most important for large thicknesses (x or ρx). This factor is listed for lead in Appendix 4.5. (see page 348).

9.6 INTERACTION OF CHARGED PARTICLES WITH MATTER

Many instruments and analytical methods involve the interaction of charged particles with matter. These include β-ray thickness monitors, smoke detectors, and tritium fuelled light sources. Analytical tools include electron beams in transmission electron microscopy and heavy particles in Rutherford Backscattering (Chapter 11). Such interactions are also important in sputtering (page 189) and for ion implantation in semiconductor manufacturing.

Heavy Particles (α, p, d, ions)

Interaction occurs primarily through Coulomb forces between the particle and electrons and nuclei in the absorber. The dominant interaction is with electrons in the target through ionization and the creation of electron hole pairs.

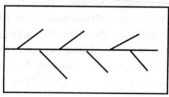

Fig. 9.12 Path of a heavy charged particle.

Simple kinetics show that the maximum change in energy ΔE of a particle of mass m and incident energy E in a head-on collision with a second mass M is $\Delta E = 4(m/M)\ E$. When heavy particles such as protons or alpha particles interact with electrons in a target, $m/M < 1800$, and a large number of collisions are required to bring the incident particle to rest. The rate of this loss is described by the specific energy loss or stopping power.

Stopping Power

The stopping power is the rate at which a particle loses energy in traversing a distance Δx.

$$ S = \frac{\Delta E}{\Delta x} = \frac{-dE}{dx} $$

ΔE is the energy loss in distance Δx, while $(-dE/dx)$ is the **specific energy loss**. The **range** R is the distance required to bring the particle to rest. For charged particles the relationship between S and R is complex since S changes as the particle slows down. Both S and R may be defined in terms of a linear distance x (cm) or a mass thickness ρx (g/cm^2). The units are MeV/cm or MeV-cm^2/g respectively.

For heavy particles, the momentum transfer per collision is small and the path of the particle within the target is generally along a straight line in the incident particle direction. The excited valume is roughly cylindrical. In general, the width of the distribution of particle energies increases and the direction of motion changes as the energy decreases with penetration.

For incident electrons, $m_0/M \sim 1$, large angle scattering can occur and the path of the particle and the analysis of the scattering process is generally more complex. The excited volume is roughly hemispherical with the base of the hemisphere towards the surface. The high density of excited electrons at the surface promotes the emission of secondary electrons (Fig. 3.21).

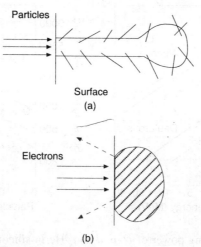

Fig. 9.13 The excited volume due to: (a) Heavy particles. (b) Electrons.

For alpha particles, initially $dE/dx \propto 1/E$. At the surface, the particle is mono-energetic and highly directional. Initially $dE/dx \propto 1/E$ but due to the statistical nature of the loss process, near the end of its path, the average energy varies widely between particles. This is known as **straggling.** The energy transfer to the material and hence the potential radiation damage is large in this region. The range of 6.3 MeV α-particles in air is about 5 cm. The stopping power for various particles in Si is shown as a function of energy in Fig. 9.14(a). Figure 9.14(b) shows how the stopping of alpha particles varies with the atomic number of the absorber. Plastics such as polyethylene $(CH_2)_n$ have a stopping power close to that of helium.

The calculation of range for a given incident energy has to take into account the change in energy with penetration. Calculations are simplest for thin targets for which S is constant. The thickness Δx for energy loss ΔE is then $\Delta E/S$.

For bulk materials, the range may be estimated by considering the path as a sequence of slabs in which energy ΔE is lost. The thickness of the m'th slab is $\Delta E/S(E_m)$ where $E_m = E_o - (m - 1)\Delta E$ is the energy at the beginning of the slab. The range is the sum of all the slabs in the sequence. Such calculations can be conveniently done by spreadsheet.

Question (vi)   Use the above approach to estimate the penetration of 5.486 MeV alpha particles from a ^{241}Am source into: (a) Al (Z = 13) and (b) polyethylene (mean Z of about 3). Use Fig 9.14 to estimate the stopping power for the material over the energy range. In practice R(Al) = 24μm and R(polyethylene) = 35μm..

The penetration of 5 MeV α-particles into plastics and light metals is about 20–40 μm. The penetration of 1 MeV protons and deuterons is 10–15 μm. This small penetration but high energy loss of heavy ions implies that the exitation density is high and severe damage or atom displacement will occur in surface regions of the target. For this reason, α-particles absorbed on the skin or ingested into the body produce very severe radiation damage and are extremely dangerous.

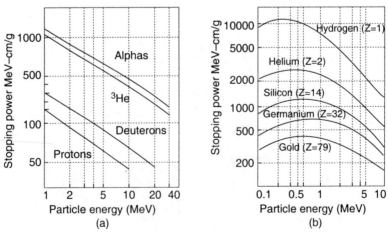

Fig. 9.14   (a) Stopping power of α, p, d and ^{3}He in silicon Z = 14. (b) Stopping power of alpha particles in materials of different Z.

Light Charged Particles—Electrons and β^--rays

The penetration of electron beams is of interest in electron microscopy and for analytical techniques such as the electron beam microprobe. β-ray thickness monitoring of polymers and thin metals is also a well known industrial tool. The incident electron is comparable in mass to the target electrons and large angle scattering occurs. The path is complex and the stopping power is high. The approximately hemispherical volume of the excited volume is shown on Fig. 9.13. The stopping power for initially monoenergetic electrons as a function of energy is graphed in Appendix 4.4. A minimum value of dE/dx occurs between 1 and 3 MeV depending on Z. For many materials this minimum value is about 1.5 MV-cm^2/g. For instrumentation and electron beam analysis lower energies in the > 10 keV range are of importance. An empirical fit of the stopping power of Appendix 4.5 for $E > 100$ keV shows that the electron stopping power varies as

$$S = E^{-0.68} Z^{0.29}$$

Using a similar spreadsheet analysis as for α-particles (Fig. 9.14) electron range values to 20% can be calculated by a simple spreadsheet calculation. This is sufficient for consideration of the depths of excitation achieved in an electron beam microprobe.

The range of β^--particles from isotopes such as ^{90}Sr is of interest for thickness measurement and in smoke detectors. In principle such attenuation has to take into account the energy spread of the β^- particles emitted in a β-decay (Fig. 9.3) up to the endpoint energy E_β. Experimentally it is found that β-radiation is attenuated up to a maximum range according to the simple expression:

$$I(x) = I(0) \, e^{-\mu(\beta)x}$$

where $\mu(\beta)$ is an absorption coefficient. This relationship is empirical and results from the combined effects of the incident energy distribution and electron scattering as a function of energy. For a number of sources incident on light metals such as Al, the attenuation coefficient $\mu(\beta)/\rho$ is related to the end-point energy of the β-source E_p (MeV) by the relationship (R.D. Evans).

$$\mu(\beta)/\rho = 17/Z \, E_\beta^{1.14} \ \text{cm}^2/\text{g}$$

Atoms within the target may be ionized or otherwise excited by the incident electrons. For example, **cathodoluminescence** which is the production of light in a material by an electron beam is fundamental to the television cathode ray tube. Radiation is also emitted due to Bremsstrahlung energy-loss (Fig. 9.6) when high energy electrons are decelerated within a target material.

Backscattering and Secondary Electron Emission

Particularly at low energies and for high Z absorbers, the path of an electron within an absorber is complex. As noted earlier (Fig. 9.13) the shape of the excitation volume is roughly a hemisphere with its base adjacent to the surface. Thus some **primary** electrons may re-emerge or scatter back to the surface with substantially all the incident energy. The **back-scatter** reflection coefficient η for absorbers is shown in Fig. 9.15.

$$\eta = \frac{\text{Reflected intensity}}{\text{Incident intensity}} = \frac{I_R}{I_0}$$

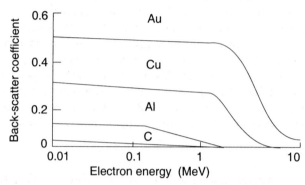

Fig. 9.15 Reflection coefficient η for back-scattered electrons.

This effect is important in that it may lead to erroneous measurements of the flux of β-particles and electron beams collected by a target. Collection of the back-scattered electrons is also used as a means of back-scatter imaging of a surface in scanning electron microscopy (Fig. 11.6). A related effect is the loss through the surface, of low energy (50 eV) secondary electrons from the high density of localized electrons excited by the incident particle (Fig. 3.22). The **secondary electrons** are utilized in photomultipliers (Fig. 3.21) and in scanning electron microscopy (Sec. 11.2). In insulators the emission of more secondary electrons than are incident in the primary beam means that a surface charges during electron bombardment. To avoid this insulators must be metallized to provide a conducting path.

9.7 THE INTERACTION OF NEUTRONS WITH MATTER

Neutrons are uncharged and therefore interact only through the nuclear force in classical collisions, or through absorption by nuclei.

Fast Neutrons

These are slowed through collision processes. The maximum transfer of energy and momentum occurs when the masses of the colliding bodies are the same. Light nuclei having a small probability of fast neutron absorption are termed **moderators.** Important nuclei for this propose are carbon and deuterium (^{2}H) in the form of heavy water and hydrogen. Inelastic scattering of neutrons with more than a few MeV energy from heavy nuclei is an important mechanism for energy loss. The target nucleus is left in an excited state which rapidly decays by emission of γ-rays. Inelastic scattering plays an important role in the design of shielding for high energy neutrons.

Slow Neutrons

These have roughly thermal energies (1/40 eV) and are usually absorbed by target nuclei. The probability of absorption can be described by an absorption cross-section in a manner similar to that for photons and the attenuation is given by

$$I(x) = I(0) \exp(-\Sigma x)$$

where

$\Sigma = N\sigma = $ **macroscopic cross-section** (cm^{-1}).

$N = $ Number of target nuclei (cm^{-3}).

$\sigma = $ Absorption or capture cross-section for the nucleus (cm^2) (1 barn = 10^{-24} cm^2)

If the target contains i different nuclei having concentration N_i nuclei/cm^3, $\Sigma_{\text{TOT}} = \Sigma_I N_i \sigma_i$.

Neutron Activation for Isotope Production

An important application of neutron sources is to activate or form a radioisotope of an element (Sec. 11.5). The element to be activated has, say thickness L and density ρ, then the rate of isotope production $R = \sigma N I$, where I is the incident neutron flux ($cm^{-2}s^{-1}$).

If N is the number of isotopic nuclei produced which subsequently decay with a decay constant λ, then

$$\frac{dN}{dt} = R - \lambda N; \quad N(t) = \frac{\sigma_t N_t I}{\lambda}\left(1 - e^{-\lambda t}\right)$$

The actual number of nuclei that are produced has to take into account the decrease in the neutron density as the beam of initial flux I_{in} ($cm^{-2}s^{-1}$) penetrates the material.

- For a **thin** target the neutron beam is not attenuated, $I = I_{in}$.
- For a **thick** target $I(x)$ varies with depth as neutrons are absorbed. The total flux has to be integrated over the number of neutrons absorbed in thickness L. The equivalent flux is

$$I(L) = I_0\, e^{-\sigma N L} = I_{in}\, e^{-\Sigma L}$$

where

$\Sigma\ (cm^{-1}) = \sigma N$ is the macroscopic cross-section

$N = $ the number of nuclei (cm^{-3}).

$= \rho N_0/A$, where N_0 is Avogadro's Number.

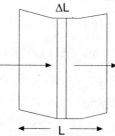

Fig. 9.16 Absorption of neutrons in a section of a bulk target.

REFERENCES AND FURTHER INFORMATION

1. Glenn F. Knoll, *Radiation Detection and Measurement,* 2nd ed., John Wiley and Sons, Inc., New York, 1989.

2. R.D. Evans, *The Atomic Nucleus,* McGraw Hill, New York, 1955.

3. *Atomic Mass Data: Atomic Data and Nuclear Data Tables,* Academic Press, New York, 1977–1998

4. Table of Isotopes: Handbook of Chemistry and Physics, Any volume, Chemical Rubber Company, Cleveland, Ohio.

5. B.D. Cullity, *Elements of X-Ray Diffraction,* 2nd ed., Addison-Wesley Publishing Co. Inc., Reading (Mass.), 1978.

6. Physical Review D-Particles and Fields, Part 1, Vol 54 (1996).

   This provides an excellent review of physical constants as well as nuclear physics data.

7. *Table of Isotopes,* 7th ed., X-ray emission—Appendix 3, Table 10. Michael Lederer and Virginia S. Shirley (Eds.), John Wiley and Sons, Inc., New York, 1978.

8. E. Browne and R.B. Fierstone, *Table of Radioactive Isotopes,* Virginia S. Shirley (Eds.), John Wiley and Sons, Inc., New York, 1986.

9. Earl K. Hyde, *The Nuclear Properties of the Heavy Elements III: Fission Phenomena,* Prentice Hall, Inc., Englewood Cliffs, New Jersey.

ANSWERS TO QUESTIONS

(i) (a) 2.6×10^{12} (b) $0.11 \ \mu C$

(ii) 4.8 yr, 0.54 μCi, 1.38.

(a) $^{90}_{38}\text{Sr} \rightarrow {}^{90}_{39}\text{Y} + \beta^- + \bar{v}$

(b) $^{55}_{26}\text{Fe} + e^- \rightarrow {}^{55}_{25}\text{Mn} + v$

(c) $^{229}_{88}\text{Ra} \rightarrow {}^{225}_{86}\text{Rn} + {}^{4}_{2}\text{He}$

(d) $^{137}_{55}\text{Cs} \rightarrow {}^{137}_{56}\text{Ba*} + \beta^- + \bar{v}; \ {}^{137}_{56}\text{Ba*} \rightarrow {}^{137}_{56}\text{Ba} + \gamma \, (0.663 \, \text{MeV})$

(e) Three modes of decay

$$^{64}_{29}\text{Cu} + e^- \rightarrow {}^{64}_{28}\text{Ni} + v \ (43\%)$$

$$^{64}_{29}\text{Cu} \rightarrow {}^{64}_{28}\text{Ni} + \beta^+ + v; \ 2511 \, \text{keV} \ \gamma \ (19\%)$$

$$^{64}_{29}\text{Cu} \rightarrow {}^{64}_{30}\text{Zn} + \beta^- + \bar{v} \ (39\%)$$

(iv) $x = 5.82$ cm, $\rho x = 65.8$ g/cm^2

(v) 0.052 J

(vi) Range for Al – 25.5 μm, range for polyethylene – 35.8 μm.

DESIGN PROBLEMS

9.1 A ^{137}Cs source of 663 keV γ-rays is to be used in the laboratory. What thickness of lead should be used to reduce the γ-ray intensity by a factor of 1000. (Mass absorption coefficient 0.22 cm^2/g, density of lead 11.3 gm/cm^3).

9.2 The transmission of γ-rays through a material of density 14.7 g/cm^3 is determined by a source/sample/counter arrangement. The number of counts recorded in 1 min intervals for varying thickness of absorber are as shown. The counts are uncorrected for background. Determine the value and uncertainty of the linear and mass attenuation coefficients for the material.

Thickness (cm)	0	.3	.6	.9	1.2	1.5	1.8	2.1	2.4	3.0	3.6	4.2
Counts (1 min)	1056	640	412	265	163	90	71	44	36	18	11	10

9.3 A thin coating of plastic 0.03 mm thick is deposited on a minute speck of ^{226}Ra which has a mass of 50 microgram to form a sphere of radius 0.03 mm. ^{226}Ra emits alpha particles of 4.70 MeV and has a half life of 1.6×10^3 yr. If heat is conducted away from the surface of the sphere at a rate proportional to the temperature difference between the plastic and the ambient temperature of 27°C with a coefficient of 7.7×10^{-9} W/K, what is the temperature of the plastic?

9.4 A radio-isotope power supply is fuelled by 475 g of ^{238}PuC (plutonium carbide). ^{238}Pu has a half life of 88 years and emits an alpha particle of energy 5.6 MeV per disintegration. The thermal to electrical efficiency of the system is 5.4%

(a) What is the specific activity (Ci/g) of the fuel?

(b) What is the thermal power produced per Curie?

(c) What is the total electrical power output of the supply?

9.5 An ^{241}Am source emitting γ-rays of energy 59 keV is used to monitor the thickness of Al sheet of nominal thickness 2.54 cm. Calculate the counting rate in the detector without the sheet in place so that the thickness will be determined to 0.1% in a counting time of 10 s. The mass attenuation coefficient for Al (density 2.7 g/cm^3) is 0.246 cm^2/g.

9.6 The thickness of nominal 20 cm thick plastic sheet (12.5% H$_2$ and 87.5% C by weight and density is 0.9 g/cm^3) is to be monitored by noting the attenuation of a 663 keV gamma ray parallel beam passing perpendicularly through the sheet. The detector is well shielded so that background and gamma scattering into the detector can be ignored. The detector counting rate with no sheet in place has a mean value of 10,000 per second. The counting period for the plastic sheet is 2 s.

(a) Use Appendix 4.3 and the expression for mixtures on page 235 to compute the mass attenuation coefficient of the plastic. Estimate or use uncertainties for all quantities and calculate an uncertainty in the linear attenuation coefficient μ.

(b) Calculate the value and uncertainty of the thickness ignoring the uncertainty in μ.

(c) Include the uncertainty in μ to obtain a final value for x.

9.7 The transmission of 5.9 keV X-rays is used to monitor the thickness of a plastic foil of chemical composition $(CH_2)^n$ during manufacturing. The nominal thickness is 10 mg/cm^2, and the count rate without the foil is 1×10^4 per second. What is the minimum counting time needed in order to detect a 5% change in the foil thickness? The % change in counts due to change in foil thickness should be twice that due to statistical fluctuations.

Mass absorption coeff. of 5.9 keV photon in C = 11.3 cm^2/g
Mass absorption coeff. of 5.9 keV photon in H = 0.50 cm^2/g
Atomic weight of C = 12.010 amu
Atomic weight of H = 1.008 amu

10

Radiation Detection and Measurement

Radiation is detected by its interaction with matter. This interaction may give rise to permanent change in the material which can be observed, or to transient pulses of electrical charge or visible light. This chapter examines the principles of nuclear radiation detectors. It concludes with a review of the quantitative radiation measurements used to define radiation dose.

10.1 GENERAL PRINCIPLES OF RADIATION DETECTION

1. Radiation interacts with the materials of the detector and energy is deposited within it.

2. The deposited energy produces:
   - (i) electron-ion pairs in gases and plastics
   - (ii) electron-hole pairs in semiconductors and plastics
   - (iii) excitation of molecules and atoms
   - (iv) damage to the material.

3. Radiation is measured by:
   - (i) the collection of free charge in ionization chambers and semiconductor detectors
   - (ii) the observation of visible light emitted when molecules and atoms decay—scintillation counters
   - (iii) development of a latent image in photographic emulsions
     - — tracks of energetic particles
     - — general darkening by X-rays
   - (iv) the counting of damage tracks in plastics

Static detection. Particles striking photographic film cause '*tracks*' which can be enhanced by chemical development. The effective charge and the mass of the radiation can be estimated from the ionization density (dE/dx) along the track. Similar damage effects can occur in plastics, in rock samples, or in materials in which the energy can be stored. For example, in thermoluminescent dosimeters, radiation impinges on a phosphor which emits light when it is subsequently heated

275

under controlled conditions. The integrated light emission is a measure of the radiation dose to which the material has been exposed.

Dynamic detection.   The radiation causes ionization in a gas or a material with the creation of ion-electron or electron-hole pairs. These charges may be detected directly (in an ionization chamber, a proportional counter, Geiger counter or semiconductor detector) or be observed through light emitted when the electron-hole pairs recombine (a scintillation detector).

Charge Collection

The incident radiation generates electron-hole or electron-ion pairs between a set of electrodes. These charges separate under the influence of an electric field E_f and are drawn to the electrodes creating a current in an external circuit.

Typical collection times are set by the mobility of the charge carriers and can range from ns to µs. The current in the external circuit is shown in Fig. 10.1. For complete charge collection, all carriers must reach the electrodes without recombination.

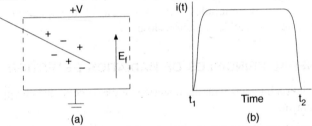

(a)   (b)

Fig. 10.1   (a) Electron-hole generation by radiation. (b) Current pulse over collection time $t_2 - t_1$.

Quantity of charge.   The energy W (eV) required to produce an electron-ion pair in a gas, or an electron-hole pair in a semiconductor is a property of the detector material. Thus an incident particle of energy E creates an average number of N charge pairs, where $N = E/W$. If all these pairs migrate to the electrodes the charge collected (C) is

$$Q = \int i(t)\,dt = qN = \frac{qE}{W}\ \text{Coulombs}$$

where q is the electronic charge (1.61×10^{-19} C).

If the electrode system is considered to be a capacitor of capacitance C (F), the charge collection generates a voltage pulse V of magnitude

$$V = \frac{Q}{C} = \frac{qE}{CW}\ \text{Volts}$$

In order to allow successive particles to be recorded by a measurable voltage step, the capacitor must be discharged. This is accomplished by a resistor R connected across the capacitor plates which discharges the capacitor with time constant RC (often chosen to be 100 µs). The detector circuit and resulting voltage pulse is shown in Fig. 10.2.

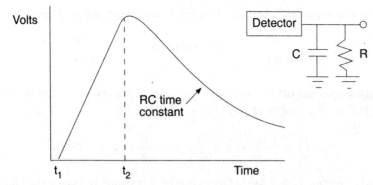

Fig. 10.2 Generation of a voltage pulse by the collection of pair charges on electrodes of capacitance $C(F)$. The inset shows the circuit used to measure the voltage.

Experimental Factors Determining Counting Rates

Statistical generation of charge. The interaction of radiation with matter is a statistical process and the number of pairs N produced by a succession of monoenergetic particles will vary. If W represents an average value for the pair creation energy this creates a range of pulse heights V for successive particles. The probability to produce a certain number of pairs can be described by Poisson statistics (page 11), where $N_0 = E/W$ is the mean value and the standard deviation σ is $\sqrt{N_0}$ or $\sqrt{E/W}$. The equivalent pulse height distribution has a mean value $V_0 = qE/WC$ with a standard deviation $\sqrt{qE/WC}$.

Full Width at Half Maximum (FWHM). In nuclear counting of large sample populations it is convenient to treat the probability distribution as a Gaussian. The standard deviation σ is then assessed through the FWHM (page 16) where FWHM $= 2.35\sigma$.

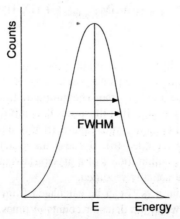

Fig. 10.3 Count distribution $N(E)$ in energy defined in terms of the FWHM. The relationship with the standard deviation σ is indicated.

For a monoenergetic particle of energy E, the voltage pulse height spectrum is

$$V_0 = \frac{qE}{WC} \text{ with a FWHM of } \Delta V = 2.35 \sqrt{\frac{qE}{WC}}$$

Fractional resolution R.   On an energy scale E this can be written in terms of the FWHM of the number of pairs produced $N = E/W$,

$$R = \frac{\Delta E}{E} = \frac{2.35\sqrt{N}}{N} = \frac{2.35}{\sqrt{N}} = 2.35\sqrt{\frac{W}{E}} \; ; \; \Delta E = 2.35\sqrt{EW}$$

For good resolution (small ΔE), a large value of N is required. Thus a good detector system requires a small pair production energy W.

Fano factor F.   Energy loss in detectors is not a completely statistical process due to details of the loss process, incomplete charge collection and other electronic effects. The energy resolution ΔE is often **smaller or 'better'** than predicted by the above equations. This is quantified empirically by the Fano Factor F.

$$F = \frac{\text{Observed variance in } N}{\text{Statistically expected variance in } N}$$

Since the variance $= \sigma^2$,

$$\text{FWHM}_{observed} = \sqrt{F}. \text{ FWHM}_{statistical}$$

On an energy scale

$$\Delta E_{observed} = 2.35 \sqrt{EWF}$$

$F < 1$ and can be as low as ~ 0.1 for a semiconductor or gas detector.

Fluctuations such as electronic noise, drift and fluctuations in the charge generation basic process contribute to the width of the distribution. If these effects are independent and statistical in nature, they add in quadrature (page 18).

$$(\text{FWHM})^2{}_{total} = (\text{FWHM})^2{}_{statistical} + (\text{FWHM})^2{}_{noise} + \ldots$$

Dead Time

In any detector system there is a minimum time that must separate two events in order that they are recorded as separate events. The RC time constant in Fig. 10.2 is one factor which influences this time. The minimum time interval between pulses is called the **dead time** τ. Figure 10.4 illustrates how this dead time has a significant impact on the measured counting rate. Figure 10.4(a) shows the actual number of pulses which should be recorded (6). Because the system does not respond to a second pulse if it occurs within time τ of a previous event, Fig. 10.4(b) shows that only four (4) pulses are actually recorded.

The **true** count rate n (s^{-1}) can be calculated from the **observed** count rate m (s^{-1}). In one second when the detector counts m times, the time the detector is not active is $m\tau$. The number of pulses which have been missed in this time is $nm\tau$. The true number of pulses is therefore $n = m + nm\tau$. Thus,

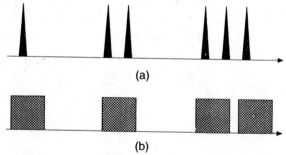

Fig. 10.4 Effect of dead time on the observed counting rate: (a) True event rate (6 pulses). (b) Measured count rate. The dead time is denoted by the shaded areas.

$$\text{True count rate (s}^{-1}) \; n = \frac{m}{1 - m\tau}$$

A **deadtime correction** must be made in all counting experiments.

Radiation events leading to a count in a detector occur within some type of source but have to interact with the detector in order to produce a recordable event. The size, geometry and possible loss of radiation through absorption will affect the number of counts recorded. This is defined by two types of efficiency factor.

Counting Efficiency

Intrinsic efficiency of a detector. For a **detector** there is a finite probability that the radiation will pass through the detector without interaction.

$$\varepsilon_{int} = \frac{\text{Number of pulses recorded}}{\text{Number of particles incident}}$$

= 100% for heavy charged particles
≤ 100% for electrons
< 100% for photons and neutrons

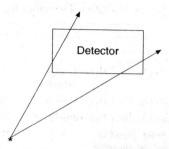

Fig. 10.5 The intrinsic efficiency defines how well a **detector** responds to radiation.

Absolute efficiency. For a source and a finite size of detector, there is a probability that emitted radiation will pass by the detector or be absorbed in the medium before it reaches the detector.

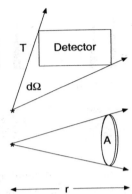

Fig. 10.6 The absolute efficiency defines the probability of interaction of radiation from a source with a detector.

$$\varepsilon_{abs} = \frac{\text{Number of pulses detected}}{\text{Number of particles emitted}}$$

$$= \varepsilon_{int} \frac{d\Omega}{4\pi} T = \varepsilon_{int} \frac{A}{4\pi r^2} T$$

$d\Omega$ = solid angle subtended by the detector

T = transmission coefficient in the medium between the source and detector.

Question (i) An air-filled gas detector requires 35.2 eV to create an ion pair. The detector capacitance is 100 pF and a 100 K discharge resistor is used. The detector is used to detect 5 MeV alpha particles. (a) Draw the output pulse as a function of voltage and time. (b) If the electronics requires that the pulse decay for a time of $3RC$ before a new event is recorded, what is the real event rate if the recorded count rate is 2×10^4/s. (c) Sketch the voltage pulse height spectrum for a Fano factor of (i) 1, (ii) 0.2.

10.2 TYPES OF RADIATION DETECTORS

Various types of radiation detectors are listed in Table 10.1.

Table 10.1 Types of Radiation Detectors for Charged Particles

Detector	Detecting medium	Radiation	Energy dispersive	Resolution at keV	%	E_0
Ionization chamber	Gas (Ar, He, CH_4)	Charged particles	Yes			
Proportional counter	Gas (Ar, He, CH_4)	Particles	Yes	0.7	12	5.9
Geiger-Mueller counter	Gas (Ar, He, CH_4)	Particles, X	No			
Scintillation counter	Solid, liquid or organic phosphor	β, γ	Yes	40	6	660
Semiconductor surface barrier	Si	X, α	Yes	12	0.25	4860
Lithium drifted semiconductor detectors	Si Ge	γ, X	Yes	0.15 1.8	2.6 0.1	5.9 1330

Gas Filled Detectors

The geometry of most gas detectors is that of a cylinder with a positive anode wire of small radius located at the centre of a negative cylinder. The electrical field in the vicinity of the anode is high.

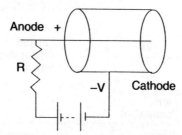

Fig. 10.7 Cylindrical geometry of a gas filled detector.

When charged particles pass through a gas, they ionize the gas to form ion-electron pairs. Photons are absorbed to generate energetic electrons which then ionize the gas. The energy absorbed to produce one ion-electron pair in a gas is typically 30–35 eV (Table 10.2).

Table 10.2 Energy Required to Create Ion-electron Pairs in a Gas. Data from S.C. Curran, "Proportional Counter Spectrometry"

Gas	Energy Dissipation per Ion-electron Pair (eV/ion pair)	
	Fast electrons	Alphas
Ar	27.0	25.9
He	32.5	31.7
H_2	38.0	37.0
N_2	35.8	36.0
Air	34.0	35.2
O_2	32.2	32.2
CH_4	30.2	29.0

The electric field produced by the applied voltage V separates the + and − free charges and collects the charge on the electrodes to measure the energy. The different regions of operation of gas filled detectors are shown in Fig. 10.8. The mechanisms which occur in each of these regions are described as follows:

Region 1: Low voltage (field). Ion-electron pairs recombine before reaching the anode. Increasing voltage separates the ions and a greater charge collection is observed. This is not a useful range of operation since the pulse voltage is a function of applied voltage.

Region 2: Ion saturation region. The electric field is strong enough to

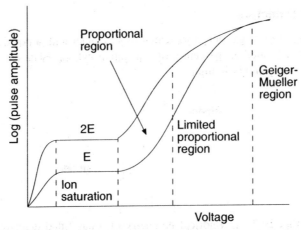

Fig. 10.8 The different voltage regions of operation of a gas filled detector. The pulse amplitude for events depositing different amounts of energy are shown.

collect all the ions produced and the charge collected is proportional to the energy absorbed. Ion-electron pairs are separated immediately on formation and all electrons are collected. The field is insufficient to cause further ionization. The magnitude of charge collected is low (for example for He, a 1 MeV pulse creates 3.07×10^4 pairs. The charge collected is 5.8×10^{-15} C). Such ionization chambers are often used as radiation monitors either by measuring the average current as a measure of background radiation flux, or by observing pulses and hence measuring the energy of the incident particles.

Region 3: Proportional region. Free electrons produced in the 'drift' volume acquire sufficient energy in the field to cause further ionization, principally as a "Townsend Avalanche' in the high field region near the fine wire anode. The multiplication factor $M \sim 10^2–10^5$ depends on the applied voltage, but is constant for a given field. The output signal is proportional to the incident energy. Proportional counters are used in X-ray spectroscopy.

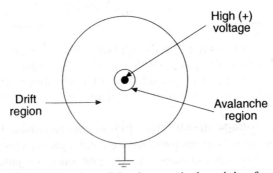

Fig. 10.9 The high field region of the detector is the origin of avalanches leading to charge multiplication and large signals. Where the output pulse voltage continues to be proportional to the incident energy.

At higher fields, the difference in the mobility of the very large numbers of positive ions and negative electrons present gives rise to space charge effects. These cause the pulse amplitude to have only a limited proportionality to the incident particle energy.

Region 4: Geiger-Mueller region.
At very high voltages, (2000 V), the ionization produced by a single incident particle induces an avalanche with causes further ionization and X-ray production. The pulse voltage becomes independent of the input energy.

The avalanche leads to a complete sheath of positive ion space charge around the anode wire which terminates the discharge. The positive ions drift radially outwards and another Geiger discharge can be generated only when the space charge has been sufficiently diffused. The Geiger counter therefore has a dead time during which the detector is insensitive to incident radiation.

When a positive ion strikes the cathode there is a small probability that a free electron will be produced. If this occurs a second Geiger pulse will be generated. To suppress this effect and to 'turn-off' the original discharge in a controlled manner, quenching agents are added to the detector gas. These are organic compounds such as ethyl alcohol or halogens such as Cl_2 or Br_2.

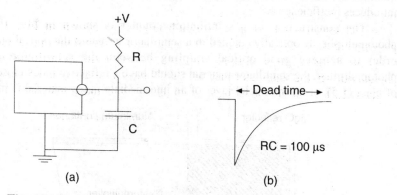

Fig. 10.10 Circuit configuration for a Geiger-Mueller counter.

Geiger-Mueller counters are operated with an applied voltage sufficient to generate reproducible avalanche behaviour, but below that which produces multiple avalanches. The counters provide no information on energy but excellent information on count rate. Geiger counters are the standard instrument used for radiation monitoring and surveying where the count rate alone is required. The output is often a series of audible pulses denoting the rate at which particles are incident on the counter.

Scintillation Counters

Organic and inorganic luminescent materials or scintillators respond to the absorption of energy by emitting light.

The mechanism normally involves the absorption of primary energy by the matrix of the material, internal energy transfer, and radiative emission at

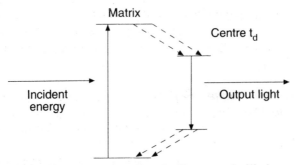

Fig. 10.11 Luminescence processes leading to scintillation counting.

'luminescence centres'. The emission process is characterized by a decay time t_d, which reflects the time over which the light is emitted.

Good **scintillator** materials have high light output for a given incident energy and the light output is proportional to the absorbed energy. The energy required to produce a photon is 3.6–5 eV. Energy conversion is rapid, the materials are transparent to their own emission, and there is a low probability for energy conversion to non-radiative thermal processes. Long-term light storage or 'phosphorescence' is also a minimum. However the measurement of light within the overall scintillation process introduces inefficiencies.

The construction of a scintillation counter is shown in Fig. 10.12. A photomultiplier is optically coupled to a scintillator to record the optical output. In order to achieve good optical coupling between the scintillator and the photomultiplier, the scintillator material should have a refractive index close to that of glass (1.5), or an interfacial layer of an intermediate index material is included.

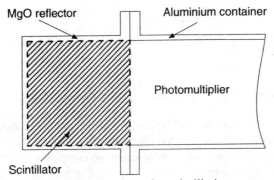

Fig. 10.12 Construction of a scintillation counter.

The magnitude of the electronic signal is determined by the radiation energy deposited in the scintillator, the luminescence conversion process, and the transfer of light to the photomultiplier, and electron generation within the potomultiplier. Interpretation of the recorded spectra must take these stages into account.

Scintillator Materials

Scintillator materials that are currently in use include— organic scintillators such as single crystal anthracene or stilbene, liquids in which other organics are dissolved, or plastics. The luminescence process characteristic of organic compounds takes place within individual organic molecules, emits in the blue region of the spectrum (400 nm), and is very fast (t_d = 5–30 ns). The refractive index of organic materials (1.3–1.6) is compatible with that of glass.

Single crystals have been used as **inorganic scintillators**, of which sodium iodide doped with 0.1% thallium (NaI(Tl)) is the most efficient. 4.69 eV is required to produce a photoelectron. NaI(Tl) has an efficiency of 4 times anthracene which is the best organic scintillator. In inorganic materials, the luminescence centre is usually associated with an impurity or defect in the host lattice. The decay time is considerably longer than for organic materials (230 ns for NaI(Tl)) and the emission spectrum can occur across the visible spectrum. The refractive index is generally higher than that of glass (1.85 for NaI(Tl)). The advantage is that the average atomic number Z of the inorganic matrix is high so that γ-ray absorption is enhanced. Bismuth germanate $Bi_4Ge_3O_{12}$ has an average value of $Z \approx 25$ and is an effective γ-ray scintillator for this reason even though its conversion efficiency is less than that of NaI(Tl).

Energy Transfer to Scintillator Materials

The luminescence process is excited most effectively by energy transfer from excited electrons. Thus the effect of other radiation is related to how it produces such electrons—by direct interaction, Compton scattering, knock-on of protons by elastic collisions and their subsequent interaction with the lattice. Excitation densities which are very large and highly localized tend to inhibit or destroy luminescence by radiation damage or non-radiative thermal de-excitation. Processes related to energy loss in the two types of scintillators are reviewed below.

Organic scintillators (particularly plastics).

    (i) Fast electrons —direct excitation of atomic electrons with low dE/dx, but high luminescence efficiency.

    (ii) γ-rays — recoil electrons produced by Compton scattering. This is favoured by the primary constituents of plastic (H, C, N, O).

    (iii) Protons and α-particles — higher dE/dx than electrons, but a higher light loss by non-luminescent thermal de-excitation. Lower luminescence efficiency.

    (iv) Neutrons — knock-on of protons, then an energy loss like protons.

In organic scintillators the shape of the light pulse as a function of time can provide information about the incident radiation. Thus α-particles tend to induce pulses which have a slower decay than γ-rays.

Pulse shape discrimination can be used to discriminate neutrons from γ-rays by measurement of the ratio

$$\frac{\text{Light in slow component}}{\text{Light in fast component}}$$

The response of an organic scintillator to γ-rays is primarily through Compton Scattering, hence the intrinsic efficiency η for a scintillator block of thickness L

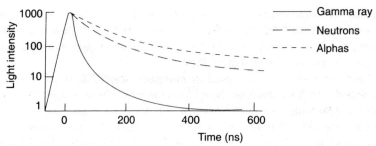

Fig. 10.13 Pulse shapes associated with excitation of an organic scintillator by α-particles, neutrons and γ-rays.

and linear absorption coefficient μ is determined by the Compton scattering cross-section. Some fraction of the scattered photons are lost from the process with a variable amount of energy being deposited in the scintillator. In the following equations A is a constant representing the geometrical factors in the experiment.

$$I_0 \;\boxed{}^{\,L}\; I = I_0 e^{-\mu L} \qquad \eta = A\,\frac{\sigma_{cs}}{\sigma_{cs} + \sigma_{pe} + \sigma_{pp}}\,(1 - e^{-\mu L})$$

The neutron response of a scintillator result from collisions between incident neutrons and protons in the material. The energy of the protons is then transferred to the scintillator. The response is related to the total energy E_n absorbed in the detector (p. 271).

$$I_0 \;\boxed{}^{\,L}\; I = I_0 e^{-\Sigma L} \qquad \eta = AE_n\,(1 - e^{-\Sigma L})$$

Inorganic scintillators. Because of the large average value of Z, the response of an inorganic scintillator to γ-rays reflects the photoelectric effect, Compton scattering and pair production at $E_\gamma - 1.02$ MeV if $E_\gamma > 1.02$ MeV. The degree to which Compton scattered photons are reabsorbed simplifies the results in large volume inorganic detectors. A response due to back-scattering of γ-rays from the surroundings may also be observed.

The response of a NaI(Tl) detector to a ^{137}Cs source which emits a 663 keV γ-ray is shown in Fig. 10.14.

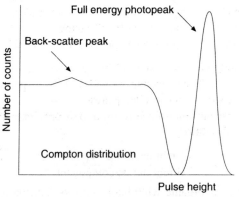

Fig. 10.14 NaI(Tl) response to a ^{137}Cs source.

The intrinsic efficiency for counts recorded in the photopeak is,

$$\eta = A \, \frac{\sigma_{pe}}{\sigma_{cs} + \sigma_{pe} + \sigma_{pp}} \, (1 - e^{-\mu L})$$

For NaI at 663 keV, in units of ($\times 10^{-24}$ cm^2) $\sigma_{pe} = 1.77$ and $\sigma_T = 13.1$. Reabsorption of scattered Compton photons in large detectors increases the relative size of the photopeak. The principal disadvantage of scintillation detectors is the large energy/photon required and the fact that detection involves a series of statistical processes. This means that the resolution is poor at 2.5-20%.

Question (ii)   What is the fractional resolution R of the photopeak in a scintillation counter for a 0.5 MeV γ-ray. The photon energy for light emission from the scintillator is 4.6 eV. The photopeak energy conversion efficiency for the scintillator is 13%, 56% of the emitted photons reach the photocathode. The quantum efficiency of the photocathode is 20%.

Semiconductor Detectors

An ideal detector would be a solid block of material which normally contains no free charges, but in which incoming radiation could create electron hole pairs in proportion to the energy of the radiation. If such charges could be collected and measured rapidly, the resulting charge or current flow would be an accurate measure of the radiation energy.

Early attempts at this ideal detector utilized charge detection in diamond, but although charges could be created, the problems of measuring these charges in an external circuit by attaching electrodes to the detector proved insurmountable. However, as discussed in Appendix 3, the depletion region of a semiconductor *pn*-junction conforms almost exactly to the requirements for the ideal detector. In thermal equilibrium, virtually no free charge exists, the *p* and *n* type regions constitute two good electrodes, and the collection time is set by the mobility of charge carriers in Ge or Si. Finally, the energy required to produce an electron/hole pair is low (3.62 eV/hole-electron pair in Si, and 2.96 eV in Ge). Thus the intrinsic fractional resolution (FWHM) of a silicon detector for a 1 MeV charged particle is 0.2 %. The development of high sensitivity, high resolution, large volume γ-ray detectors based on this concept has revolutionized many aspects of nuclear physics.

The n⁺-p or p⁺-n Junction.   Junctions are fabricated with one side doped with N_A *p*-type impurities and the other doped with N_D *n*-type impurities.

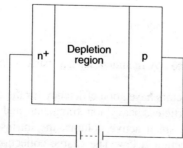

Fig. 10.15   An *n⁺p*-junction under reverse bias.

In general, one side is more heavily doped and the depletion region lies primarily in one type of material, and under reverse bias V is of width x_d,

$$x_d = \left[\frac{2V\varepsilon}{q} \frac{N_A + N_D}{N_A N_D} \right]^{1/2}$$

where ε is the permittivity of silicon, and q is the electronic charge.

For $N_D = 10^{15}$ cm^{-3}, $N_A = 10^{18}$ cm^{-3}, with $V = 1000$ V, the width of the depletion region is $x_d = 31$ µm. This width is greater than the range of charged particles in silicon, and such detectors can be made in areas of 1–5 cm^2. In principle, the collection time t_d is the time for hole (the slowest particle) to migrate across the depletion width

$$t_d = x_d/\mu_h E = x_d^2/\mu_h V = 0.2 \text{ ns}$$

where $\mu_h = 1700$ cm^2/V-s is the hole mobility in silicon. Such junctions are employed as charged particle detectors. The semiconductor physics of the p–n junction is reviewed in Appendix 3 (see Ref. 9.1, Chapters 11, 12 and 13). Methods of fabricating a junction range from diffusion of impurities, to ion implantation, to the adsorption of chemical species.

Silicon surface barrier detectors.

A thin slice of n-type Si has a contact made to its rear face and is exposed to controlled atmosphere. Surface absorption causes a p-type layer to form at the surface. A thin gold layer (~40 µg/cm^2 or 2 µm thick) is evaporated on to the surface as an ohmic contact and the slice is mounted.

Surface barrier detectors must always be operated under reverse bias, and are often kept under vacuum. The (reverse) polarity of the voltage applied to a detector should be carefully checked and the output should be monitored when the voltage is first applied. If the noise increases, stop!

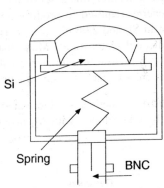

Si

Spring   BNC

Fig. 10.16   A mounted surface barrier detector.

Surface barrier detectors have good efficiency for the detection of α-particles, protons, deuterons, α-particles, heavy ion fragments and low energy electrons. However, because of the small active volume, the intrinsic efficiency for γ-ray, X-ray and β-particle detection is low. The charge collection time is around 10 ns and the resolution (FWHM) for 1 MeV α-particles is 12–20 keV or (1–2%).

Large volume Ge(Li) and Si(Li) gamma ray detectors. Large volume detectors with a depletion depth of up to 15 mm are made by the controlled diffusion of compensating impurities. These are called lithium-drifted Ge or Si detectors. Li atoms in Ge or Si are highly mobile at elevated temperatures and act as donor (*n*-type) impurities.

A layer of Li is evaporated on to the surface of a *p*-type Ge single crystal. If a voltage is applied, at about 300°C, the Li⁺ ions diffuse into the material and distribute themselves in such a fashion as to minimize the electric field in all parts of the sample.

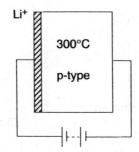

Fig. 10.17 Fabrication of large volume detectors using Li drift.

This distribution corresponds to exact compensation of the original *p*-type material, and on cooling to 77K, the material acts as an intrinsic semiconductor over virtually its entire volume. The *n*-type layer, and residual *p*-type material act as good contacts to the intrinsic volume. Planar, cylindrical or well type configurations are shown in Fig. 10.18.

The detector is maintained at 77K to minimize the thermal generation current and to inhibit the back-diffusion of Li⁺ ions from the silicon or germanium. A vacuum is required to eliminate the adsorption of surface impurities such as water vapour. The detector and its preamplifier are mounted on a copper cold finger (Fig. 4.16) with a zeolite gas absorber (Fig. 6.10). The detector is reverse-biased for use.

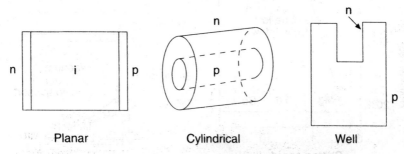

Fig. 10.18 Lithium drifted detectors having various configurations.

The type of housing is illustrated in Fig. 10.19(a). Figure 10.19(b) compares the resolution possible with a Ge(Li) detector to that of a NaI(Tl) scintillator.

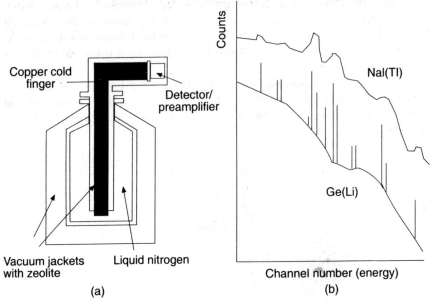

Fig. 10.19 (a) Housing for a semiconductor detector. (b) Resolution of a Ge(Li) detector compared to that for a NaI(Tl) scintillator.

The energy required to create an electron-hole pair in Ge is 2.96 eV. For γ-rays the intrinsic photopeak efficiency is η (page 287). At 663 keV for Ge $\sigma_{pe} = 0.26$ and $\sigma_T = 8.95$ cm². Because of the relatively low Z, large volume detectors are necessary to absorb the Compton scattered photon. The photopeak resolution of a germanium lithium drifted detector for a 0.663 MeV γ-ray is

$$\Delta E/E = 2.35 \sqrt{F/N} = 2.35\sqrt{2.96F/E} = 0.2\% = 1\text{--}2 \text{ keV}$$

with a Fano factor of 0.1. The improvement in γ-ray resolution achieved over scintillation counters is shown in Fig. 10.19(b).

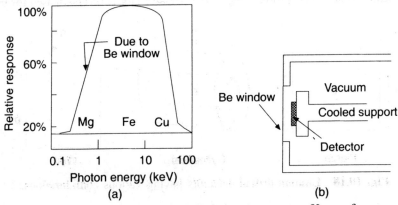

Fig. 10.20 Silicon X-ray detectors. (a) Relative response to X-rays from various elements. (b) Experimental configuration in which the beryllium window absorbs radiation from all elements below Mg.

Intrinsic Ge can be fabricated directly into large volume detectors by making ion-implanted electrodes. This eliminates the need for Li-drifting. These high purity germanium detectors are stored at room temperature but are used at 77 K. The larger atomic number of Ge compared with Si means that it is a more efficient γ-ray detector.

Si detectors for X-rays. At low energy (<40 keV) the full energy photoelectron peak efficiency for silicon is almost as high as for germanium. Such detectors are used for X-ray fluorescence experiments using a beryllium window (Fig. 10.20).

Neutron Detectors

The lack of charge of a neutron means that indirect means must be used to detect neutrons. These usually involve the results of a classical collision with protons, or by observing the end product of a nuclear reaction.

BF₃ counters. Thermal neutrons undergo the following reactions:

$$n + {}^{10}B = {}^{7}Li + \alpha \qquad Q = 2.792 \text{ MeV}$$

$$n + {}^{10}B = {}^{7}Li^{*} + \alpha' \qquad Q = 2.310 \text{ MeV}$$

^{10}B is 20% abundant. The particle recoils with roughly 1.5–1.9 MeV producing ionization in the BF_3 gas. This can be detected in a conventional proportional counter.

³He counters (thermal neutrons)

$$n + {}^{3}He = p + {}^{3}H$$

Both p and ^{3}H recoil to produce ionization.

Scattering on hydrogen (fast neutrons)

$$n + p = n' + p'$$

The excited proton is produced and detected in a plastic scintillator.

Personnel Monitors

On-going personnel monitoring of total radiation dose is essential for workers exposed to radiation. Two major procedures are current:

Calibrated film badges. These monitor accumulated dosages by the darkening of a photographic film. They are processed through a national program for radiation protection. Separate sections of the badge are appropriately shielded to allow separate estimates of α, n and charged particle exposures.

Fig. 10.21 Pocket electroscope for personnel monitoring.

Electroscope monitors. A pocket gold leaf electroscope is charged and the rate of discharge is a measure of ionization produced within the chamber.

10.3   RADIATION DOSE

The units of radiation dose reflect the nature of the interaction between radiation and organic matter and the different effects of a specific radiation on biological tissue.

Radiation causes ionization. This can not only break molecular bonds, but also lead to the production of highly reactive chemicals which may have deleterious effects in tissue. While the primary measure of a radiation field is the ionization it produces, the ionization density or specific energy loss (dE/dx) is also important. An important reference to radiation monitoring is (Ref. 10.2). International agreements are made through the International Commission on Radiation Units and Measurement (ICRU).

Units of Radiation Dose: γ- and X-rays

The **exposure** to a radiation field of γ- or X-rays is measured by the **number of ions produced per unit mass of dry air.**

$$X = \frac{\Delta q}{\Delta m} \; C/kg$$

This is measured as the sum of all the electrical charge of one sign Δq that is produced in a volume of air of mass Δm, when all the charge liberated in that volume is stopped in that volume. The S.I. unit of exposure is Coulomb/kg. This is called an SI. Non-SI units such as the Roentgen, the Rad and the rem which remain in common use are reviewed in Appendix 4.6.

Exposure rate $\dot{X}$ It is the ionization rate in SI/time (SI/s, μSI/hr, mSI/a)

Consider an exposure rate of 1 μSI/hr of γ-rays.
How many ions are produced per cm^3 of air per hour?

1 μSI/hr = 10^{-6} C/kg/hr; density of air 1.293×10^{-6} kg/cm^3
      = 1.293×10^{-12} C/cm^3 of air/hr
For γ-rays, most ions would be singly charged with a
charge of $q = 1.6 \times 10^{-19}$ C
1 μSI/hr = 8.08×10^{6} ions/cm^3/hr

Absorbed dose *D.*   The **energy absorbed per unit mass** in a body is called the Absorbed Dose. The units for γ-ray absorption are

$$\text{Gray (Gy)} = 1 \text{ Joule/kg (J/kg)}$$

Absorbed dose rate $\dot{D} = \dfrac{dD}{dt}$. It is the rate at which an absorbed dose is received. The units are Dose/time. A common unit is the yearly dose or dose/annum (Gy/a).

Question (iii) For a weakly absorbing radiation of intensity $I/cm^2/s$ in a thin absorber of thickness Δx the energy absorbed is $\mu I \Delta x$, where μ is the linear attenuation coefficient (p 81). What flux (in cm^2/s) of 663 keV γ-rays from a ^{137}Cs source is required to give a dose of 1Gy within 60 min in a thin slab of biological tissue of density 1300 kg/m^3. The linear attenuation (absorption) coefficient for this radiation in tissue is 0.41 m^{-1}.

For γ-rays the biological damage to **tissue** is a function of the absorbed energy. However, other radiations have different (dE/dx) and therefore cause different degrees of biological damage. To take account of this, the dose is weighted by a **Quality Factor** Q to give the **dose equivalent** in units of **Sievert (Sv)**.

Dose Equivalent H = Dose × Quality Factor

Dose Equivalent Rate $\dot{H} = \dot{D} \times Q$ (Sv/hr or mSv/a, etc.)

The value of Q reflects the type of radiation and its specific energy loss over a given energy range. Representative values are shown in Table 10.3 (see Refs. 10.6 and 10.9).

Table 10.3 Columns 1 and 2—Specific Energy Loss and Quality Factor Q for radiation Damage in biological tissue; Columns 3 and 4—Q for Specific Radiations and Energy Ranges

dE/dx (keV/m)	Q	Type of Radiation	Q
< 0.35	1	X-rays and γ-rays	1
7	2	β-rays E_{max} < 30 keV	1.7
23	5	β-rays E_{max} > 30 keV	1.0
53	10	α-particles < 10 MeV	20
>175	20	Heavy ion recoils	20
		Neutrons: thermal < 10 keV	5
		fast 10–100 keV	20
		100 keV–2 MeV	10
		> 2 MeV	5

Question (iv) If the 663 keV γ-ray source in the previous question is replaced with an α-particle source of the same energy and flux, compare (a) the dose and (b) the dose equivalent received by the tissue. The range of α-particles in the tissue of density 1300 kg/m^3 is about 4 μm.

Dose Rates from Natural and Man-made Sources

Radiation doses are received from naturally occurring sources of radiation: cosmic rays, radionuclides produced by cosmic rays in the earth's atmosphere, and radionuclides originating in the earth's crust. The dose an individual receives varies with location and lifestyle. A person who flies receives additional radiation from cosmic rays (the dose from a 10 h flight is 0.02 mSv), while locations near to specific radionuclides may increase dose rates. A convenient unit is the mSv/a.

Table 10.4 shows dose rates from natural background radiation at various locations. The variations reflect the prevalence of radionuclides such as radon and its daughter products, potassium-40 (^{40}K), as well as atmospheric ^{14}C. For example, the higher value for Southern Ontario reflects the geological nature of the rocks of the Canadian Shield. Table 10.4 also compares equivalent dose rates from man-made sources (see Refs. 10.2 and 10.9).

Table 10.4 Dose Rates and Contributions to Dose Rate from Various Sources. Data from *Environmental Impact Statement on the Concept for Disposal of Canada's Nuclear Fuel Waste*, Atomic Energy of Canada Ltd., Lamarsh

Equivalent dose rate (mSv/a)	Location	Equivalent dose rate (mSv/a)	Source
Natural sources		**Natural sources** (whole populations)	
0.4 to 5	World (local values)	Cosmic rays	0.28
2.4	World (average)	terrestrial γ	0.26
2.6	Canada (average)	Heavy elements	0.08
3.0	Southern Ontario	^{40}K	0.19
3.6	United States	^{14}C	0.007
Man-made sources	Typical dose	**Man-made sources**	Typical dose
		X-ray diagnosis	0.39
0.03	Dental X-ray	Nuclear medicine	0.14
0.10	Chest X-ray	Nuclear power	0.0005
		Occupational	0.008

Effects of Radiation: Permissible Limits

The effects of radiation are not well understood. They depend to some extent on the nature of the exposure—whether it is an acute short-term exposure in a nuclear explosion or in radiation therapy, or it is low-level but continuous. General categories may be defined as given in Table 10.5.

Table 10.5 General Categories of Exposure to Radiation and Permitted Limits

Acute:	Large dose, short time, not repeated	0.1–0.3 Sv
Medium:	Intermediate dose, short time, repeated	0.01–0.1 Sv
Low:	Low level, continuous	0.01–0.1 mSv

The current Canadian and American Atomic Energy Control Board limit on whole body dose rate for atomic radiation workers is 50 mSv/a. The United Kingston has a limit of 15 mSv/a. The current AECB limit on whole body dose rate for the general public is 5 mSv/a. The lethal whole body dose from penetrating ionizing radiation (50% probability of death in 30 days) is 2.5–3.0 Gy.

The maximum permissible limits for radiation dose (MPD) are set out by various national control bodies for various occupations and situations. The regulations take into account the type of exposure and the occupation of the individual, and whether specific organs have been affected (Table 10.6 (Ref. 10.9)). For example, an intake of radioactive iodine is automatically concentrated by the body into the thyroid.

Table 10.6 Regulations for Maximum Permissible Dose (MPD)

Type of Exposure	Maximum dose mSv/a
Occupational exposure	
Annual limit	50
Long-term accumulation to N years of age	5(N-18)
Skin	150
Hands	500
Other organs	150
Fertile women	5
Public or occasionally exposed	
Students	1
Individuals	5
Emergency—life saving	
Whole body	1000
Extremities	2000

The medical effects of radiation are well established. In the case of acute or short-term doses, the observed clinical effects are the following (Table 10.7).

Table 10.7 Clinical Effects of Acute Radiation Dose

Acute dose (Sv)	Probable clinical effects
0–0.5	No observable short-term effects
0.5–1.0	Slight blood change
1–2	Moderate blood change, vomiting, fatigue, recovery in a few weeks
2–6	Vomiting, severe blood change, haemorrhage and infection, loss of hair
6–10	Recovery from 1 month to 1 year at low dose end, 20% of survival at high end. Vomiting, loss of hair, 80–100% likely death within two months

Question (v) (a) Estimate the radiation dose equivalent that a 80 kg worker would receive in a daily period of 4 hours at a distance of 1.5 m from an unshielded 200 µCi ^{60}Co source if it emits one 1.33 MeV and one 1.17 MeV γ-ray for each decay. Treat the source as a point emitter. Assume that the worker has approximate area 0.37 m^2 and that 30% of any impinging radiation is absorbed by his body. (b) If he is working in North America on a 5 day work week. Is he under any hazard in this situation?

Calculation of Exposure and Dose

In order to compute the dose received by a body it is necessary to determine the energy transferred to the material.

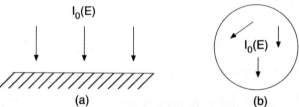

Fig. 10.22   Two modes for radiation damage: (a) External radiation incident on a body. (b) Internal or ingested sources.

External sources: γ or X-ray.   The absorption of γ-rays or X-rays of energy E and flux I_0 photons/cm^2/s falling on the surface of a medium is determined by the absorption coefficient $\mu_a(E)$. The amount of radiation that is absorbed by a body is determined by its size L relative to the absorption length for the radiation $1/\mu_a$.

For γ-rays, generally $L < 1/\mu_a$

Since $I(x) = I(0)\ e^{-\mu x}$; $I_{\text{absorbed}} = \mu\ I(x)\ dx$; If x is small

$$\boxed{I(x) \sim I(0)}$$

The exciting intensity changes little within the absorber. The energy absorbed is $(\mu I L)$ (see Problem 10.3).

For X-rays, or for large bodies where $L \geq 1/\mu_a$

$$I_{\text{absorbed}} = \int_0^L I(0)\ e^{-\mu x}\ dx = \frac{I(0)}{\mu_a}\ (1 - e^{-\mu L});$$

$$= \frac{I(0)}{\mu_a}\ \text{if}\ L \to \infty$$

The absorbed energy is a maximum independent of the size of the body.

Exposure rate for γ-rays.   The exposure rate $\dot{X}$ for $I(0)$ photons/cm^2-s of γ-rays of energy E is the ionization created per second per kg of air (C/kg/s). Because the absorption of γ-rays is small, the energy is absorbed in a 'thin' slab of thickness Δx and area 1 cm^2. $\mu_a(E)^{\text{air}}$ is the linear absorption coefficient for γ-rays of energy E in air in cm^{-1}. $q = 1.6 \times 10^{-19}$ C is the electronic charge.

$$\dot{X} = q\ \frac{I(0)\ E}{W_{\text{air}}}\ \frac{\mu_a(E)^{\text{air}}\ \Delta x}{\rho \Delta x}\ \text{SI/s}$$

If E is in MeV and the ionization energy for air $W_{air} = 34$ eV/ion-pair, this expression can be written in terms of the mass absorption coefficient for air $(\mu_a(E)/\rho)^{air}$ cm²/g (Appendix 4.7).

$$\dot{X} = 4.71 \times 10^{-12} \, I(0) \, E \left(\frac{\mu_a(E)}{\rho} \right)^{air} \text{SI/s}$$

Dose rate in air. The dose rate $\dot{D}$ from γ-rays in air is the energy (J/s) absorbed by a unit area of a 'thin' slab of thickness Δx in one second divided by the mass of the slab $(\rho \, \Delta x)$. The equation in terms of the mass absorption coefficient $(\mu_a(E)/\rho)^{air}$ (Appendix 4.7), for photon energy E in Joules, can be written in the following manner.

$$\dot{D} = \frac{\text{Absorbed energy (J/s)}}{\text{Mass (kg)}} = \frac{EI\mu_a(E)^{air} \, \Delta x}{\rho \, \Delta x} = EI \left(\frac{\mu_a(E)}{\rho} \right)^{air} \text{Gy/s}$$

Dose rate for air or media for which $L \geq 1/\mu_a$ For γ-rays in larger volumes of air, for X-rays, or for other media in which the absorption of γ-rays is larger, the total rate of energy absorption in the mass contained within the relevant volume must be calculated.

Dose rate for radiation in tissue. In radiation protection the important quantity is the dose rate $\dot{D}$ to biological tissue. This is the rate of energy absorption per unit mass of tissue. $(\mu_a(E)/\rho)^{tissue}$ cm²/g is the mass absorption coefficient for photons of energy E in tissue—Appendix 4.7. For γ-rays the quality factor $Q = 1$ and the dose rate $\dot{D}$ is the same as the dose equivalent rate $\dot{H}$. For other radiation, the appropriate quality factor must be included. E is measured in MeV. For biological tissue a convenient unit in magnitude is mSv and time 1 hr = 3600 s .

$$\dot{H} = 5.76 \times 10^{-4} \, Q \, I(0) \, E \left(\frac{\mu_a(E)}{\rho} \right)^{tissue} \text{mSv/hr}$$

Dose Rates for Charged Particles

Heavy charged particles including β-rays have a very short range in both air and tissue, so that radiation is generally not a problem *unless the sources are internal (ingested) or are deposited on the skin*. In this case, severe local damage may result because of the high dE/dx. The action of the body can also concentrate a source within a particular organ so that severe damage results. Calculations for this purpose take into account the activity $C(t)$ of the source in Bequerel and the mass of the organ in grams affected by the radiation.

$$\text{Dose equivalent rate } \dot{H} = 5.76 \times 10^{-4} \, \frac{C(t)EQ}{M} \text{ mSv/hr}$$

where
  $C(t)$ = Source strength in the organ (Bq) M = Mass of organ in g
  E = Average energy of charged particles Q = Quality factor
      emitted in MeV

Many sources decay with β- and γ-emission as well as charged particles. If a source is swallowed or inhaled all of these decay products will cause radiation damage. A parameter ζ, the effective energy equivalent is used to approximate the overall effect of a given source on an organ. This can be estimated from,

$$\zeta = \sum_i f_i F_i Q_i E_i$$

where

f_i is the fraction of radiation emitted of type i
F_i is the fraction retained within the organ
Q_i is the quality factor for the ith component
E_i is the energy in MeV

Representative values for ζ and Q for important sources are shown in Table 10.8.

$$\text{Dose equivalent rate to an organ of mass } M \text{ kg } \dot{H} = \frac{5.76 \times 10^{-4} \, C(t) \, \zeta}{M} \text{ mSv/hr}$$

Several factors must be considered when dealing with the effects of ingested sources:

(a) the radioactivity of the source decays with decay constant λ.

(b) the body may take up only a fraction q of the source material which has been ingested.

(c) the body can excrete sources. To a good approximation for many sources the excretion can be described as an exponential decay with a time constant λ_b. To allow for biological excretion, an effective total decay constant λ_{eff} is defined. This implies that the radioactivity decays **faster** than expected from the physical half-life of the radionuclide.

$$\lambda_{\text{eff}} = \lambda + \lambda_b$$

The total dose **after time T** to an organ from a source of initial strength $C(0)$ which decays as $C(t) = C(0) \, e^{-\lambda t}$ is

$$H = \int_0^T \dot{H} \, e^{-\lambda_{\text{eff}} t} \, dt = \frac{1.38 \times 10^{-5} \, C(0) \, \zeta Q \, (1 - e^{-\lambda T})}{M \, \lambda_{\text{eff}}} \text{ Sv}$$

where λ_{eff} is measured in day^{-1} and $\lambda_{\text{eff}} = 0.693/T_{1/2}$. This information is given in Table 10.8.

Question (vi)   Three days after an accidental exposure by inhalation to ^{131}I a radiation worker has a thyroid burden of 0.10 µCi of ^{131}I. Calculate the total radiation dose to his thyroid assuming that iodine is not excreted from the organ. The mass of the thyroid is 20 g. See Table 10.8 for the properties of ^{131}I.

Shielding of Radioactive Sources

When shipping and storing radioactive sources they must be shielded to reduce the strength of the surrounding radiation field. Such sources may be treated as point sources of radiation.

Table 10.8 Physical and Biological Data for Radionuclides

Nucleus	Name	Half-life (days) Radioactivity	Biological	Organ	ζ MeV	Q ingestion	Q inhalation
^{3}H	Tritium	4500	11.9	Total body	0.01	1	1
^{14}C	Carbon-14	210000	10	Total body	0.054	1	0.75
		0		Bone	0.27	1	0.02
^{24}Na	Sodium-24	0.63	11	Total body	2.7	1	0.02
^{60}Co	Cobalt-60	1900	9.5	Total body	1.5	0.3	0.4
^{90}Sr	Strontium-90	11000		Total body	1.1	0.3	0.4
^{131}I and ^{131}Xe		8.04	138	Total body Thyroid	0.44	1	0.75
^{137}Cs and ^{137}Ba		11000	70	Total body	0.59	1	0.75
			140	Bone	1.4	0.04	0.03
^{236}Ra and daughters		590000		Bone	110	0.04	0.03
^{235}U	Uranium-235		300	Bone	230	0.000011	0.028
^{239}Pu	Plutonium-239	890000		Bone	270	0.000024	0.2
		0					
^{87}Kr	Krypton-87	0.053		Submersion†	2.8		
^{41}Ar	Argon-41	0.076		Submersion†	1.8		

†Submersion implies the value for ζ for a person surrounded by a cloud of gas.

At distance R from a source of intensity $I(E)$ the γ flux/unit area of photon energy E is

$$\frac{I(E)}{4\pi R^2} \quad \text{where } I(E) \text{ is in Bq (1 decay/s) or Ci } (3.7 \times 10^{10} \text{ decay/s})$$

The exposure $\dot X$ and dose equivalent rate $\dot H$ is given by

$$\dot X(R) = \frac{4.71 \times 10^{-12}}{4\pi R^2} \sum_i I_i\,(E_i)\, E_i \left(\frac{\mu_a(E_i)}{\rho}\right)^{\text{air}} \text{SI/s}$$

$$\dot H(R) = \frac{5.76 \times 10^{-4}}{4\pi R^2} \sum_i I_i\,(E_i)\, Q_i\, E_i \left(\frac{\mu_a(E_i)}{\rho}\right)^{\text{air}} \text{mSv/hr}$$

where the summation takes into account i γ-rays of source strength I_i and energy E_i.

The strength of the source is now reduced by surrounding it with a shield of material of thickness t and linear attenuation coefficient $\mu(E)$. However, this only takes into account the γ-rays that pass through the shielding without interaction. There are γ-rays of lower energy due to Compton scattering inside the shielding, and since after the scattering event they may proceed in a direction which differs from the incident direction, some of these scattered γ-rays may reach the region of interest to produce radiation damage (see page 266). To take account of these effects, the **build-up factor** $B(\mu(E_i),t) > 1$ defined on page 266 and listed in Appendix 4.5 is determined for a γ-ray of energy E_i that has traversed an absorber of thickness t and linear attenuation coefficient $\mu(E_i)$. This factor multiplies the expressions given above.

$$\dot{X}_s(R') = \dot{X}(R)\, e^{-\mu(E)t}\, B(\mu,\, t)$$

$$\dot{H}_s(R') = \dot{H}(R)\, e^{-\mu s(E)t}\, B(\mu,\, t)$$

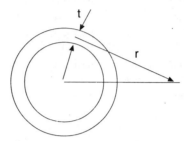

Fig. 10.23 Enhanced radiation exposure within a spherical shell due to build-up.

REFERENCES AND FURTHER INFORMATION

1. Glenn F. Knoll, *Radiation Detection and Measurement,* 2nd ed., John Wiley and Sons Ltd, New York, 1989.

2. J.R. Lamarsh, *Introduction to Nuclear Engineering,* 2nd ed., Addison-Wesley, Reading (Mass.), 1989, Chapter 9.

3. *Instruments and Systems for Nuclear Spectroscopy,* EE&G Ortec Nuclear Instruments, & G Ortec Inc., 100 Midland Road, Oak Ridge, Tennessee 37830-9912 USA

4. Canberra Nuclear Inc., *Research Parkway,* Meriden, Connecticut 06450 USA.

5. *Source for scintillators and scintillation counters,* Bicron Corporation, 12345 Kinsman Rd, Newbury, Ohio 44065-9677 USA.

6. Atomic Energy Control Act (1974): *Canada Gazette, Part II,* Vol. 108, 1783, Amendments (1978): Canada Gazette, Part II, Vol. 112, 406.

7. *Radioisotopes and Transportation Division:* Atomic Energy Control Board, P.O. Box 1046, Ottawa, K1P 5S9 Canada.

8. G. Margaritondo, *Introduction to Synchrotron Radiation,* Oxford Press, New York, 1988.

9. Physical Review D: Particles and Fields: *Review of Particle Physics 54,* 150 (1996) (Section 25: Radioactivity and Radiation Protection).

10. ICRP Publication 60, 1990, *Recommendations of the International Commission on Radiological Protection,* Pergamon Press, Oxford, 1991.

ANSWERS TO QUESTIONS

(i) (a) Pulse shape: Amplitude 227 μV

Decay $RC = 10$ μs

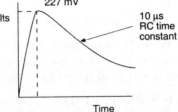

(b) Real counting rate: 5×10^4/s

(c) Gaussion distribution with mean value 227 μV resolution (i) FWHM = 35.6 μV (ii) FWHM = 15.9 μV

(ii) 1582 counts should be recorded: fractional resolution 5.9%

(iii) 8.25×10^8 γ-rays/cm²/s

(iv) (a) 60.8 Gy   (b) 1216 Sv

(v) (a) 0.002 mSv, (b) no.

(vi) 12.0 mSv

DESIGN PROBLEMS

10.1 Describe briefly the operation of a scintillation counter and a semiconductor counter for γ-ray spectroscopy. Include in your discussion the complete system used to accumulate energy spectra. Describe and explain (i) the pulse height distribution, (ii) the resolution.

10.2 (a) ^{137}Cs emits a γ-ray of energy 663 keV. Describe the characteristic features of the spectrum recorded by

(i) a NaI-Tl scintillation counter having a resolution of 15.3%

(ii) a solid state lithium drifted Ge detector of 2% resolution.

(b) Estimate the size of the unamplified voltage pulse corresponding to the photopeak in the two cases. The Ge detector/preamplifier has a capacitance of 7.5 pF. In addition to the intrinsic energy conversion, only 10% of photons generated in the scintillation process lead to a photoelectron. A 10 stage photomultiplier has a gain of 2.7×10^5 with an output capacitance of 75 pF.

(c) Discuss the reasons for the improvement in resolution obtained by use of the solid state counter, making particular reference to the mechanisms through which statistical fluctuations arise in the two systems.

10.3 A 3 mm thick Si(Li) detector is isolated from the atmosphere by a 0.025 mm thick Be window. To a good approximation the photoelectric attenuation coefficients may be written as $\mu = C \times En$ (where μ is in cm-1 and E is in keV). With the values of C and n given below estimate the efficiency of the detector system for mono-energetic X-rays in the range 2–100 keV. Explain what influence the Compton effect and finite electron path lengths could have on the result.

Data: For Be, $C = 1.5 \times 10^3$, $n = -3.35$; for Si, $C = 1.01 \times 10^5$, $n = -3.08$

10.4 Sketch and describe the energy spectrum of 2.6 MeV gamma rays obtained by an intrinsic Ge detector. If the electronic noise of the system is 0.9 keV, discuss the resolution (FWHM) of the peaks in the spectrum.

10.5 A dentist places the nose cone of a 100 kV X-ray machine against the face of a patient to obtain an X-ray picture of the teeth. Without filtration, considerable low energy (assume 20 keV) X-rays are present in the beam. If the intervening tissue is equivalent to 5 mm of water,

(a) Calculate the relative fraction of 20 keV and 100 keV X-rays absorbed by the skin tissue.

(b) What thickness of aluminium filter would be required to reduce the 20 keV radiation exposure by a factor of 10?

See the following diagram:

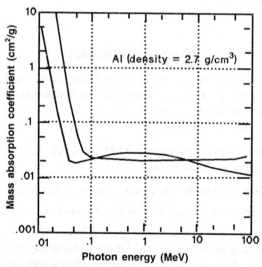

10.6 A MBq ^{137}Cs source (half life = 30 yr), uniformly distributed over a 0.5 cm^2 area, is placed in a shirt pocket by mistake. The source emits a 663 keV γ-ray per decay. If the shirt is worn for 10 hrs., calculate the total skin dose received.

μ_a/ρ for 663 keV γ-rays in tissue = 0.0315 cm^2/g

μ/ρ for 663 keV γ-rays in tissue = 0.0834 cm^2/g

10.7 The operators of a nuclear power station wish to release under controlled conditions 30,000 Ci of ^{85}Kr (an inert gas) which is trapped in the containment vessel. They want to do this because it is feared that an accidental release under adverse atmospheric conditions will give rise to high doses to the local population. Two extreme scenarios may be imagined:

Scenario I.   Under ideal conditions the ^{85}Kr is released at the top of a 20 m high stack and proceeds downwind to a safe disposal area as a well defined cloud 20 m above the ground with a speed of 5.0 m/s. The cloud can effectively be modelled as a point source.

Scenario II. Under poor atmospheric conditions the ^{85}Kr is assumed to mix uniformly with the atmosphere to a height of 100 m and over a circular area of radius 0.3 km. This condition persists for 12 hours.

(a) Estimate the total body dose to a member of the public standing on the ground some way from the release site directly below the path of the cloud in scenario I. Such a person may have a mass of 90 kg and an effective area of 0.15 m^2.

(b) What is the dose to lung tissue for someone in the uniform mixing zone in scenario II?

Data: ^{85}Kr $T_{1/2}$ = 10.7 yr, Average β-energy per decay 0.22 MeV, A γ-ray of energy 514 keV is produced in 0.5% of ^{85}Kr β-decays. Mass of lungs = 1.1 kg, Volume of lungs = 3.5 l, μ_{air} = 0.030 cm^2/g, $\mu(E)_{tissue}$ = 0.032 cm^2/g at E_γ = 514 keV. A person breathes about every 5s.

10.8   In most chemical forms, iodine is quickly taken up by the thyroid, whether the iodine is inhaled or ingested or injected. Shortly after an injection of iodine, the retention function for the thyroid is approximately

$$R(t) = 0.312\, e^{-0.00384t}$$

where t is in hours, and $R(t)$ is the fraction of the injected (stable) iodine. Suppose a patient is given an injection of 1.0 µCi of radioactive iodine for diagnosis of a thyroid condition:

(a) If ^{131}I is used, what would be the total dose, in Sv, received by the thyroid?

(b) If ^{123}I is used, what would be the total dose, in Sv, received by the thyroid?

The half-life for ^{131}I = 8.05 day^{c} and the effective energy equivalent for ^{131}I = 0.23 MeV for the thyroid. The half-life for ^{123}I = 13.2 hours and the effective energy equivalent for ^{123}I = 0.034 MeV for the thyroid. Assume a thyroid mass of 25 g.

10.9   A leak in the moderator heat exchange system at a nuclear power station, caused heavy water (D$_2$O) to be lost at the rate of 25 l/day to the cooling water flow and discharged into an adjacent lake. The total cooling water flow was 5 × 10^6 l/hr; and the heavy water lost had been in an average neutron flux of 6 × 10^{13} n/cm^2/s since the start up of the reactor 7.5 yr earlier. The reaction which created tritium $^2H + n = {}^3H$ gives rise to an activity $A_\lambda = \sigma\, N_x\, \Phi\, [1 - e^{-\lambda t}]$, where N_x is the number of target nuclei and ϕ is the neutron flux.

(a) What is the tritium activity in the discharged water in Bq/l?

(b) The discharge water after further dilution by a factor of 10 was picked up by the local water supply system. If the leak persisted for three days, estimate the typical radiation dose to an 80 kg person at the end of this time receiving water from the supply if they ingested approximately 5 l/day over the period of three days.

(c) Do you think this incident was a significant health hazard?

Data: Physical half life of tritium = 12.3 yr
Biological half life of tritium = 10 days
Tritium ^3H decays by β^- emission with a mean β^- energy = 0.01 MeV.
Neutron capture cross section for deuterium = 5.3×10^{-28} cm^2.

10.10 The mineral water at Bad Gastein Spas in Austria contain 4×10^{12} ^{226}Ra atoms per litre. The natural decay half-life and the biological half life are 1602 yr and 44.9 yr respectively. The mean energy for the decay of ^{226}Ra and its daughters is 110 MeV and the body retains 4% of all Ra ingested. A 20 year old man on a single visit to Bad Gastein drinks 2 l of the mineral water. Assuming that the Ra is uniformly distributed in his body and that his body weight remains constant at 80 kg, calculate the total dose that he will have received from this Ra when he will be 80 years old.

═══ 11 ═══

Analytical Instrumentation

Electrons, ion and particle beams are capable of imaging materials and analyzing their surfaces. Some important imaging and analytical techniques are described in this section.

11.1 TRANSMISSION ELECTRON MICROSCOPY (TEM)

Electron microscopy is an important analytical tool. The resolution of any microscope or optical instrument is limited by diffraction (Fig. 7.5). This is important when the size of the object being observed approaches the wavelength of the observing radiation. For optical microscopes this implies a limit of about one third of the wavelength of visible light (150 nm). However, the de Broglie wavelength of a 60 keV electron $\lambda = h/p = h/\sqrt{2mE} = 0.03$ Å, so that the theoretical resolving power of an electron microscope is 0.01 Å, although various aberrations tend to restrict electron microscopes to resolutions of 5–25 Å.

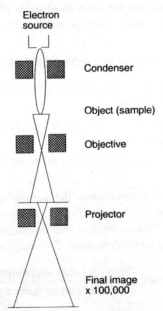

Fig. 11.1   The electron optics of a transmission electron microscope.

305

An electron microscope has a ray path very similar to its optical analog. Electrons are focused to pass through a thin object. The emerging beam enters magnetic lenses which magnify the image for projection on a fluorescent screen or photographic plate.

Electron Lenses

Most electron-optic lenses are magnetic.

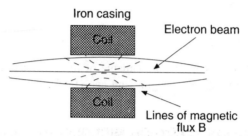

Fig. 11.2   Electron lens.

The major difficulty with TEM is that of specimen preparation. Specimens have to be thinned by grinding and then by sputtering or ion beam, chemical or electrochemical etching to a fraction of a µm to allow transmission of the electron beam.

The **magnification** is set by the ray optics, the effective size of the electron source, and the focal lengths of the lenses. The highly focused beam of a transmission electron microscope can also be used for the study of crystal structures using electron diffraction patterns, and for analysis of areas of $< µm^2$ by the excitation of X-ray fluorescence in the electron beam microprobe—Fig. 8.11.

11.2   SCANNING ELECTRON MICROSCOPY (SEM)

The SEM utilizes the principle of electron beam scanning to form an image on a television or oscilloscope monitor.

- An *X-Y* voltage generator scans the electron beam in the microscope over the sample in a raster pattern.

- The same voltages are used to deflect the beam of an oscilloscope over the full picture area.

- The electron beam interacts with the sample and causes phenomena to occur such as the back-scattering of electrons, secondary emission, luminescence etc. This is detected and its magnitude is used to modulate the intensity of the display monitor. The image is viewed on the monitor.

$$\text{Magnification} = \frac{\text{Size of image on monitor screen}}{\text{Raster pattern size on sample}}$$

Magnifications from 100 to 25,000 are available on most SEM's.

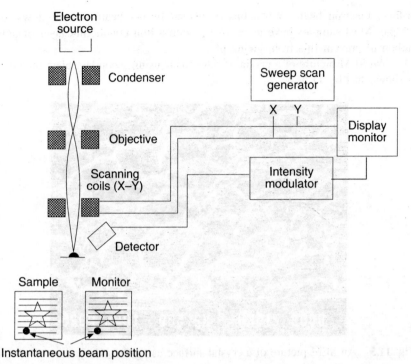

Fig. 11.3 The electron optics and operating principle of a scanning electron microscope.

Detection methods. The most frequently detected parameter in a conventional SEM is that of secondary electron emission. The secondary electron coefficient depends on:

- the specimen material
- the angle of an element of the surface to the beam.

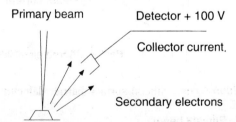

Fig. 11.4 Collection of secondary electron emission.

The current collected by the Faraday cage detector at any instant is proportional to the total of the secondary emission current from a given point on the sample, plus any elastically back-scattered primaries.

Since the number of secondaries per primary > 1, if the target is not a good conductor it will charge up and effectively reach the acceleration voltage of the primary electron beam. When this occurs no further beam current flows to the sample. Most samples have to be coated with a thin conducting layer of gold or carbon to prevent this from taking place.

An SEM picture of a crystal surface made using secondary electron emission is shown in Fig. 11.5.

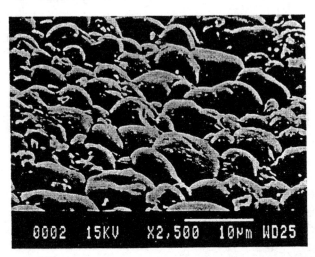

Fig. 11.5   An SEM picture of a crystal surface using secondary electron emission.

Other phenomena stimulated by the beam can be observed using different detectors, with the detector output used to control the display intensity. These include:

(a) *Back-scattered primaries*: negatively biased collector (Fig. 9.15)

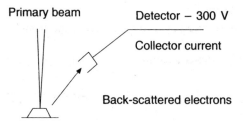

(b) *X-ray fluorescence*: silicon surface barrier detector (Fig. 10.16)

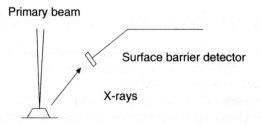

(c) *Cathodoluminescence:* photomultiplier.

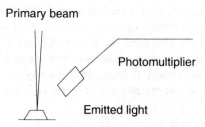

(d) *Specimen current*: electrical leakage current to ground (Problem 4.8)

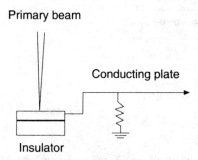

Fig. 11.6 Detection techniques.used to modulate the display intensity in a scanning electron microscope.

11.3 ENVIRONMENTAL SCANNING ELECTRON MICROSCOPE (ESEM)

A difficulty for all electron beam instruments is that the high vacuum required in the electron beam column makes the examination of hydrated samples, or of samples that degas in vacuum difficult.

The environmental scanning electron microscope avoids this problem. An electron beam from a conventional high vacuum column exits into a sample chamber through a small hole. The sample chamber contains an ionizable gas (often water vapour) at pressures of the order of 3-4 mtorr. The electron beam ionizes some of the gas to form a positive ion 'envelope' around the primary beam. This positive ion cloud discharges the build-up of negative ion charge on uncoated sample surfaces, while the high pressure of the sample chamber is not affected by sample outgassing. The ESEM is often used for the study of cements, ceramics, polymers and biological specimens.

11.4 SURFACE ANALYTICAL METHODS

Surfaces play a fundamental role in the properties of materials and systems. A number of important analytical techniques have been developed to obtain chemical and physical information about surfaces.

Auger Electron Spectroscopy

Auger electrons are emitted from a surface whenever incident radiation—photons, electrons, ions or neutral atoms—interacts with an atom with sufficient energy to remove an inner shell electron from the K, L or M shells (Fig. 8.4). When higher level electrons fall into the empty state two things can happen—a characteristic X-ray is emitted, or the energy is transferred directly to a neighbouring electron which is released as an Auger electron. Figure 11.7 compares the mechanisms involved.

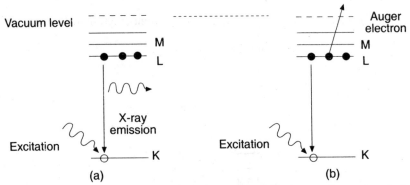

Fig. 11.7   The mechanism of: (a) X-ray emission. (b) Auger electron emission.

Since electron beams in the 1 to 30 keV range of energy can easily be produced and focused, this is the excitation of choice for Auger electron spectroscopy.

The kinetic energy of an Auger electron resulting from a transition involving the K and L levels is

$$E_{KE} = E_K - 2E_L - \phi$$

where ϕ, the work function for the surface is the kinetic energy which is lost by the electron in escaping from the sample surface into vacuum. The low energy of the Auger electrons means that only electrons in the top 0.5 to 3 nm (2–10 monolayers) escape from the surface. The energy of these electrons is analyzed by some form of electron energy analyzer.

Figure 11.8(a) shows a chart of the principal Auger electron energies. Figure 11.8(b) is a representative AES spectrum from silver. The sensitivity of the measurement is enhanced by calculating the differential of the simple count rate N versus E curve (page 114).

Information available from Auger electron spectroscopy includes:

(a) electron energies characteristic of transitions within the source atoms.

(b) shifts in the energies related to the chemical bonding in the surface layers.

X-ray Photoelectron Spectroscopy (XPS)

The excitation of core electrons by X-ray photons of frequency v can release photoelectrons from the surface of a solid. Their kinetic energy upon release is

$$KE = hv - BE + \phi$$

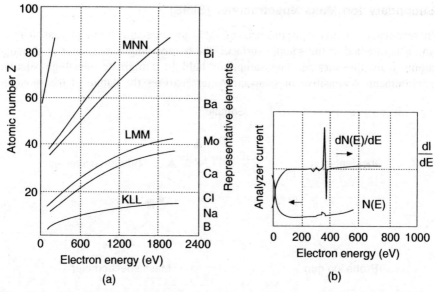

Fig. 11.8 (a) Chart of the principal Auger electron energies. (b) Experimental AES data from silver.

Information is therefore available on the individual levels plus changes in work function ϕ due to variations in chemical composition and environment at the surface. The photo-emitted electrons have low energy and therefore scattering ensures that only electrons within the top few atomic layers are emitted.

An XPS spectrum is shown in Fig. 11.9.

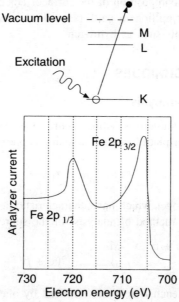

Fig. 11.9 XPS spectra for iron (Fe).

Secondary Ion Mass Spectrometer (SIMS)

In secondary ion mass spectrometer (SIMS) an energetic beam of ions such as argon is directed at the sample surface. The bombarding ions sputter or dislodge atoms from the surface. The sample is held in a high or ultra-high vacuum environment. A sensitive mass spectrometer analyzes the nature of the sputtered

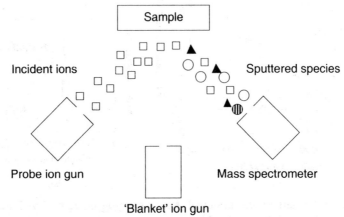

Fig. 11.10   Schematic representation of the principles of SIMS.

atoms. Modern argon ion guns provide well controlled high sputtering rates for good sensitivity and a full surface analysis is achieved of the top 5 nm of the surface. Depth profiling is undertaken by use of a large area gun to sputter the sample uniformly over considerable distances. An analysis gun and ion analyzer then measures the local composition of the surface. Trace element analysis can be achieved with parts per million to parts per billion sensitivity. The technique is sensitive to light elements such as hydrogen.

11.5   NUCLEAR TECHNIQUES

Neutron Activation Analysis

A nuclear reactor or a radium-beryllium source (see page 261) is a source of thermal neutrons. Nuclei placed in such a source capture neutrons and undergo a reaction of the form

$$_A^Z X + n \rightarrow _{A+1}^Z Y$$

to produce unstable isotopes which have characteristic decay schemes (page 271). These can be used as a method of analysis for the source nucleus.

For example $^{55}Mn + n \rightarrow {}^{56}Mn$

$$\text{\rule{0.4pt}{12pt}\rule{40pt}{0.4pt}}{}^{56}Fe + \beta^- + \upsilon$$

Because of the decays, each β-decay is followed by one 0.847 γ-ray decay. This emission is detected in the experiment.

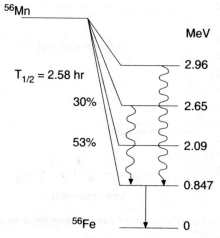

Fig. 11.11 Decay scheme for ^{56}Mn created by irradiation by neutrons.

The rate of production of atom Y at any time t after the sample X containing N_X nuclei is placed in a neutron flux ϕ is

$$\frac{d\,N_Y(t)}{dt} = \text{rate of production} - \text{rate of decay}$$

$$= \sigma N_X \phi - \lambda N_Y(t)$$

where σ is the reaction cross-section and λ is the decay constant for the product nucleus N_Y. Assuming that $N_Y = 0$ at $t = 0$, and that the rate of reaction is such that over the course of a reaction the number of starting nuclei N_X is essentially unchanged, then

$$N_Y(t) = \frac{\sigma N_X \phi}{\lambda} [1 - e^{-\lambda t}]$$

$$= \frac{R}{\lambda} [1 - e^{-\lambda t}]$$

where $R = \sigma N_X \phi$ is the rate of production of N_Y at time $t = 0$. The activity A_Y resulting from the formation of N_Y is,

$$A_Y(t) = \sigma N_X \phi [1 - e^{-\lambda t}]$$

These equations are valid during the time t_R the target nucleus is being activated by the neutron source. If the material is irradiated for time t_R, N_Y and A approach saturation as $t_R > 1/\lambda$ at values of $N_{Y_0} = (\sigma N_X \phi)/\lambda$ and $A_{Y_0} = \sigma N_X \phi$ respectively. The activity attains 95% of the saturation value after a time $t = 3/\lambda$.

On removal from the source at time t_R, the activity decays at time t'

$$N_Y(t') = N_{Y_0}\, e^{-\lambda t'}$$

$$A_Y(t') = A_{Y_0}\, e^{-\lambda t'}$$

The activity $A(t)$ as a function of time is shown in Figure 11.12.

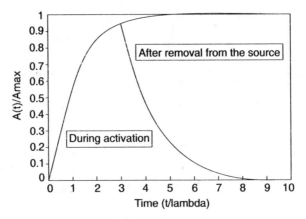

Fig. 11.12 Activity $A_\gamma(t)/A_t$ resulting from a neutron activated reaction. Time in units of $1/\lambda$.

Experimental practice. Absolute measurement of the above quantities requires (page 280) knowledge of the solid angle subtended by the source at the detector $d\Omega$, the detector efficiency ε_{int}, and the transmission probability T for γ-rays to reach the detector. The branching ratio (BR) is the probability that the nucleus decays according to a particular mode of decay (Sec. 9.3). The number of γ-rays detected in a time t' after a period of excitation in the reactor of t_R is

$$N = T\,\varepsilon_{int}\,\frac{d\Omega}{4\pi}\,(BR)\,\frac{\sigma N_X \phi}{\lambda}\,(1 - e^{-\lambda t})\,e^{-\lambda t'}$$

In practice, measurements are often made by a comparison technique in which a standard sample containing a known number of nuclei N_{Xs} (for example a sample of pure Mn) is irradiated for the same time as the test sample N_{Xt}. If the two samples are then counted by the same detector and experimental arrangement for identical time intervals Δt after removal from the neutron flux, many of the above parameters cancel out. The transmission factor T takes into account the effects of different absorption of the emitted radiation in the two paths.

X-ray Fluorescence

X-ray fluorescence is the analysis of materials by the excitation of characteristic X-rays. As the primary excitation source, an electron beam (page 237) or a radioactive source such as ^{125}I in which a 27.4 keV X-ray is emitted through electron capture is used (Fig. 11.13(a)).

Sources are mounted in a collimator so that the incident X-rays enter the sample, excite characteristic X-rays, which are then detected by a Si(Li) detector (Fig. 11.13(b)). The advantage of XRF is that it can detect X-rays from many different atoms in the target simultaneously. Quantitative analysis calculates the relative numbers of different atoms in the target from the areas under the peaks corresponding to the K_α and K_β emissions.

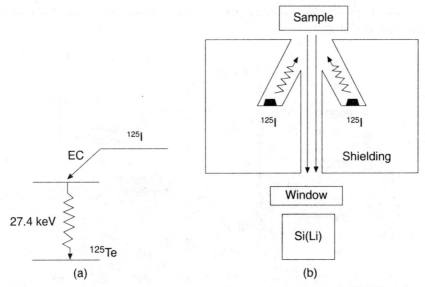

Fig. 11.13 (a) Decay scheme leading to a 27.4 keV Te X-ray from [125] I. (b) A collimator system mounted above a Si(Li) detector.

The sample holder is as shown. The incident X-rays enter the sample, excite characteristic X-rays which are then detected by the Si(Li) detector. This technique is excellent for detecting the presence of elements. The calculation of a **quantitative** materials analysis relating the area under a given characteristic X-ray peak to the number and chemical identity of atoms in the target is relatively complex. It has to take into account the factors illustrated in Figure 11.14, some of which require a preliminary estimate of the approximate composition. For comparative purposes, the intrinsic efficiency for all X-rays is assumed to be the same. The measurement parameters include:

1. The mass attenuation coefficient for the incident X-rays in the target material (page 263)

2. The fluorescence yield ω_α for production of a characteristic K_α X-ray from a particular atom (page 260, Appendix 4.2).

3. The attenuation coefficient for exciting X-rays (page 263).

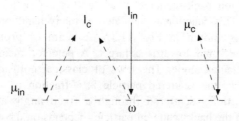

Fig. 11.14 Factors determining the X-ray flux from an XRF experiment.

Figure 11.15 shows an XRF spectrum for stainless steel—an alloy of iron, chromium and nickel.

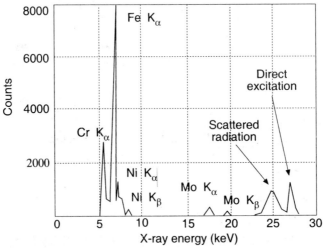

Fig. 11.15   An XRF spectrum for commercial stainless steel excited using an ^{125}I source.

Most stainless steel also contains a small amount of molybdenum which has a relatively high fluorescence yield. This acts as a good 'marker' for stainless steel with respect to iron. The narrow peak at 27.4 keV is due to direct excitation of the detector by incident X-rays, while the lower broader peak is from Compton scattered X-rays in the collimator and detector structures.

For operation of XRF sources in air, a restriction on the lightest elements that can be detected is set by absorption in the beryllium window covering the detector. This limits XRF to elements with Z greater than silicon. This limitation is less severe for the electron beam microprobe where both sample and detector are held in vacuum. The highest energy is determined by the energy of the source.

Rutherford Backscattering

When a mono-energetic beam of ions of energy E and mass m_1 is incident on a target nucleus of mass m_2 the ions undergo elastic 'billiard ball' collisions. The energy of an ion back-scattered at angle θ is a function of the parameters $K = f(m_1, m_2, \theta)$. This function is tabulated in many handbooks [see Ref. 11.4]. The experiment is shown in Fig. 11.16. A beam of protons, deuterons or α-particles is accelerated towards a target. A detector measures the scattered ion flux at a particular angle. The laws of classical collisions shows that the maximum energy of the scattered particle is a fraction of the incident energy (page 267). If particles scattered through a fixed angle are considered, the energy spectrum of the back-scattered particles depends on the physical form of the sample.

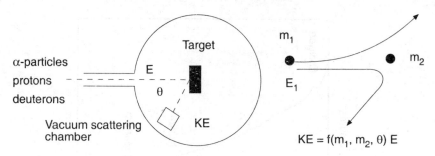

Fig. 11.16 Rutherford backscattering: experiment and theory.

For a **thin film** there is a minimum energy loss in the primary beam as the particles enter and leave the sample, and the scattered spectrum shows **one peak** for each element in the target. The peak occurs exactly at the energy predicted by the kinetic factor. Each line spectrum is broadened by the resolution of the detector. The area under the peak is a measure of the number of target ions of a particular species which is present. The RBS spectrum for a sample using a high resolution detector is illustrated in Fig. 11.17.

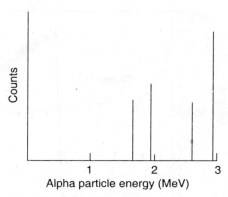

Fig. 11.17 RBS spectrum from a thin film sample containing different elements. 3 MeV alpha particles scattered through angle $\theta = 155°$.

For a **bulk material** the primary beam loses energy as it enters the target and the backscattered particles have progressively lower energies depending on how far a particular particle in the primary beam penetrated before interacting. The spectrum is broadened into a series of steps with the high energy limit of the step being characteristic of an element. The height of the step is a measure of the concentration of the element. The scattering cross-section for RBS varies roughly as $1/E^2$ (slower particles scatter more) which accounts for the increase in scattering at low E.

Rutherford backscattering allows elemental concentrations to be measured as a function of depth from the surface. If a layer of a certain thickness is observed, RBS will show a distinct feature depending on the dE/dx and the film thickness. Thermal diffusion of an element in the surface and hence of its distribution within the sample can be observed by measurements made as a function of temperature.

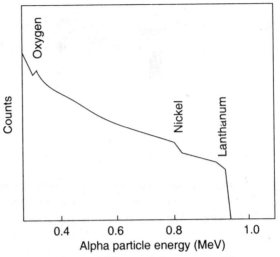

Fig. 11.18 RBS spectrum for 1 MeV α-particles scattered from ceramic La_2NiO_4; the position of the steps for each element are noted.

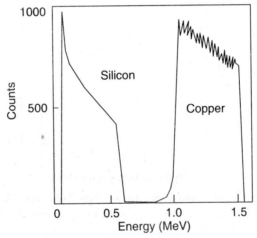

Fig. 11.19 RBS spectrum from a 500 nm thick film of copper on silicon. The high energy feature is due to the copper film; the step to the silicon. Merging of these features on heating would show that the copper was diffusing into the silicon.

11.6 ATOMIC FORCE AND TUNNELING SCANNING MICROSCOPES

If a fine metal probe is placed near to a conducting surface, when the distance is sufficiently close, a quantum mechanical tunneling current flows between the probe and surface. The magnitude of current is a sensitive measure of the spacing. An extension of this effect is the observation of mechanical and chemical forces between a probe and a surface.

The development of highly sensitive piezoelectric actuators which allow fine movements (sub-angstrom) of a probe with respect to a surface have created a new family of scanning tunneling and force microscopes.

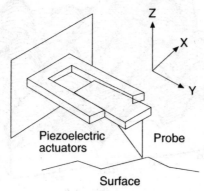

Fig. 11.20   Piezoelectric scanner for fine movements of a tip close to a surface.

A piezoelectric scanner is driven along an X-Y plane by voltages which are in synchronism with comparable voltages controlling the position of the beam of a video monitor. This is analogous to the mechanism of a scanning electron microscope.

A third actuator adjusts the probe tip in the Z direction. In one mode of operation, the tunneling current or force between the probe and surface is maintained constant by a voltage applied to the actuator. The magnitude of this voltage is used to modulate the brightness of the monitor beam. The resulting picture is a measure of the surface morphology or mechanical interaction.

Figure 11.21 is an example of the surface morphology of a conducting oxide deposited by sputtering. The surface morphology reflects the grains within the film.

Scanning microscopy is critically dependent on signal to noise processing. This includes the elimination of mechanical vibrations in the probe and supports, high quality signal processing, and careful image generation.

The types of interaction include tunneling current, mechanical forces, electrochemical effects, and even biological phenomena.

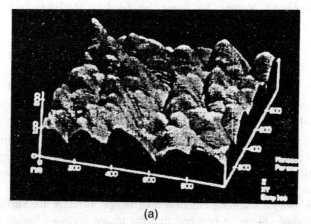

(a)

Fig. 11.21   Cont.

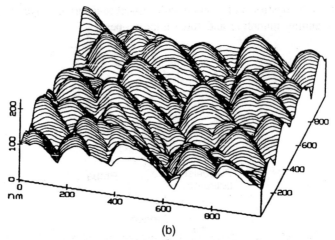

(b)

Fig. 11.21   (a) Atomic force image of the surface morphology of a sputter deposited and oxidized ruthenium oxide (D. McIntyre). (b) A three-dimensional plot of the surface contours; the vertical scale is approximately 200 nm.

REFERENCES AND FURTHER INFORMATION

1. J.W.S Hearle, *The Use of the Scanning Electron Microscope*, Oxford University Press, New York, 1972.

2. Rigaku Canada, 34 Bersczy St. Aurora, Ontario L4G 1W9 Canada.

3. John B. Hudson, *Surface Science: An Introduction*, Butterworth-Heinmann, Boston, 1992.

4. L.C. Feldman and J.W. Mayer, *Fundamentals of Surface and Thin Film Analysis,* North Holland, New York, 1986.

5. J.R. Bird and J.S. Williams, *Ion Beams for Materials Analysis*, Academic Press, Australia, 1989.

6. J.W. Mayer and E. Rimini, *Ion Beam Handbook for Materials Analysis*, Academic Press, New York, 1977.

DESIGN PROBLEMS

11.1 (a) A scanning electron microscope has a magnification of 2500 with a monitor screen size of 20 cm. What is size of the raster pattern on the sample?

  (b) If the surface of a metal is unpolished with a fine grain structure, which would be the best technique to image the surface morphology—secondary electrons or elastically back-scattered electrons. Give reasons for your answer.

11.2 When using a scanning electron microscope to image the surface of a piece of intrinsic silicon, it is noted that the image becomes highly distorted and occasionally disappears from the monitor screen. Using the information on secondary electron emission provided on pages 75 and 270, give a quantitative account of why this happens. How would you resolve the problem in practice? If the secondary emission coefficient decreases with electron energy, is there any way you could operate the machine without modifying the sample?

11.3 A scanning electron microscope with a display monitor 20 cm × 20 cm is used in light-detecting mode to identify luminescent defects on the surface of a crystal which are about 5 μm in diameter. If the microscope is operated at a magnification of 25,000, and the beam moves with a sweep frequency of 200 Hz, what will the current pulse recorded by the photomultiplier look like? Recommend a value of the time constant of the *RC* circuit to be attached to the photomultiplier anode (page 77), if this pulse is to be recorded as a voltage without distortion.

11.4 Discuss an appropriate way to quantify the variations in surface morphology shown in Fig. 11.21.

What would be the advantages and disadvantages of using atomic **force** microscopy compared with **tunnelling** microscopy?

11.5 In coal-treatment facilities it can be important to determine the trace elements in coals and their by-products in order to avoid pollution of the environment.

A sample of coal weighing 1 g is to be analyzed by neutron activation analysis for the trace element arsenic. It is placed in a reactor whose flux is $5 \times 10^{13} n/cm^2$-s. If the half-life of ^{76}As is 26.4 hr, (a) how long must the sample be left in the reactor to reach 90% of its saturation activity? (b) If the sample is irradiated for one week and the measured activity of ^{76}As is 100 μCi five hr after removal from the reactor, calculate the weight of arsenic in 1 g of this coal. The cross section for As = 4.3 b.

11.6 (a) A steel lathe cutting tool of mass 10g is irradiated in a reactor in order to produce ^{59}Fe ($T_{1/2} = 45$ days). After the tool has been used on a production line, a 2% efficient Ge(Li) γ-ray detector is used to measure γ-rays from the decay of ^{59}Fe both from the tool and from the cutting oil used with the tool. The following observations were made:

(1) Counting rate for the tool alone = 1.8×10^6 per minute.

(2) Counting rate for the cutting fluid = 20 per minute.

Calculate the amount of tool wear that has taken place assuming that all the cutting oil was collected. Discuss the importance of self absorption and radioactive decay in your calculations. The mass attenuation coefficient for γ-rays in Fe is 0.60 cm^2/g.

(b) If the reactor used for the activation had a thermal neutron flux of 1.0×10^{12} n/cm^2/s, for how long was the tool bit irradiated. (The steel was 80% Fe, 0.33% of Fe is ^{58}Fe and the thermal neutron capture cross-section σ_c for ^{58}Fe is 1.0×10^{-24} cm^2, and the half life $T_{1/2} = 2.58$ hr).

11.7 Discuss the principles of elemental analysis using Rutherford Backscattering. Describe how isotopes are identified and how spatial distributions can be deduced. Discuss the limitations of this technique for thick targets.

————12————

Occupational Health and Safety

As developers, promoters and users of science and technology, engineers and scientists have important responsibilities to the general public. Incidents such as the nuclear reactor accidents at Chernobyl or Three Mile Island or the chemical emissions at Bhopal, occurred through bad design, bad practice or scientific incompetence. Such disasters make it imperative that all engineers, scientists and business people consider safety to be as important as economics or technical advantage. These considerations extend from the design and human factors inherent in an instrument panel, to the environmental acceptability of a new process, and the disposal of waste created in an industrial process. The legal and moral consequences of incorrect actions can be personally severe. They range from substantial fines to imprisonment, to living with the stigma of a long-term personal responsibility for the injury of innocent people, or of permanently damaging the world in which we live.

All persons in a workplace—a university, an industry or a government laboratory, are bound by significant legislation. Under various circumstances a person can function as an employer, a supervisor or a worker. As a professional, additional considerations must be taken into account. For example, for an engineer, the following legislation is applicable.

> "The practice of professional engineering" means any act of designing, composing of plans and specifications, evaluating, advising, reporting, directing or supervising **wherein the safeguarding of life, health, property or the public welfare is concerned** and that requires the application of engineering principles;

Legislation covering scientists is generally less well defined, but the theme of ethical and general responsibility for the safety of the public is the basis for the activities for any responsible professional.

This chapter reviews representative legislation covering the occupational safety of workers, chemicals, radiation safety and safe electrical and laboratory practices. While specific legislation varies around the world, the common theme is to define a safe work place. The summary is not comprehensive or legally complete: in a specific situation where potential hazards are identified, the safety officer at one's place of employment should be contacted.

12.1 OCCUPATIONAL HEALTH AND SAFETY

As an example of legislation, a representative Occupational Health and Safety Act and the associated Regulations for Industrial Establishments are reviewed. The legislation covers the responsibilities of employers, supervisors and workers who work in the Canadian province of Ontario. All persons in these three categories who contravene the provisions of the Act may be subject to a fine of up to $25,000 and/ or be subject to imprisonment for a period of up to 12 months. The wording of relevant sections of the Act read as follows:

An **employer** shall ensure:

- that equipment, materials and protective devices as prescribed are provided and are maintained in good condition.
- that the measures and procedures prescribed are carried out in the workplace, and the equipment, materials and protective devices provided are used as prescribed.
- that a floor, roof, wall, pillar, support or other part of the workplace is capable of supporting all loads to which it may be subjected without causing the materials therein to be stressed beyond allowable unit stresses.

Without limiting the strict duty imposed by the above, the employer shall:

- provide information, instruction and supervision to a worker to protect the health and safety of the worker.
- when appointing a supervisor, appoint a competent person.
- acquaint a worker or a person in authority over a worker with any hazard in the work and in the handling, storage, use, disposal and transport of any article, device, equipment or biological, chemical or physical agent.

A **supervisor** shall ensure that a worker:

- works in the manner and with the protective devices, measures and procedures required by the Act and the regulations; and
- uses or wears the equipment, protective devices or clothing that his or her employer requires to be used or worn.

Without limiting the duty imposed by the above; a supervisor shall,

- advise a worker of the existence of any potential or actual danger to the health or safety of the worker of which the supervisor is aware.
- where so prescribed, provide a worker with written instructions as to the measures and procedures to be taken for protection of the worker.
- take every protection reasonable in the circumstances for the protection of a worker.

A **worker** shall

- work in compliance with the provisions of this Act and the regulations.

- use or wear the equipment, protective devices or clothing that the employer wants to be used or worn by the employees.

- report to the employer or supervisor the absence of or any defect in any equipment or protective device of which the worker is aware and which may endanger the worker or another worker.

- report to his or her employer or supervisor any contravention of this Act, or of the regulations, or the existence of any hazard of which he or she knows.

- where so prescribed, have, at the expense of the employer, such medical examinations, tests or X-rays, at such time or times and at such place or places as prescribed.

No **worker** shall

- remove or make ineffective any protective device required by the regulations or by his or her employer without providing an adequate temporary protective device, and when the need for removing or making ineffective the protective device has ceased, the protective device shall be replaced immediately.

- use or operate any equipment, machine, device or work in a manner that may endanger himself or herself or any other worker.

- engage in any prank, contest, feat of strength, unnecessary running or rough and boisterous conduct.

In a given workplace, the employer shall establish a joint health and safety committee consisting of at least two persons of whom at least half shall be workers who do not exercise managerial functions. It is the function of the committee, and it has the power, to:

- identify situations that may be source of danger or hazard to workers.

- make recommendations to the employer and the workers for the improvement of the health and safety of workers.

- recommend to the employer and to the workers the establishment, maintenance and monitoring of programs, measures and procedures with respect to the health and safety of workers.

- obtain information from the employer regarding:

  - the identification of potential or existing hazards of materials, processes or equipment.

  - health and safety experience and work practices and standards in similar or other industries of which the employer has knowledge.

Application to practice. Obvious breaches of the above legislation occur in an industrial plant if saws, conveyors or other machinery are operated after removal of guard rails or protective enclosures, if protective equipment such as eyeglasses are not used near lasers, or if gloves are either not available or are not used near furnaces. In case of injury, the worker, his or her supervisor, and the employer may be held liable. Less obvious breaches of the Act occur by eating in an area where poisonous materials are present, and there is inadequate hygiene or poor ventilation.

In a laboratory setting where prototype apparatus is developed or used, the safety practices still apply. Common breaches are poor electrical wiring where live electrical leads are exposed, or where ineffective grounding procedures are used on custom built equipment. Inadequate solvent ventilation, storage or handling; or poor mechanical construction of a large structure such as an antenna or a test rig have to be guarded against. A common but regrettable attitude is that the researcher knows where he or she has to be careful in using a piece of custom equipment. However, others work in the same area or may borrow the equipment. A consequent accident may be held to be the responsibility of the first worker, or the supervisor of the group, or the employer.

> Health and Safety regulations almost always lead to better science and engineering, and if the appropriate attitude is developed early in a career, a major technological accident may be avoided at a later stage.

Safe practices are an essential form of insurance. In an increasingly litigious and environmentally conscious age, a careless act can cost a career, lead to corporate and personal bankruptcy, and decimate an industry. In the aftermath of Chernobyl and Three Mile Island, the nuclear industry has learned this to its cost. On the other hand, technology can lead to great benefits. A careful, well documented and public assessment of the benefits and disadvantages of a technical development is not a frill.

12.2 CHEMICAL SUBSTANCES

Within the general context of health and safety legislation, there are increasingly stringent requirements for the handling of chemicals. These particularly concern the effects of chemicals on workers and the public, and the safe handling, transport and disposal of all chemicals.

Chemical Disposal

In conjunction with national environmental legislation, municipalities generally enact by-laws to control waste discharges to municipal sewers. These by-laws are comprehensive. They severely limit the amount and nature of chemical materials that can be discharged into sewers from institutions and from industry. They provide for legal proceedings from the municipality and the Ministry of the Environment and substantial fines when release of prohibited materials exceeds the limits set in the by-laws.

Designated Substances in the Workplace

Substances which are identified as hazardous are placed on a 'designated' list. This requires that Health and Safety authorities be notified of the use of such chemicals, and an on-going program of inspections. In Canada, regulations exist for lead, mercury, asbestos, vinyl chloride, coke oven emissions, isocyanates, silica, benzene and acrylonitrile. Notice of possible designation have been published to regulate noise, arsenic, formaldehyde, cadmium, chromium, ethylene oxide, styrene, nickel and coal tar products.

In practice, the effect of designation is to make the substance more expensive to use. Additional costs include safe handling facilities and protective equipment for workers, the requirement for official approvals for use and for on-going inspection, and the potential liability inherent in products utilizing the substance. Product liability considerations have been significant for asbestos and urea formaldehyde foam insulation (UFFI). Engineering design must take into account such environmental and chemical factors.

Even for non-designated substances and products, environmental legislation makes it increasingly difficult to transport and dispose of all kinds of liquid and solid wastes. The design of industrial processes **should** take this into account. For example, if it is required to create a protective coating of nickel on an industrial part it may be technically and economically attractive to use a chemical method involving the electroless deposition of nickel. The alternative is a capital-intensive technique such as vacuum deposition. However, a full economic analysis may show that the operating costs of disposal of the chemical residues eliminates the economic advantage of the chemical method.

A new factor is the need to consider the ease with which a product or material may be recycled. For example, an aluminium alloy of superior properties may be eliminated for a particular purpose because it contains an element which is detrimental to the recycling of aluminum cans. The economic cost of separating the alloy by recyclers may be greater than the benefits provided by its properties. In the automotive industry the practice is growing of labelling plastic components with the type of plastic in order to simplify recycling.

Workplace Hazardous Materials Information System (WHMIS)

National regulations implemented by a combination of federal and provincial/territorial legislation affect all those who sell, buy, transport, store or use chemicals. These regulations contain elements which are coming into use internationally. The primary impetus is 'right-to-know' legislation setting up a hazard communication system whereby chemical information is provided to workers in three ways:

1. **Labels**—comprehensive labelling of hazardous materials using internationally recognized symbols.

2. **Material Safety Data Sheets (MSDSs)**—provision of material safety data sheets for hazardous materials.

3. **Training**—effective training in the use of hazardous materials.

Suppliers must classify hazardous materials which they supply to workplaces according to standard hazard classes. All hazardous materials are required to bear a standardized label and the supplier must develop and distribute material safety data sheets with detailed health and safety information. Employers are responsible for ensuring that hazardous materials are properly labelled and that materials safety data sheets are readily available. Employers are required to train their workers to understand the information on the labels and the MSDSs, and to apply the information in their work practices.

The standard hazard symbols and the format of a material safety data sheet are shown in Figs. 12.1 and 12.2. Individual manufacturers may have a different layout of this sheet, but all of the information must be present.

HAZARD SYMBOLS: WHMIS

Column I Classes and Divisions	Column II Hazard Symbols
Class A—Compressed Gas	
Class B—Flammable and Combustible	
Class C—Oxidizing Material	
Class D—Poisonous and Infectious	
(1) Materials Causing Immediate and serious Toxic Effects	
(2) Material causing other Toxic Effects	
(3) Biohazardous Infectious Material	
Class E—Corrosive Material	
Class F—Dangerously Reactive Material	

Fig. 12.1 Workplace Hazardous Materials Information System (WHMIS) symbols.

MATERIAL SAFETY DATA SHEET (MSDS)

An MSDS sheet must be available for all chemicals used in a laboratory or available for commercial sale. The information is provided in specific categories.

Section I: Material Identification and Use Section II: Hazardous Ingredients or Material	Section VI: Toxicological Properties of Product Section VII: Preventive Measures
Section III: Physical Data for Material Section IV: Fire and Explosion Hazard of Material Section V: Reactivity Data	Section VIII: First Aid Measures Section IX: Preparation Date of MSDS Person Responsible:

Fig. 12.2 Format of a Materials Safety Data Sheet (MSDS).

12.3 RADIATION SAFETY

Electromagnetic Emissions

Radiation hazards arise over a wide frequency spectrum. Within the electromagnetic spectrum, they potentially include low frequency electromagnetic fields associated with high current power lines, radio-frequency or microwave emissions from generating equipment. Radio-frequency power supplies, microwave ovens, transmitting equipment, and X-ray emissions from video display terminals can also be a hazard if the equipment is not correctly designed and maintained. Safety in a laboratory normally is ensured if equipment is properly installed according to manufacturer's instructions, and that interlocks and safety features are circumvented rarely and are handled with care during maintenance.

In the case of custom equipment, say, in an equipment that utilizes radio-frequency or microwave power, openings should be covered or taped with metal foil and substantial ground connections should be maintained. In any uncertainty, the organization, university or company safety office should be consulted.

Radioactivity and Nuclear Sources

The most common use of radioactivity in a university or industrial environment is through radioisotopes (page 253). These may give rise to γ-, β- or α-radiation. The use, storage and eventual disposal of such sources must be rigorously controlled. Careless handling creates a hazard not only to present personnel, but to future generations.

The use of radioisotopes is permitted under a radioisotope user permit issued through the appropriate organizational radiation committee. This body is set up subject to the regulations of the Atomic Energy Control Board. There are standard conditions associated with such a permit, and various practices are essential.

1. A copy of the AECB poster 'Rules for Working with Radioisotopes in a Basic Laboratory' must be displayed. The permit holder or designate is responsible that these rules are followed.

2. A copy of the Radioisotope User Permit must be displayed in each laboratory approved in this permit. A current list of personnel authorized to work with radioisotopes must be pasted beside it. Only authorized personnel should handle radioisotopes.

3. Students working in a teaching or research laboratory must receive proper instruction in the safe handling of radioisotopes.

4. Gloves should be used whenever there is a chance of contamination.

5. Laboratory aprons should be worn when working with radioisotopes.

6. A dosimeter must be used if it is required as per the licence. When it is not being worn, it should be stored away from radioactive materials.

7. Dry powders or volatile substances must be handled in a fumehood.

8. Disposable absorbent liners should be used on trays or other surfaces.

9. Glassware used for radioactive work must not be used for other purposes.

10. On a regular basis (at least weekly), equipment, trays, floor, and working surfaces should be decontaminated. A record should be kept of such actions.

11. Hands must be washed before leaving the laboratory.

Radioactive Storage and Waste Disposal

1. Storage facilities for radioactive materials must be marked with a radiation warning symbol.

2. Radioactive material must be stored in a secure area.

3. Radioactive waste must not be disposed of without approval from the person in charge.

Accidents

1. Local building emergency procedures should be known and complied with.

2. Immediate steps to confine contamination should be taken and the entry of persons restricted.

3. The person in charge must be notified.

The conditions for a university permit require that radiation badge monitors be worn by all technicians and demonstrators in the laboratory. All laboratory areas where open source radioisotopes are used must be monitored weekly for removable surface contamination, and records must be kept of the results. If the contamination exceeds

0.5 Bq/cm^2 the surfaces must be decontaminated. All sealed sources containing 1 mCi or more of radioisotope must be checked for leakage at least every 6 months using a wipe test. Records must be kept. All devices containing a sealed source must be labelled to indicate this fact. Sealed sources must be disposed of through the Department of Occupational Health and Safety or other responsible authority.

12.4   GENERAL ELECTRICAL AND TESTING STANDARDS—CSA APPROVAL

Manufacturers and users of products have a need to define the capability of a product against a standard. For a new product or technology this involves reaching an industry-wide agreement on which characteristics or parameters need to be measured. In North America, the major organization which facilitates this is the American Society for Testing and Measurement (ASTM). ASTM committees consider product areas and issue an ASTM standard. Compliance with such standards may be voluntary or compulsory. Standard practice in, say, electrical or building construction is enforced through National Electrical and Building Codes.

In Canada the Canadian Standards Association (CSA) is an independent testing organization whose label 'CSA Approved' denotes that a product meets standards outlined in specific documents related to the product area. A comparable organization in the United States is the 'Underwriters Laboratories'. The Canadian Electrical Code covering all building construction and power distribution systems is CSA Standard C22.1—1986, while CSA Standard C22.2 1—M1961 covers Consumer and Commercial Products.

The Canadian Standards Association maintains independent testing laboratories around the country to which prototype products are submitted. CSA tests do not provide information on the quality of the performance of an instrument in terms of its particular function, but evaluates whether it meets safety criteria defined within the relevant CSA standard. Such criteria include:

(a) Case impact resistance,

(b) Exposed parts having a potential difference of > 30 V which can be contacted with both hands at the same time shall not be capable of supplying > 0.5 mA through a 1500 ohm resistance,

(c) Fuses must be connected on the live (load) side,

(d) Cabinet openings must be such as to eliminate the possibility of contact with live internal circuitry,

(e) Live (ac) and low voltage (ac or dc) leads must be separated. It is this requirement that puts many computer power switches on the rear of the cabinet,

(f) CSA approved components should be used in construction.

CSA approval simplifies the sale and approval of consumer or technical products to the public. It is good practice to be aware of the general outlines of CSA requirements in all laboratory construction.

Safe Electrical Practice in the Laboratory

A major part of research and development is the construction of prototype instrumentation. It is **never** a saving of time and effort to 'rough' wire a circuit together. If the wiring is done properly the first time, it is likely to work, and will be safe and capable of being translated into industrial practice. If any doubt regarding the Electrical Code requirements for low voltage (Class 2) circuits arises, a qualified electronics technician or technologist should be consulted.

1. Use appropriate circuit boards, connectors and boxes.

2. Wire neatly and with logic. If it looks neat, it is likely to work well.

3. Ensure that the AC main connections are correct. For single phase 115 V circuits this means checking the three terminals—black or red (live), white or blue (neutral), green (ground) of the standard 3 conductor wall outlet as shown in Fig. 12.3.

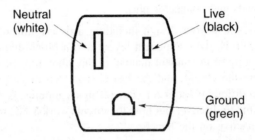

Fig. 12.3   Connections for a standard 115 V AC outlet.

4. Ensure that all cabinets, switches, etc. are securely grounded by connection with the green wire.

5. The wire gauge requirements should be recognized. If a long length of wire (more than 3 m) is used, the next heavier gauge should be used.

Current	Gauge (AWG)
400 mA	24
0.5 – 1 A	22
1 – 5 A	18
5 – 10 A	16
10 – 15 A	14

6. Run custom heaters through isolation transformers. Note that a 'Variac' autotransformer provides a direct connection to the mains and does not satisfy this criterion.

7. Disable interlocks rarely and only with great care if servicing is required. Make sure that a temporary protective device (for example a suitable fuse or breaker) is in place. Replace it as soon as possible.

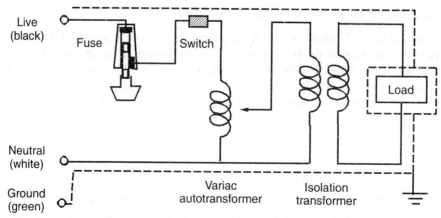

Fig. 12.4 Power control for an AC load.

8. Label leads and connector pins.

9. Fully document your design including a description of the function, the panel controls, the component layout and a block diagram of the circuit. Store a copy of instrument manual in the laboratory files. Today **you** may remember the circuit, and the last change you made to make it work. It will be a different matter for yourself in six months, or for your successor at any time to discover out how the circuit works. The result will be a box sitting inactive on the bench.

> Do not do things in a hurry or when tired.

12.5 GENERAL LABORATORY AND WORKSHOP PRACTICE

In addition to keeping all laboratory and workshop areas clean and tidy, certain practices are standard in a laboratory and workshop setting.

1. Gas cylinders must always be secured to a wall or solid object.

2. Large or heavy structures must be appropriately secured. The possibility of the 'worst case' leading to an accident must always be considered.

3. In the workshop:

    (a) wear appropriate dress—no shorts or sandals; secure ties, loose hair and any other item of personal attire which may become trapped in a machine.

    (b) always use protective glasses.

    (c) when drilling or sawing, firmly anchor the workpiece and proceed with caution.

    (d) use only those machines which you are authorized to use and for which you have received training from a competent person.

    (e) **never** remove or disable any protective device.

    (f) unplug electrical appliances before making repairs.

4. While working with furnaces or hot objects:

   (a) use protective gloves. Never assume that something is cool.

   (b) maintain a clear space around all furnaces, do not litter the area with flammable material which may ignite after a period of drying, or when a hot object is removed from the furnace.

   (c) construct a brick or insulating hearth upon which hot objects can be placed on removal from the furnace.

   (d) always check that the circuits which supply power to a furnace are 'fail safe', and ensure that when the temperature sensing element fails, or if power is removed from the furnace and then re-applied, then the power control either turns off or re-establishes itself to its control setting. If this is not the case, it is a fact of life that full power will be applied to the furnace in the middle of the night—resulting in the ruin of your sample and experiment, and potential damage to the furnace, its surroundings, and your credibility.

REFERENCES AND FURTHER INFORMATION

1. **Province of Ontario**: *Occupational Health and Safety Act and Regulations for Industrial Establishments,* May 1995.
   All workplaces have a copy of this Act in the form of a small green book.

2. **WHMIS Regulations**
   Provincial WHMIS Regulations
   Ontario Government Bookstore, Toronto (416) 326-5122

   Canadian Centre for Occupational Health and Safety
   Hamilton (416) 572-2981

   MSDS—A Basic Guide
   MSDS—An Explanation of Common Terms

   WHMIS courses are run at most organizations on a periodic basis, and particularly in the early summer.

3. **Atomic Energy Control Board Regulations**
   Atomic Energy Control Act (1974)
   Canada Gazette Part II, Vol 108, 1783
   Amendments (1978): Canada Gazette, Part II, Vol 112, 406

   Radioisotopes and Transportation Division
   Atomic Energy Control Board
   Ottawa, Canada

4. Canadian Standards Association
   178 Rexdale Boulevard Rexdale, Ontario, M9W 1R3 Canada (416) 747-4044

5. **Directory of Accredited Testing Organizations**
   Standards Council of Canada
   350 Sparks Street, Ottawa, Ontario K1R 7S8 Canada (613) 238-3222

DESIGN PROBLEMS

12.1 Describe the main provisions of the Occupational Health and Safety Act relative to your location.

   You notice a badly frayed electrical cable in a workplace. What do you do, and what are the consequences of not doing something, if you are: (a) a worker, (b) a supervisor, and (c) an employer.

12.2 Describe the principal features and requirements of the WHMIS safety legislation regarding the handling of chemicals in the workplace.

12.3 A safety guard has been removed from a machine you wish to use. What are your responsibilities under the Health and Safety legislation if you are an employer, a supervisor, or a worker respectively.

12.4 Examine an instrument, for example a computer, which has a label indicating that it has CSA approval. Write down the features of the design which makes it conform to CSA regulations.

12.5 What are your responsibilities as a **worker** under the Occupational Health and Safety Act of your locality.

12.6 Under a regular maintenance program you change the oil in a large rotary vacuum pump using as a backing pump in a vacuum evaporator for metallizations in electronic microcircuits. How do you dispose of 10 l of contaminated oil? What information do you provide to the disposal agency?

Appendix 1

The SI System of Units

Base units

Metre	m	Length of path travelled in vacuum by light in 1/299,792,458 s.
Kilogram	kg	Mass of the international prototype of the kilogram.
Second	s	Duration of 9192631770 periods of the radiation arising from the transition between two hyperfine levels of the ground state of ^{133}Cs
Ampere	A	Current in two infinite lengths of parallel conductors placed 1 mm apart in vacuum which gives rise to a mutual force of 2×10^{-7} N/m.
Kelvin	K	Temperature equal to 1/273.16 of the thermodynamic temperature of the triple point of water.
Candela	cd	Luminous intensity in a given direction of a source of radiation of frequency 540×10^{12} Hz when the radiant intensity of the source is 1/683 W/steradian.
Mole	mol	An amount of substance which contains as many elementary entities as there are atoms in 0.012 kg of carbon-12.

Supplementary units

Plane angle   rad   Solid angle   sr

Derived units

			Definition
Hertz	Hz	Frequency	s^{-1}
Joule	J	Energy	$m^2 \, kg \, s^{-2}$
Newton	N	Force	$m \, kg \, s^{-1}$
Pascal	Pa	Pressure	$N \, m^{-2}$
Watt	W	Power	$J \, s^{-1}$
Coulomb	C	Electric charge	$s \, A$
Volt	V	Electric potential difference	$J \, A^{-1} \, s^{-1}$

335

Derived units (Cont.)			Definition
Ohm	Ω	Electric resistance	$V\ A^{-1}$
Siemens	S	Electric conductance	$A\ V^{-1} = \Omega^{-1}$
Farad	F	Electric capacitance	$A\ s\ V^{-1}$
Weber	Wb	Magnetic flux	$V\ s$
Henry	H	Inductance	$V\ A^{-1}\ s$
Tesla	T	Magnetic flux density	$V\ s\ m^{-2}$
Lumen	lm	Luminous flux	$cd\ sr$
Lux	lx	Illuminance	$m^{-2}\ cd\ sr$
Degree celsius	°C	Celsius temperature	$°C = K - 273$
Becquerel	Bq	Activity of a radioactive source	s^{-1}
Gray	Gy	Absorbed dose ionizing radiation	$J\ kg^{-1}$
Sievert	Sv	Dose equivalent	$J\ kg^{-1}$

Appendix 2
Fourier Content of Common Periodic Waveforms

The Fourier spectrum of a periodic waveform carries information with repect to the form of the waveform—lower frequencies define the overall periodicity, while higher frequencies reflect the details—edges, corners and spikes. A knowledge of the spectrum then allows the bandwidth of the processing circuitry to be designed such that the waveform can be reproduced with a reasonable degree of accuracy.

The voltage magnitude of the Fourier content of fine common periodic waveforms each of which have a peak of $E(V)$, out to the seventh harmonic is given below. The table was adapted from General Radio:

Waveform	Name	Fund	2nd	3rd	4th	5th	6th	7th
	Square wave	$\dfrac{4E}{\pi}$	0	$\dfrac{4E}{3\pi}$	0	$\dfrac{4E}{5\pi}$	0	$\dfrac{4E}{7\pi}$
		(127%)	(0%)	(42%)	(0%)	(25%)	(0%)	(18%)
	Triangular	$\dfrac{8E}{\pi^2}$	0	$\dfrac{8E}{9\pi^2}$	0	$\dfrac{8E}{25\pi^2}$	0	$\dfrac{8E}{49\pi^2}$
		(81%)	(0%)	(9%)	(0)	(3.2%)	(0%)	(1.6%)
	Sawtooth	$\dfrac{2E}{\pi}$	$\dfrac{1E}{\pi}$	$\dfrac{2E}{3\pi}$	$\dfrac{1E}{2\pi}$	$\dfrac{2E}{5\pi}$	$\dfrac{1E}{3\pi}$	$\dfrac{2E}{7\pi}$
		(64%)	(32%)	(21%)	(16%)	(13%)	(11%)	(9%)
	Half-wave rectifier	$\dfrac{1E}{\pi}$	$\dfrac{2E}{3\pi}$	0	$\dfrac{2E}{15\pi}$	0	$\dfrac{2E}{35\pi}$	0
		(32%)	(21%)	(0%)	(4%)	(0%)	(1.8%)	(0%)
	Full-wave rectifier	$\dfrac{2E}{\pi}$	$\dfrac{4E}{3\pi}$	0	$\dfrac{4E}{15\pi}$	0	$\dfrac{4E}{35\pi}$	0
		(64%)	(42%)	(0%)	(8%)	(0%)	(4%)	(0%)

Appendix 3

The p–n Junction in Pictures

The Basic Material

The crystal is electrically neutral, and the number of negative and positive charges per unit volume are equal.

Charge picture

Energy band picture

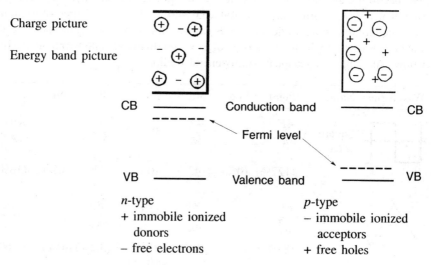

n-type	p-type
+ immobile ionized donors	− immobile ionized acceptors
− free electrons	+ free holes

The Junction in Thermal Equilibrium: Charge picture

When the junction is formed, electrons flow to p-type, holes flow to n-type. Recombination occurs at the junction.

The immobile charged impurity concentration plus the change in the free carrier concentrations caused by recombination results in two layers of impurity charge being 'uncovered' on either side of the junction.

The electric field associated with the charge layers opposes further the free carrier movement. The final result is a region around the junction where the number of free carriers is severely depleted.

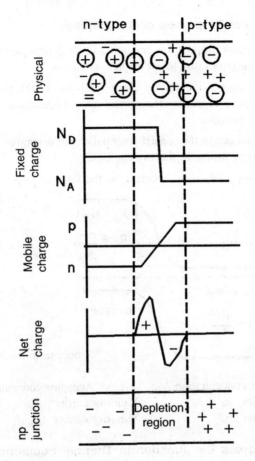

The Junction in Thermal Equilibrium: Potential energy picture

From the point of view of an electron, the transfer of an electron from *n* to *p* raises the potential of the *p*-type materials and makes it more difficult for subsequent electrons to transfer, i.e the energy of the conduction electrons in the *p*-region is raised with respect to those in the *n*-region.

An equilibrium is established in which the attractive energy between electrons and holes is just balanced by the difference in the potential energy of the two regions. Thermodynamically this occurs when the Fermi level in the two materials attains the same energy.

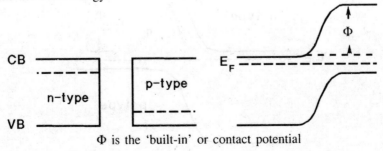

Φ is the 'built-in' or contact potential

Expected Physical Properties of the Junction

(a) Current flow through the junction will be determined by the number of charge carriers close to the depletion region.

(b) The two layers of impurity charge associated with the junction form a parallel plate capacitor arrangement and a capacitance may be associated with the junction.

(c) The magnitude of the contact potential Φ will depend on the concentration of impurity atoms in the two regions.

(d) $n_n \cdot p_n = p_p \cdot n_p = n_i^2$, where n_i is the intrinsic concentration

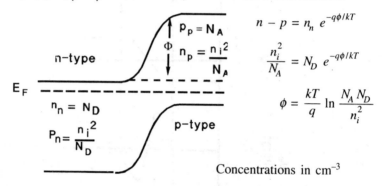

$$n - p = n_n \, e^{-q\phi/kT}$$

$$\frac{n_i^2}{N_A} = N_D \, e^{-q\phi/kT}$$

$$\phi = \frac{kT}{q} \ln \frac{N_A N_D}{n_i^2}$$

Concentrations in cm^{-3}

n-type: Donor concentration N_D *p*-type: Acceptor concentration N_A
Majority carrier n_n Majority carrier p_p
minority carrier p_n Minority carrier n_p

Charge Flow Across the Junction in Thermal Equilibrium (no applied voltage)

Two contributions;

Forward current i_f due to the number of electrons that manage to surmount the potential barrier.

At temperature *T*. $i_f \sim e^{-q\phi/kT}$

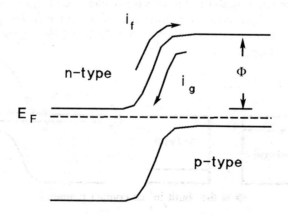

Reverse current i_g. At any temperature T minority carriers—electrons in p-type, holes in n-type, will be created in the regions close to the junction. These may migrate to the junction and fall down the potential hill. In equilibrium, no net current flows through the junction $(i_f + i_g = 0)$

$$\text{Hence } i_f = i_g$$

Voltage Applied to the Junction: Energy band picture

Consider the p-type material made positive with respect to the n-type. This makes it easier for majority carriers to cross the junction, i.e. lowers the height of the barrier.

Forward current i_f will be changed to $i_f(V)$

$$\text{where } \frac{i_f(V)}{i_f} = \frac{e^{-q(\phi-v)/kT}}{e^{-q\phi/kT}} = e^{+qV/kT}$$

Reverse current i_g still only depends on the rate of thermal generation of minority carriers and will be essentially unchanged.

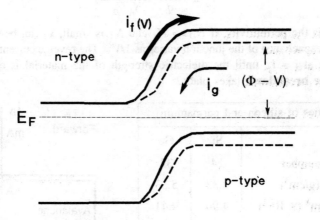

Total current through junction $i = i_f(V) - i_g = i_f e^{qV/kT} - i_g$, and since $i_f = i_g$

> **Forward voltage current characteristic** $i = i_g(e^{qV/kT} - 1)$

Thermal Generation Current i_g

Holes produced in the n-type, and electrons produced in the p-type materials can diffuse to the junction and transfer across the depletion layer. The number doing so will depend on the number present (determined by the temperature) and what number do actually reach the junction. Since the recombination is occurring along with diffusion, the average distance that a minority carrier drifts before recombination is the diffusion length $L = \sqrt{D\tau}$, all carriers within one diffusion length may reach the junction. Contributions to the average diffusion current are therefore $qD_p dp/dx = qD_p p_n/L_p$ for holes and $qD_n n_p/L_n$ for electrons.

The diode current is therefore

$$i = \left(\frac{qD_p P_n}{L_p} + \frac{qD_n n_p}{L_n} \right) (e^{+qV/kT} - 1)$$

Substituting $n \cdot p = A\, e^{-Eg/kT}$, where E_g is the band gap energy. $i = i'\exp\left(\frac{q(V - E_g)}{kT} \right)$

Voltage Applied to the Junction: Charge picture

Forward direction (p positive, n negative).   Additional holes are attracted into the n-type material, and extra holes enter the p-type material. This constitutes 'minority carrier injection'.

Reverse direction (p negative, n positive).   The majority carrier concentrations are repelled from the junction and the width of the depletion layer increases. The junction width as a function of voltage is given by

$$x_d = \left[\frac{2V\varepsilon}{q} \frac{N_A + N_D}{N_A N_D} \right]^{1/2}$$

where ε is the permittivity. If $N_D \gg N_A$ and N_A is small, x_d can be large.
    The capacitance of the junction varies as $1/V^2$. The reverse current will saturate with bias at $i = i_g$, until the dielectric strength of the material is exceeded and **avalanche breakdown** takes place.

Properties of silicon and germanium		
	Si	Ge
Atomic number	14	32
Density (g/cm³)	2.33	5.33
Atoms/cm³ ($\times 10^{22}$)	4.96	4.41
Relative permittivity	12	16
Energy gap (eV)	1.16	0.67
Intrinsic n_i ($\times 10^{10}$)	1.5	2400
Mobility (cm²/V-s)		
Hole	480	1900
Electron	1350	3900
eV/electron-hole pair	3.62	2.96

Appendix 4
Selected Atomic and Nuclear Data

A4.1 Commonly Used Laboratory Nuclear Sources

Radiation	Source	Type of decay	Half-life	Particle energy emission		Photon energy emission		Purpose
				MeV	Prob. %	MeV	Prob. %	
Alpha α	$^{241}_{95}$Am	α	432.7 yr	5.443	13	0.060	36	Range expt.
				5.486	85	Np L X-ray 38		
	$^{244}_{96}$Cm	α	18.11 yr	5.763	24	Pu L X-ray 9		
α	$^{252}_{98}$Cf	α	2.645 yr	6.076	15			
+				6.118	82			
Fission		Fission		20 γ/f		80 < 1 MeV		
				4 n/f		E_n = 2.14 MeV		
Beta⁻ β⁻	$^{90}_{38}$Sr	β⁻	28.5 yr	0.546	100			Attenuation
	$^{90}_{39}$Y	β⁻		2.283	100			
	$^{137}_{55}$Cs	β⁻	30.2 yr	0.514	94	0.663	85	
				1.176	6			
Beta⁺ β⁺	$^{22}_{11}$Na	β⁺, EC	2.603 yr	0.545	90	0.511 Annih γ		
						1.275	90	
Gamma γ#	^{60}Co	β⁻	5.271 yr	0.316	100	1.173	100	Attenuation
						1.333	100	
γ#	$^{137}_{55}$Cs	β⁻	30.2 yr	0.514	94	0.663	85	
				1.176	6			
X-ray	$^{55}_{26}$Fe	EC	2.73 yr			Mn K X-ray		Mossbauer
						0.00589 25		
						0.00649 3.4		
	$^{125}_{53}$I	EC	60 days			Te K X-ray		X-ray
						0.00274 100		Fluorescence

These sources are convenient for laboratory experimentation. The γ-sources labelled # have the particle which is emitted removed by the encapsulation.

Reference: Table 26.1 *Phys. Rev.* D 54,154 (1996)

A4.2 Atomic and Nuclear Data for the Elements

Atomic weight and density values taken from *Handbook of Chemistry and Physics,* 66th ed., CRC Press, Ohio.

Emission data taken from Montenegro, Baptista and Duarte, Atomic Data and Nuclear Data Tables 22, 2 1978.

Mass absorption data for 30 keV X-rays from various sources including Cullity (Ref. 8.2).

Fluorescence yield ω_K data from W. Bambynek, et al, *Rev. Mod. Phys.,* 44,716 (1972).

Element	Z	A	ρ (g/cm^3) (30 keV)	μ/ρ (cm^2/g)	K_α	K_β	ω_K
H	1	1.00797	0.0899 g/l	.390			
He	2	4.0026	0.177 g/l				
Li	3	6.939	0.534	.180	0.05		
Be	4	9.0122	1.848	.185	0.11		
B	5	10.811	2.34	.198	0.18		
C	6	12.011	1.8–2.1	.256	0.28		
N	7	14.0067	1.251 g/l	.350	0.39		
O	8	15.999	1.429 g/l	.372	0.52		
F	9	18.998	1.696 g/l		0.68		
Ne	10	20.183	0.90 g/l	.580	0.848		
Na	11	22.989	0.971	.750	1.052	1.067	
Mg	12	24.312	1.738	.940	1.253	1.302	
Al	13	26.981	2.699	1.17	1.486	1.557	.0357
Si	14	28.086	2.33	1.36	1.739	1.836	.0470
P	15	30.973	1.82–2.25	1.67	2.012	2.139	.0604
S	16	32.064	2.07	2.10	2.306	2.464	.0761
Cl	17	35.453	3.214	2.47	2.622	2.817	.0942
Ar	18	39.948	1.7837 g/l	2.95	2.957	3.191	0.115
K	19	39.102	0.862	3.50	3.312	3.589	0.138
Ca	20	40.08	1.55	3.97	3.691	4.012	0.163
Sc	21	44.956	2.992	4.65	4.088	4.459	0.190
Ti	22	47.90	4.54	5.32	4.509	4.931	0.219
V	23	50.942	6.11	6.04	4.949	5.427	0.250
Cr	24	51.996	7.18	6.81	5.411	5.947	0.282
Mn	25	54.938	7.21	7.62	5.895	6.492	0.314
Fe	26	55.847	7.874	8.45	6.404	7.059	0.347
Co	27	58.933	8.9	9.40	6.925	7.649	0.381
Ni	28	58.71	8.902	10.5	7.472	8.265	0.414
Cu	29	63.546	8.96	11.45	8.041	8.907	0.445
Zn	30	65.37	7.13	12.3	8.631	9.572	0.479
Ga	31	69.72	5.907		9.243	10.263	0.510

Element	Z	A	ρ (g/cm^3)	μ/ρ (cm^2/g) (30 keV)	K_α	K_β	ω_K
Ge	32	72.59	5.323		9.876	10.983	0.540
As	33	74.921	1.97		10.532	10.729	0.567
Se	34	78.96	4.79		11.211	12.502	0.596
Br	35	79.904	7.59 g/l	19.0	11.907	13.299	0.622
Kr	36	83.80	3.73 g/l		12.631	14.126	0.646
Rb	37	85.47	1.532		13.375	14.979	0.669
Sr	38	87.62	2.54	24.0	14.142	15.859	0.691
Y	39	88.905	4.45		14.933	16.766	0.722
Zr	40	91.22	5.53		15.746	17.701	0.730
Nb	41	92.906	8.57		16.584	18.661	0.748
Mo	42	95.94	10.22	30.0	17.443	19.648	0.764
Tc	43	97	11.5		18.327	20.663	0.779
Ru	44	101.07	12.41		19.234	21.705	0.793
Rh	45	102.905	12.41		20.167	22.778	0.807
Pd	46	106.4	12.02		21.122	23.876	0.819
Ag	47	107.868	10.50	41	22.103	25.007	0.830
Cd	48	112.40	8.65		23.108	26.165	0.840
In	49	114.82	7.31		24.138	27.192	0.850
Sn	50	118.69	5.75	45	25.192	28.571	0.859
Sb	51	121.75	6.691		26.271	29.721	0.867
Te	52	127.60	6.24		27.377	31.000	0.875
I	53	126.90	4.93		28.508	32.289	0.882
Xe	54	131.30	5.887 g/l		29.666	33.619	0.889
Cs	55	132.90	1.873		30.851	34.981	0.895
Ba	56	137.34	3.5		32.062	36.372	0.901
La	57	138.91	5.98–6.18		33.299	37.795	0.906
Ce	58	148.12	6.67–8.23		34.566	39.251	0.911
Pr	59	140.91	6.782		35.860	40.741	0.915
Nd	60	144.24	6.8		37.182	42.264	0.920
Pm	61	145			38.532	43.818	0.924
Sm	62	150.35	7.536		39.911	45.405	0.928
Eu	63	154.96	5.259		41.320	47.030	0.931
Gd	64	157.23	7.895		42.757	48.688	0.934
Tb	65	158.92	8.272		44.226	50.373	0.937
Dy	66	162.50	8.536		45.724	52.110	0.940
Ho	67	164.93	8.803		47.253	53.868	0.943
Er	68	167.26	9.051		48.813	55.672	0.945
Tm	69	168.93	9.332		50.406	57.506	0.948
Yb	70	173.04	6.97		52.030	59.356	0.950
Lu	71	174.97	9.872		53.687	61.272	0.952
Hf	72	178.49	13.29		55.382	63.222	0.954
Ta	73	180.95	16.6	21.5	57.098	65.212	0.956
W	74	183.85	19.3	22.5	58.856	67.233	0.957
Re	75	186.2	21.02		60.648	69.298	0.959
Os	76	190.2	22.57		62.477	71.401	0.961

Element	Z	A	ρ (g/cm^3)	μ/ρ (cm^2/g) (30 keV)	K_α	K_β	ω_K
Ir	77	192.2	22.42		64.339	73.548	0.962
Pt	78	195.967	21.45	27.4	66.241	75.735	0.963
Au	79	196.967	19.32	28.4	68.177	77.971	0.964
Hg	80	200.59	13.546		70.154	80.240	0.966
Tl	81	204.37	11.85		71.167	82.562	0.967
Pb	82	207.19	11.35	32	74.956	84.922	0.968
Bi	83	209	9.747		76.315	87.328	0.969
Po	84	209	9.32		78.452	89.781	0.970
At	85	210			80.624	92.287	0.971
Rn	86	222	9.73 g/l		82.843	94.850	0.972
Fr	87	223			85.110	97.460	0.972
Ra	88	226	5		87.419	100.113	0.973
Ac	89	227	10.07		89.773	102.829	0.974
Th	90	232.038	11.66		92.22	105.61	0.975
Pa	91	231	15.37		94.67	108.43	0.976
U	92	238.03	18.95		97.18	111.30	0.977
Np	93	237	18.20		101.07	114.24	0.977
Pu	94	244	19.84		103.76	117.26	0.978

Notes: The emission energies quoted for K_α and K_β are weighted averages of the fine
structure $K_{\alpha 1}$ and $K_{\alpha 2}$ etc. (page 233). Details of the weighting factors and individual
line intensity are found in Ref. 8.4.

The uncertainty in the calculations of fluorescent yield are as follows:

A4.3   Mass Attenuation and Absorption Coefficients of γ rays

Mass Attenuation Coefficient

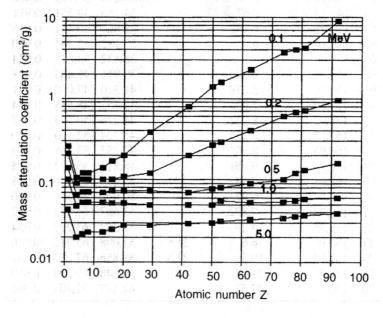

Mass Absorption Coefficent

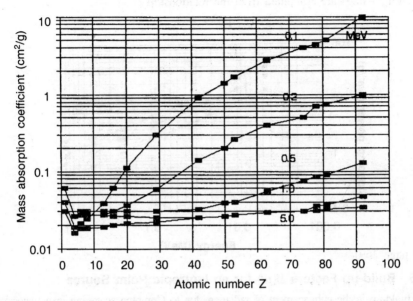

(a) Mass attenuation coefficients and (b) Mass absorption coefficients for γ-rays having energies $0.1 < E < 5$ MeV for elements across the periodic table. Adapted from L.T. Templin, Editor, Reactor Physics Constants ANL-5800 (1963).

A4.4 Total Stopping Power for Electrons in Air, Water, Aluminum and Lead

From W. Heitler, *The Quantum Theory of Radiation*, Oxford University Press 1954.

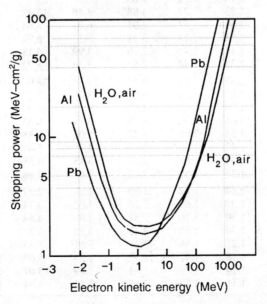

Stopping power for electrons in air, water and Pb at low energies (<100 keV). The straight lines are computed from the relationship

$$S = E^{-0.68} Z^{0.29}$$

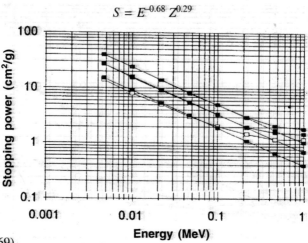

(See page 269)

A4.5 Build-up Factors $B(\mu t)$ for an Isotropic Point Source

The build-up is the enhancement of radiation due to Compton scattering in a material of thickness t. For this table, the emitter is assumed to be a point source. The linear attenuation coefficient for the radiation of energy E in the material is μ.

Material	E				μt			
	MeV	1	2	4	7	10	15	20
Water	0.255	33.09	7.14	23	72.9	166	456	982
	0.5	2.52	5.14	14.3	38.8	77.6	178	334
	1	2.13	3.71	7.68	16.2	27.1	50.4	82.2
	2	1.83	2.77	4.88	8.46	12.4	19.5	27.7
	3	1.69	2.42	3.91	6.23	8.63	12.8	17
	4	1.58	2.17	3.34	5.13	6.94	9.97	12.9
	6	1.46	1.91	2.76	3.99	5.18	7.09	8.85
	8	1.38	1.74	2.4	3.34	3.72	4.9	5.98
Aluminum	0.5	2.37	4.24	9.47	21.5	38.9	80.8	141
	1	2.02	3.31	6.57	13.1	21.2	37.9	58.5
	2	1.75	2.61	4.62	8.05	11.9	18.7	26.3
	3	1.64	2.32	3.78	6.14	8.65	13	17.7
	4	1.53	2.08	3.22	5.01	6.88	10.1	13.4
	6	1.42	1.85	2.7	4.06	5.49	7.97	10.4
	8	1.34	1.68	2.37	3.45	4.58	6.56	8.52
	10	1.28	1.55	2.12	3.01	3.96	5.63	7.32
Lead	0.5	1.24	1.42	1.69	2	2.27	2.65	2.73
	1	1.37	1.69	2.26	3.02	3.74	4.81	5.86
	2	1.39	1.76	2.51	3.66	4.84	6.87	9
	3	1.34	1.68	2.43	3.75	5.3	8.44	12.3
	4	1.27	1.56	2.25	3.61	5.44	9.8	16.3
	6	1.18	1.4	1.97	3.34	5.69	13.8	32.7
	8	1.14	1.3	1.74	2.89	5.07	14.1	44.8
	10	1.11	1.23	1.58	2.52	4.34	12.5	39.2

H. Goldstein, *Fundamental Aspects of Reactor Shielding*, Addison-Wesley (1959)

A4.6 Non-SI Radiation Units

Exposure: (SI)

1 Roentgen (R) = Amount of radiation which will create 1 esu of charge per cm^3 of dry air at 1 atm and 273 K.

1 esu = 3.33×10^{-10} C; 1 cm^3 air = 1.293 mg

1 Roentgen (R) = 2.58×10^{-4} C/kg (SI)

Dose: (Gy = 1 J/kg)

1 Rad = 100 erg/g delivered to a material

1 Rad = 0.01Gy

Dose Equivalent: (Sv = Gy × Q)

1 rem = 1 Rad × Q = 0.01 Sv

$$X = 1.83 \times 10^{-8} \, I(E) \, \frac{E\mu_a(E)^{air}}{\rho} \, R/s$$

A4.7 Mass Absorption Coefficients for Photons in Air and Tissue

(After Lamarsh—Ref. 9.2)

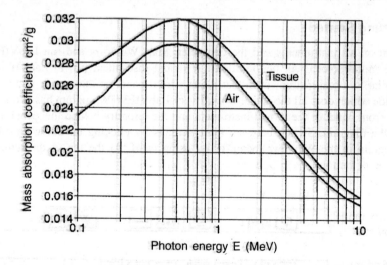

Photon energy E (MeV)

Appendix 5

LabVIEW™ Instrumentation Software—An Introduction

LabVIEW is a windows-based specialized programming language developed by National Instruments to simplify interfacing a computer with instrumentation. It also provides a substantial portfolio of functions. It is entirely graphical. Functions are represented as icons and these are dragged around the screen with a mouse and 'wired' together to form programs. There are many useful built-in functions and a number of different types of loops. LabVIEW™ allows a graphical interface to be quickly set up with a variety of instruments. This appendix introduces LabVIEW™ procedures with a particular emphasis on the IEEE488 bus. Permission to use the LabView™ icons has been granted by National Instruments Inc.

Getting Started

Turn on all instruments and the computer. Start Windows and run LabVIEW™. The initial screen will be "untitled". A virtual instrument (vi) in LabVIEW™ is similar to a real instrument in that it has a front panel on the outside and a circuit inside which does all the actual work. Every LabVIEW™ program has two screens, the front panel of the virtual instrument and the circuitry behind this panel, known as the diagram. Switch between these screens by clicking on Show Panel/Show Diagram in the Windows menu, or by using ctrl-f. At the top of the screen is a series of buttons.

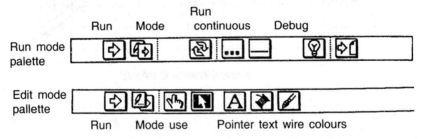

The set of buttons depends on the mode. Switch modes by clicking the mode button. When the pencil is on top, the mode is 'edit'; when the arrow is on top the mode is 'run'.

Creating a Program—The Front Panel

Decide on the readouts and controls needed on the front panel of the virtual instrument. An example of a pull-down menu is shown. The panel may include input variables, digital outputs, graphs, and even switches. Either click on the spot where the control is to be placed, then choose the appropriate control from the control menu, or simply right-click where it is needed and a menu will pop up. Choose controls from the sub-menus. Passing the pointer over an icon will give a brief description at the bottom of the menu box. Once a control is located the label for this control may be typed, followed by 'enter'. If a label is not desired the label will disappear. Right clicking on an existing control gives a list of options specific to that type of control or readout.

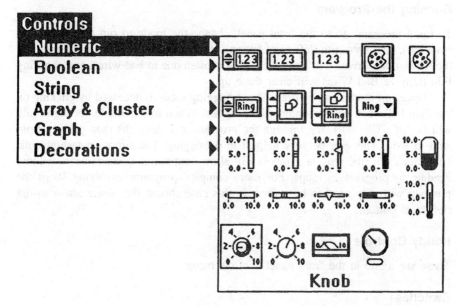

If it is later desired to label a control, 'show label' is an option in the control's pop-up menu. Controls can be resized and moved about at any time. Scales on graphs and knobs can be adjusted by clicking on the scale with the user tool and typing in a new value. All values between the maximum and minimum will automatically rescale.

Creating a Program—The Wiring Diagram

Move to the diagram screen by choosing Windows—Show Diagram (the front panel can be modified at any time). Icons for the front panel controls and readouts can be moved about without affecting the front panel layout. Necessary functions can now be added to the program from the Functions menu (this is done in the same manner as adding controls to the front panel). These functions may include arithmetic functions, input/output functions, or other graphical and analytic functions designed to fit a wide variety of needs.

Icons and the Help Screen

Functions appear as icons in the diagram. Icons are connected by creating a path for the data to follow. This is known as 'wiring', and is done with the wiring tool. Some functions have many inputs and/or outputs. These are accessed by wiring to different regions of the icon. The regions will blink when the wiring tool is moved over them. Ctrl-h will bring up a help screen. Moving the pointer over an icon will indicate the different regions that can be 'wired to' and what each region represents. A successful wiring job will leave a coloured wire. An unsuccessful job will leave a black, dashed line. Bad wires can be removed by choosing 'Remove Bad Wires' from the Edit menu (ctrl-b).

Running the Program

To run a program, go to the front panel, change the mode to run (second button from left), and press run (left-most button). If there is a problem with the program the run button's arrow will be broken. This is often due to bad wires, and choosing Edit-Remove Bad Wires will clear these up.

Three run modes can be used. The debugging mode is accessed by clicking on the light bulb icon before running the program. When in this mode, the light bulb will be 'lit'. This will step through the program and show the data passing from icon to icon. This can be very helpful in debugging. For simple measurements, continuous run (the recycle symbol) will allow continuous measurements without needing to program in loops. For more complex programs involving loops the program will only need to be started. In this case choose the single arrow to run (left-most button).

Handy Controls

These are found in the front panel controls menu:

Switches

Found in Boolean

Returns 'True' for on, and 'False' for off. Useful for controlling While Loops. Can be switched by the user while the virtual instrument is running.

Waveform Graphs

Found in Graphs

These make strip graphs of your data vs. time, similar to an oscilloscope readout.

XY Graph

Found in Graphs

This must receive clusters of arrays of data. This is usually used in conjunction with a For or While Loop. Data leaving the For Loop will do so as arrays. These should be bundled (with the Bundler function). The array that goes into the top input of the Bundler forms the x-data of the graph. The graph icon should be placed outside of the For or While Loop. If it is inside the For Loop it will plot only one point at a time, erasing previous points. The plot does not occur until the program completes the loop.

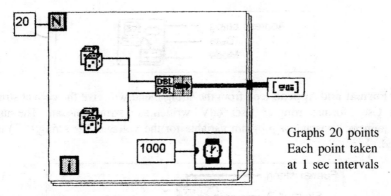

Graphs 20 points
Each point taken
at 1 sec intervals

FUNCTIONS YOU MAY NEED

The GPIB Bus

Reading data from an instrument. In order to read an instrument through the GPIB, it must be addressed by its device number, the number of bytes to be sent, and adequate time to operate must be given. Subsequently, the string number which is returned must be converted to a decimal number. Most conventional instruments use traditional GPIB.

GPIB Read

Let us take an example: a device number for one multimeter may be 16, and that for a second multimeter is 21. The byte count is 30 in both cases.

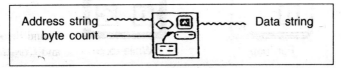

The string returned has the form NCVD+0.0013+1. This must be converted into a decimal using the following string icon. An offset of 4 strips off the first 4 digits of the string (removing NCVD).

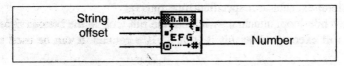

Writing information to an instrument. In order to address an instrument to set ranges etc., its number must be set as a string variable, and a second string variable used to induce the instrument to do something. The actual format must be looked up in the IEEE488 manual for the instrument. For example, a power supply may have a device number of 14. It requires a command of the form "setv 5V" to set the output to 5V. "3" connected to the 'mode' terminal passes a CR/LF to execute the command.

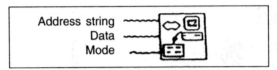

The **Format and Append** icon from the string menu will give the correct string as data. Use a format string of "vset %dV" written as a string constant. The number is a numeric constant or program variable for the voltage. The string (" ") is not wired.

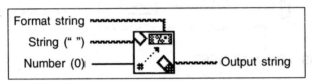

Found in Structures and Constants.

Loops. These are standard fare in most programs. While (if something is 'true') or For (N times) Loops are available in LabVIEW, as well as a number of other structures. Loops are necessary if the program is required to repeat itself or parts of itself. For simple programs this can also be accomplished using the Run Repeatedly option, but for programs involving data logging to file or functions that need to accumulate data loops are required.

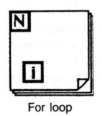

For loop

While loop

Found in Structures and Constants

For a For loop, the number of times to execute should be wired outside the loop to the N icon in the top left corner. All the data that leaves the loop will do so all at once (as an array) after the loop terminates. Thus the graph will appear all at once at the end of the loop's iterations. Include a wait icon inside the loop to make the loop execute at a specified time interval.

For a While loop, attach a 'switch' to the run icon in the bottom right corner. The loop will execute while this is 'true'. I is a counter. It can be used to 'step' voltages.

Sequences. LabVIEW does not execute the program from top to bottom or left to right. It instead executes everything in a pseudo-parallel fashion. This is usually to be desired, but under some circumstances it may be necessary to perform operations in a specified order. This is done by using Sequences. This is important if the program is long and/or complex.

Sequences look like frames from a film. Frames are layered on top of each other so that only one is visible at a time. Each frame is numbered and they execute in sequence.

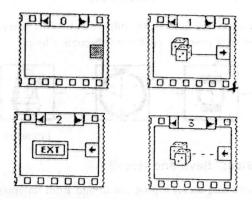

Thermolinearizer. This function will translate a voltage to a temperature. The calibration number for chromel-alumel is 3. This must be wired on. This function takes its zero point to be zero Celsius. 20 must therefore be added to the output temperature.

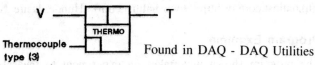

Found in DAQ - DAQ Utilities

Math functions. There are a variety of math functions available. Arguments are wired into the left, the answer comes out the right.

Bundler. This bundles data together into packets called clusters for activities such as making *x-y* plots. Resizing this icon will increase the number of inputs (i.e. how many data will be bundled together into one cluster).

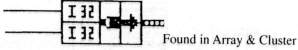

Found in Array & Cluster

Write to spreadsheet file. The program writes to a tab delimited file which can easily be parsed into a spreadsheet program. You will need to give a file name and also specify that the data should be concatenated. This is done by wiring a boolean (true) variable to the icon. If this variable is not set, the program will over-write the last data point whenever it takes a new one. See the sample program at the end of this document for an example.

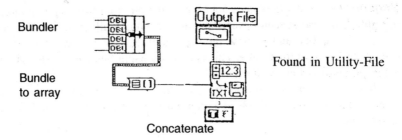

Found in Utility-File

Concatenate

Wait. Waits for specified number of milliseconds. This prevents too many data points being taken in a program controlled through a loop.

Found in Time & Dialogue

Program Libraries: Save and Recall

This implies to save and recall a program to and from an appropriate library.

Application builder.   This allows an executable LabVIEW file to be constructed which can be transferred to any Windows-based computer. See the appropriate Release Notes.

It is easy to download a function for a specific instrument from the internet (ftp.natinst.com or http://www.natinst.com). Hundreds are available free of charge.

Program Example

The program shown undertakes an experiment to measure resistance R versus temperature T as recorded by a chromel-alumel thermocouple. The data is read from digital voltmeters. The temperature is varied as an increasing series of steps by the voltage delivered to a heater by an IEEE488 controlled power supply.

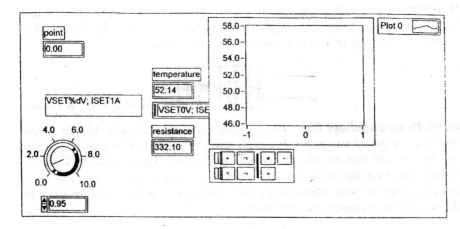

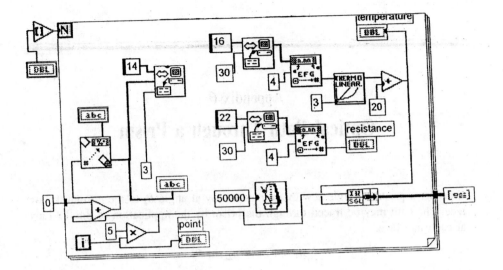

Appendix 6

Optical Path Through a Prism

Consider a prism of angle ϕ with an incident ray at angle θ_1 to the normal to one face. The path may be traced through the prism by the application of Snell's Law at each interface.

$$n_1 \sin \theta_1 = n_2 \sin \theta_2$$

where θ_2 is the angle of refraction within the glass.

The angle of deviation between the incident and emerging ray is ψ.

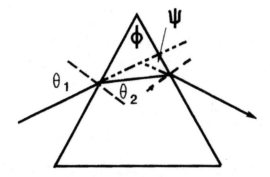

The angle of deviation varies with the angle of incidence. When it becomes minimum known as the condition of minimum deviation ψ_m, the refracted ray within the prism lies parallel to the base.

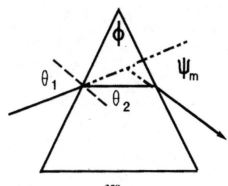

Under these conditions

$$\theta_2 = 1/2 \, \phi$$

$$\psi_m = 2 \, (\theta_1 - \theta_2)$$

From Snell's Law

$$\sin \frac{(\psi + \phi)}{2} = n_2 \sin \phi/2$$

This expression provides an effective way of measuring the refractive index of a material.

It is difficult to derive an analytical expression for ψ as a general function of the angle of incidence. A computer simulation for a prism of angle $60°$ and a refractive index $n = 1.5$ is shown.

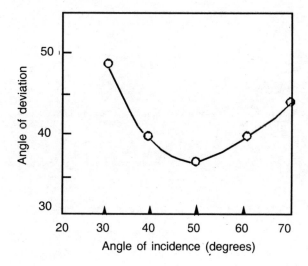

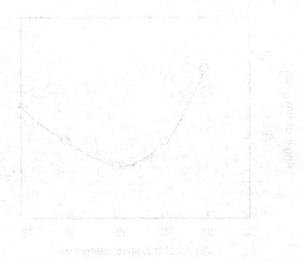

Index